RILEM State-of-the-Art Reports

RILEM STATE-OF-THE-ART REPORTS
Volume 35

RILEM, The International Union of Laboratories and Experts in Construction Materials, Systems and Structures, founded in 1947, is a non-governmental scientific association whose goal is to contribute to progress in the construction sciences, techniques and industries, essentially by means of the communication it fosters between research and practice. RILEM's focus is on construction materials and their use in building and civil engineering structures, covering all phases of the building process from manufacture to use and recycling of materials. More information on RILEM and its previous publications can be found on www.RILEM.net.

The RILEM State-of-the-Art Reports (STAR) are produced by the Technical Committees. They represent one of the most important outputs that RILEM generates – high level scientific and engineering reports that provide cutting edge knowledge in a given field. The work of the TCs is one of RILEM's key functions.

Members of a TC are experts in their field and give their time freely to share their expertise. As a result, the broader scientific community benefits greatly from RILEM's activities.

RILEM's stated objective is to disseminate this information as widely as possible to the scientific community. RILEM therefore considers the STAR reports of its TCs as of highest importance, and encourages their publication whenever possible.

The information in this and similar reports is mostly pre-normative in the sense that it provides the underlying scientific fundamentals on which standards and codes of practice are based. Without such a solid scientific basis, construction practice will be less than efficient or economical.

It is RILEM's hope that this information will be of wide use to the scientific community.

Indexed in SCOPUS, Google Scholar and SpringerLink.

More information about this series at https://link.springer.com/bookseries/8780

Antonin Fabbri · Jean-Claude Morel ·
Jean-Emmanuel Aubert · Quoc-Bao Bui ·
Domenico Gallipoli · B. V. Venkatarama Reddy
Editors

Testing and Characterisation of Earth-based Building Materials and Elements

State-of-the-Art Report of the RILEM TC 274-TCE

Editors
Antonin Fabbri
Laboratoire de Tribologie et Dynamique des Systèmes
École Nationale des Travaux Publics de l'Etat
Vaulx-en-Velin, France

Jean-Emmanuel Aubert
Laboratoire Matériaux et Durabilité des Constructions
University of Toulouse
Toulouse, France

Domenico Gallipoli
Institut Supérieur Aquitain du Bâtiment et Travaux Publics
University of Pau and Pays de l'Adour
Anglet, France

Università degli Studi di Genova
Genova, Italy

Jean-Claude Morel
Faculty of Engineering, Environment and Computing, Centre for the Built and Natural Environment
Coventry University
Conventry, UK

Laboratoire de Tribologie et Dynamique des Systèmes
École Nationale des Travaux Publics de l'Etat
Vaulx-en-Velin, France

Quoc-Bao Bui
Faculty of Civil Engineering
Ton Duc Thang University
Ho Chi Minh, Vietnam

B. V. Venkatarama Reddy
Department of Civil Engineering
Indian Institute of Science Bangalore
Bengaluru, Karnataka, India

ISSN 2213-204X ISSN 2213-2031 (electronic)
RILEM State-of-the-Art Reports
ISBN 978-3-030-83299-5 ISBN 978-3-030-83297-1 (eBook)
https://doi.org/10.1007/978-3-030-83297-1

This Springer imprint is published by the registered company Springer Nature Switzerland AG
The registered company address is: Gewerbestrasse 11, 6330 Cham, Switzerland

RILEM Technical Committee 274-TCE

Jean-Emmanuel Aubert
Christopher T. S. Beckett
Ana Armada Bras
Agostino W. Bruno
Quoc-Bao Bui
Bogdan Cazacliu
Antonin Fabbri
Paulina Faria
Domenico Gallipoli
Anne-Cecile Grillet
Guillaume Habert
Erwan Hamard
Rogiros Illampas
Ioannis Ioannou
Emmanuel Keita
Thibaut Lecompte
Pascal Maillard
Fionn McGregor
Jean-Claude Morel
Daniel V. Oliveira
Kouka Amed Jérémy Ouedraogo
Claudiane Ouellet-Plamondon
Céline Perlot-Bascoulès
Noemie Prime
Elodie Prud'homme
B. V. Venkatarama Reddy
Abbie Romano
Rui Silva

RILEM Publications

The following list is presenting the global offer of RILEM Publications, sorted by series. Each publication is available in printed version and/or in online version.

RILEM Proceedings (PRO)

PRO 1: Durability of High Performance Concrete (ISBN: 2-912143-03-9; e-ISBN: 2-351580-12-5; e-ISBN: 2351580125); *Ed. H. Sommer*

PRO 2: Chloride Penetration into Concrete (ISBN: 2-912143-00-04; e-ISBN: 2912143454); *Eds. L.-O. Nilsson and J.-P. Ollivier*

PRO 3: Evaluation and Strengthening of Existing Masonry Structures (ISBN: 2-912143-02-0; e-ISBN: 2351580141); *Eds. L. Binda and C. Modena*

PRO 4: Concrete: From Material to Structure (ISBN: 2-912143-04-7; e-ISBN: 2351580206); *Eds. J.-P. Bournazel and Y. Malier*

PRO 5: The Role of Admixtures in High Performance Concrete (ISBN: 2-912143-05-5; e-ISBN: 2351580214); *Eds. J. G. Cabrera and R. Rivera-Villarreal*

PRO 6: High Performance Fiber Reinforced Cement Composites—HPFRCC 3 (ISBN: 2-912143-06-3; e-ISBN: 2351580222); *Eds. H. W. Reinhardt and A. E. Naaman*

PRO 7: 1st International RILEM Symposium on Self-Compacting Concrete (ISBN: 2-912143-09-8; e-ISBN: 2912143721); *Eds. Å. Skarendahl and Ö. Petersson*

PRO 8: International RILEM Symposium on Timber Engineering (ISBN: 2-912143-10-1; e-ISBN: 2351580230); *Ed. L. Boström*

PRO 9: 2nd International RILEM Symposium on Adhesion between Polymers and Concrete ISAP '99 (ISBN: 2-912143-11-X; e-ISBN: 2351580249); *Eds. Y. Ohama and M. Puterman*

PRO 10: 3rd International RILEM Symposium on Durability of Building and Construction Sealants (ISBN: 2-912143-13-6; e-ISBN: 2351580257); *Ed. A. T. Wolf*

PRO 11: 4th International RILEM Conference on Reflective Cracking in Pavements (ISBN: 2-912143-14-4; e-ISBN: 2351580265); *Eds. A. O. Abd El Halim, D. A. Taylor and El H. H. Mohamed*

PRO 12: International RILEM Workshop on Historic Mortars: Characteristics and Tests (ISBN: 2-912143-15-2; e-ISBN: 2351580273); *Eds. P. Bartos, C. Groot and J. J. Hughes*

PRO 13: 2nd International RILEM Symposium on Hydration and Setting (ISBN: 2-912143-16-0; e-ISBN: 2351580281); *Ed. A. Nonat*

PRO 14: Integrated Life-Cycle Design of Materials and Structures—ILCDES 2000 (ISBN: 951-758-408-3; e-ISBN: 235158029X); (ISSN: 0356-9403); *Ed. S. Sarja*

PRO 15: Fifth RILEM Symposium on Fibre-Reinforced Concretes (FRC)—BEFIB'2000 (ISBN: 2-912143-18-7; e-ISBN: 291214373X); *Eds. P. Rossi and G. Chanvillard*

PRO 16: Life Prediction and Management of Concrete Structures (ISBN: 2-912143-19-5; e-ISBN: 2351580303); *Ed. D. Naus*

PRO 17: Shrinkage of Concrete—Shrinkage 2000 (ISBN: 2-912143-20-9; e-ISBN: 2351580311); *Eds. V. Baroghel-Bouny and P.-C. Aïtcin*

PRO 18: Measurement and Interpretation of the On-Site Corrosion Rate (ISBN: 2-912143-21-7; e-ISBN: 235158032X); *Eds. C. Andrade, C. Alonso, J. Fullea, J. Polimon and J. Rodriguez*

PRO 19: Testing and Modelling the Chloride Ingress into Concrete (ISBN: 2-912143-22-5; e-ISBN: 2351580338); *Eds. C. Andrade and J. Kropp*

PRO 20: 1st International RILEM Workshop on Microbial Impacts on Building Materials (CD 02) (e-ISBN 978-2-35158-013-4); *Ed. M. Ribas Silva*

PRO 21: International RILEM Symposium on Connections between Steel and Concrete (ISBN: 2-912143-25-X; e-ISBN: 2351580346); *Ed. R. Eligehausen*

PRO 22: International RILEM Symposium on Joints in Timber Structures (ISBN: 2-912143-28-4; e-ISBN: 2351580354); *Eds. S. Aicher and H.-W. Reinhardt*

PRO 23: International RILEM Conference on Early Age Cracking in Cementitious Systems (ISBN: 2-912143-29-2; e-ISBN: 2351580362); *Eds. K. Kovler and A. Bentur*

PRO 24: 2nd International RILEM Workshop on Frost Resistance of Concrete (ISBN: 2-912143-30-6; e-ISBN: 2351580370); *Eds. M. J. Setzer, R. Auberg and H.-J. Keck*

PRO 25: International RILEM Workshop on Frost Damage in Concrete (ISBN: 2-912143-31-4; e-ISBN: 2351580389); *Eds. D. J. Janssen, M. J. Setzer and M. B. Snyder*

PRO 26: International RILEM Workshop on On-Site Control and Evaluation of Masonry Structures (ISBN: 2-912143-34-9; e-ISBN: 2351580141); *Eds. L. Binda and R. C. de Vekey*

PRO 27: International RILEM Symposium on Building Joint Sealants (CD03; e-ISBN: 235158015X); *Ed. A. T. Wolf*

PRO 28: 6th International RILEM Symposium on Performance Testing and Evaluation of Bituminous Materials—PTEBM'03 (ISBN: 2-912143-35-7; e-ISBN: 978-2-912143-77-8); *Ed. M. N. Partl*

PRO 29: 2nd International RILEM Workshop on Life Prediction and Ageing Management of Concrete Structures (ISBN: 2-912143-36-5; e-ISBN: 2912143780); *Ed. D. J. Naus*

PRO 30: 4th International RILEM Workshop on High Performance Fiber Reinforced Cement Composites—HPFRCC 4 (ISBN: 2-912143-37-3; e-ISBN: 2912143799); *Eds. A. E. Naaman and H. W. Reinhardt*

PRO 31: International RILEM Workshop on Test and Design Methods for Steel Fibre Reinforced Concrete: Background and Experiences (ISBN: 2-912143-38-1; e-ISBN: 2351580168); *Eds. B. Schnütgen and L. Vandewalle*

PRO 32: International Conference on Advances in Concrete and Structures 2 vol. (ISBN (set): 2-912143-41-1; e-ISBN: 2351580176); *Eds. Ying-shu Yuan, Surendra P. Shah and Heng-lin Lü*

PRO 33: 3rd International Symposium on Self-Compacting Concrete (ISBN: 2-912143-42-X; e-ISBN: 2912143713); *Eds. Ó. Wallevik and I. Níelsson*

PRO 34: International RILEM Conference on Microbial Impact on Building Materials (ISBN: 2-912143-43-8; e-ISBN: 2351580184); *Ed. M. Ribas Silva*

PRO 35: International RILEM TC 186-ISA on Internal Sulfate Attack and Delayed Ettringite Formation (ISBN: 2-912143-44-6; e-ISBN: 2912143802); *Eds. K. Scrivener and J. Skalny*

PRO 36: International RILEM Symposium on Concrete Science and Engineering—A Tribute to Arnon Bentur (ISBN: 2-912143-46-2; e-ISBN: 2912143586); *Eds. K. Kovler, J. Marchand, S. Mindess and J. Weiss*

PRO 37: 5th International RILEM Conference on Cracking in Pavements—Mitigation, Risk Assessment and Prevention (ISBN: 2-912143-47-0; e-ISBN: 2912143764); *Eds. C. Petit, I. Al-Qadi and A. Millien*

PRO 38: 3rd International RILEM Workshop on Testing and Modelling the Chloride Ingress into Concrete (ISBN: 2-912143-48-9; e-ISBN: 2912143578); *Eds. C. Andrade and J. Kropp*

PRO 39: 6th International RILEM Symposium on Fibre-Reinforced Concretes—BEFIB 2004 (ISBN: 2-912143-51-9; e-ISBN: 2912143748); *Eds. M. Di Prisco, R. Felicetti and G. A. Plizzari*

PRO 40: International RILEM Conference on the Use of Recycled Materials in Buildings and Structures (ISBN: 2-912143-52-7; e-ISBN: 2912143756); *Eds. E. Vázquez, Ch. F. Hendriks and G. M. T. Janssen*

PRO 41: RILEM International Symposium on Environment-Conscious Materials and Systems for Sustainable Development (ISBN: 2-912143-55-1; e-ISBN: 2912143640); *Eds. N. Kashino and Y. Ohama*

PRO 42: SCC'2005—China: 1st International Symposium on Design, Performance and Use of Self-Consolidating Concrete (ISBN: 2-912143-61-6; e-ISBN: 2912143624); *Eds. Zhiwu Yu, Caijun Shi, Kamal Henri Khayat and Youjun Xie*

PRO 43: International RILEM Workshop on Bonded Concrete Overlays (e-ISBN: 2-912143-83-7); *Eds. J. L. Granju and J. Silfwerbrand*

PRO 44: 2nd International RILEM Workshop on Microbial Impacts on Building Materials (CD11) (e-ISBN: 2-912143-84-5); *Ed. M. Ribas Silva*

PRO 45: 2nd International Symposium on Nanotechnology in Construction, Bilbao (ISBN: 2-912143-87-X; e-ISBN: 2912143888); *Eds. Peter J. M. Bartos, Yolanda de Miguel and Antonio Porro*

PRO 46: Concrete Life'06—International RILEM-JCI Seminar on Concrete Durability and Service Life Planning: Curing, Crack Control, Performance in Harsh Environments (ISBN: 2-912143-89-6; e-ISBN: 291214390X); *Ed. K. Kovler*

PRO 47: International RILEM Workshop on Performance Based Evaluation and Indicators for Concrete Durability (ISBN: 978-2-912143-95-2; e-ISBN: 9782912143969); *Eds. V. Baroghel-Bouny, C. Andrade, R. Torrent and K. Scrivener*

PRO 48: 1st International RILEM Symposium on Advances in Concrete through Science and Engineering (e-ISBN: 2-912143-92-6); *Eds. J. Weiss, K. Kovler, J. Marchand, and S. Mindess*

PRO 49: International RILEM Workshop on High Performance Fiber Reinforced Cementitious Composites in Structural Applications (ISBN: 2-912143-93-4; e-ISBN: 2912143942); *Eds. G. Fischer and V. C. Li*

PRO 50: 1st International RILEM Symposium on Textile Reinforced Concrete (ISBN: 2-912143-97-7; e-ISBN: 2351580087); *Eds. Josef Hegger, Wolfgang Brameshuber and Norbert Will*

PRO 51: 2nd International Symposium on Advances in Concrete through Science and Engineering (ISBN: 2-35158-003-6; e-ISBN: 2-35158-002-8); *Eds. J. Marchand, B. Bissonnette, R. Gagné, M. Jolin and F. Paradis*

PRO 52: Volume Changes of Hardening Concrete: Testing and Mitigation (ISBN: 2-35158-004-4; e-ISBN: 2-35158-005-2); *Eds. O. M. Jensen, P. Lura and K. Kovler*

PRO 53: High Performance Fiber Reinforced Cement Composites—HPFRCC5 (ISBN: 978-2-35158-046-2; e-ISBN: 978-2-35158-089-9); *Eds. H. W. Reinhardt and A. E. Naaman*

PRO 54: 5th International RILEM Symposium on Self-Compacting Concrete (ISBN: 978-2-35158-047-9; e-ISBN: 978-2-35158-088-2); *Eds. G. De Schutter and V. Boel*

PRO 55: International RILEM Symposium Photocatalysis, Environment and Construction Materials (ISBN: 978-2-35158-056-1; e-ISBN: 978-2-35158-057-8); *Eds. P. Baglioni and L. Cassar*

PRO 56: International RILEM Workshop on Integral Service Life Modelling of Concrete Structures (ISBN 978-2-35158-058-5; e-ISBN: 978-2-35158-090-5); *Eds. R. M. Ferreira, J. Gulikers and C. Andrade*

PRO 57: RILEM Workshop on Performance of cement-based materials in aggressive aqueous environments (e-ISBN: 978-2-35158-059-2); *Ed. N. De Belie*

PRO 58: International RILEM Symposium on Concrete Modelling—CONMOD'08 (ISBN: 978-2-35158-060-8; e-ISBN: 978-2-35158-076-9); *Eds. E. Schlangen and G. De Schutter*

PRO 59: International RILEM Conference on On Site Assessment of Concrete, Masonry and Timber Structures—SACoMaTiS 2008 (ISBN set: 978-2-35158-061-5; e-ISBN: 978-2-35158-075-2); *Eds. L. Binda, M. di Prisco and R. Felicetti*

PRO 60: Seventh RILEM International Symposium on Fibre Reinforced Concrete: Design and Applications—BEFIB 2008 (ISBN: 978-2-35158-064-6; e-ISBN: 978-2-35158-086-8); *Ed. R. Gettu*

PRO 61: 1st International Conference on Microstructure Related Durability of Cementitious Composites 2 vol., (ISBN: 978-2-35158-065-3; e-ISBN: 978-2-35158-084-4); *Eds. W. Sun, K. van Breugel, C. Miao, G. Ye and H. Chen*

PRO 62: NSF/ RILEM Workshop: In-situ Evaluation of Historic Wood and Masonry Structures (e-ISBN: 978-2-35158-068-4); *Eds. B. Kasal, R. Anthony and M. Drdácký*

PRO 63: Concrete in Aggressive Aqueous Environments: Performance, Testing and Modelling, 2 vol., (ISBN: 978-2-35158-071-4; e-ISBN: 978-2-35158-082-0); *Eds. M. G. Alexander and A. Bertron*

PRO 64: Long Term Performance of Cementitious Barriers and Reinforced Concrete in Nuclear Power Plants and Waste Management—NUCPERF 2009 (ISBN: 978-2-35158-072-1; e-ISBN: 978-2-35158-087-5); *Eds. V. L'Hostis, R. Gens and C. Gallé*

PRO 65: Design Performance and Use of Self-consolidating Concrete—SCC'2009 (ISBN: 978-2-35158-073-8; e-ISBN: 978-2-35158-093-6); *Eds. C. Shi, Z. Yu, K. H. Khayat and P. Yan*

PRO 66: 2nd International RILEM Workshop on Concrete Durability and Service Life Planning—ConcreteLife'09 (ISBN: 978-2-35158-074-5; ISBN: 978-2-35158-074-5); *Ed. K. Kovler*

PRO 67: Repairs Mortars for Historic Masonry (e-ISBN: 978-2-35158-083-7); *Ed. C. Groot*

PRO 68: Proceedings of the 3rd International RILEM Symposium on 'Rheology of Cement Suspensions such as Fresh Concrete (ISBN 978-2-35158-091-2; e-ISBN: 978-2-35158-092-9); *Eds. O. H. Wallevik, S. Kubens and S. Oesterheld*

PRO 69: 3rd International PhD Student Workshop on 'Modelling the Durability of Reinforced Concrete (ISBN: 978-2-35158-095-0); *Eds. R. M. Ferreira, J. Gulikers and C. Andrade*

PRO 70: 2nd International Conference on 'Service Life Design for Infrastructure' (ISBN set: 978-2-35158-096-7, e-ISBN: 978-2-35158-097-4); *Eds. K. van Breugel, G. Ye and Y. Yuan*

PRO 71: Advances in Civil Engineering Materials—The 50-year Teaching Anniversary of Prof. Sun Wei' (ISBN: 978-2-35158-098-1; e-ISBN: 978-2-35158-099-8); *Eds. C. Miao, G. Ye and H. Chen*

PRO 72: First International Conference on 'Advances in Chemically-Activated Materials—CAM'2010' (2010), 264 pp., ISBN: 978-2-35158-101-8; e-ISBN: 978-2-35158-115-5; *Eds. Caijun Shi and Xiaodong Shen*

PRO 73: 2nd International Conference on 'Waste Engineering and Management—ICWEM 2010' (2010), 894 pp., ISBN: 978-2-35158-102-5; e-ISBN: 978-2-35158-103-2, *Eds. J. Zh. Xiao, Y. Zhang, M. S. Cheung and R. Chu*

PRO 74: International RILEM Conference on 'Use of Superabsorbent Polymers and Other New Additives in Concrete' (2010) 374 pp., ISBN: 978-2-35158-104-9; e-ISBN: 978-2-35158-105-6; *Eds. O.M. Jensen, M.T. Hasholt, and S. Laustsen*

PRO 75: International Conference on 'Material Science—2nd ICTRC—Textile Reinforced Concrete—Theme 1' (2010) 436 pp., ISBN: 978-2-35158-106-3; e-ISBN: 978-2-35158-107-0; *Ed. W. Brameshuber*

PRO 76: International Conference on 'Material Science—HetMat—Modelling of Heterogeneous Materials—Theme 2' (2010) 255 pp., ISBN: 978-2-35158-108-7; e-ISBN: 978-2-35158-109-4; *Ed. W. Brameshuber*

PRO 77: International Conference on 'Material Science—AdIPoC—Additions Improving Properties of Concrete—Theme 3' (2010) 459 pp., ISBN: 978-2-35158-110-0; e-ISBN: 978-2-35158-111-7; *Ed. W. Brameshuber*

PRO 78: 2nd Historic Mortars Conference and RILEM TC 203-RHM Final Workshop—HMC2010 (2010) 1416 pp., e-ISBN: 978-2-35158-112-4; *Eds. J. Válek, C. Groot and J. J. Hughes*

PRO 79: International RILEM Conference on Advances in Construction Materials Through Science and Engineering (2011) 213 pp., ISBN: 978-2-35158-116-2, e-ISBN: 978-2-35158-117-9; *Eds. Christopher Leung and K.T. Wan*

PRO 80: 2nd International RILEM Conference on Concrete Spalling due to Fire Exposure (2011) 453 pp., ISBN: 978-2-35158-118-6; e-ISBN: 978-2-35158-119-3; *Eds. E.A.B. Koenders and F. Dehn*

PRO 81: 2nd International RILEM Conference on Strain Hardening Cementitious Composites (SHCC2-Rio) (2011) 451 pp., ISBN: 978-2-35158-120-9; e-ISBN: 978-2-35158-121-6; *Eds. R.D. Toledo Filho, F.A. Silva, E.A.B. Koenders and E.M.R. Fairbairn*

PRO 82: 2nd International RILEM Conference on Progress of Recycling in the Built Environment (2011) 507 pp., e-ISBN: 978-2-35158-122-3; *Eds. V.M. John, E. Vazquez, S.C. Angulo and C. Ulsen*

PRO 83: 2nd International Conference on Microstructural-related Durability of Cementitious Composites (2012) 250 pp., ISBN: 978-2-35158-129-2; e-ISBN: 978-2-35158-123-0; *Eds. G. Ye, K. van Breugel, W. Sun and C. Miao*

PRO 84: CONSEC13—Seventh International Conference on Concrete under Severe Conditions—Environment and Loading (2013) 1930 pp., ISBN: 978-2-35158-124-7; e-ISBN: 978-2- 35158-134-6; *Eds. Z.J. Li, W. Sun, C.W. Miao, K. Sakai, O.E. Gjorv and N. Banthia*

PRO 85: RILEM-JCI International Workshop on Crack Control of Mass Concrete and Related issues concerning Early-Age of Concrete Structures—ConCrack 3—Control of Cracking in Concrete Structures 3 (2012) 237 pp., ISBN: 978-2-35158-125-4; e-ISBN: 978-2-35158-126-1; *Eds. F. Toutlemonde and J.-M. Torrenti*

PRO 86: International Symposium on Life Cycle Assessment and Construction (2012) 414 pp., ISBN: 978-2-35158-127-8, e-ISBN: 978-2-35158-128-5; *Eds. A. Ventura and C. de la Roche*

PRO 87: UHPFRC 2013—RILEM-fib-AFGC International Symposium on Ultra-High Performance Fibre-Reinforced Concrete (2013), ISBN: 978-2-35158-130-8, e-ISBN: 978-2-35158-131-5; *Eds. F. Toutlemonde*

PRO 88: 8th RILEM International Symposium on Fibre Reinforced Concrete (2012) 344 pp., ISBN: 978-2-35158-132-2; e-ISBN: 978-2-35158-133-9; *Eds. Joaquim A.O. Barros*

PRO 89: RILEM International workshop on performance-based specification and control of concrete durability (2014) 678 pp., ISBN: 978-2-35158-135-3; e-ISBN: 978-2-35158-136-0; *Eds. D. Bjegović, H. Beushausen and M. Serdar*

PRO 90: 7th RILEM International Conference on Self-Compacting Concrete and of the 1st RILEM International Conference on Rheology and Processing of Construction Materials (2013) 396 pp., ISBN: 978-2-35158-137-7; e-ISBN: 978-2-35158-138-4; *Eds. Nicolas Roussel and Hela Bessaies-Bey*

PRO 91: CONMOD 2014—RILEM International Symposium on Concrete Modelling (2014), ISBN: 978-2-35158-139-1; e-ISBN: 978-2-35158-140-7; *Eds. Kefei Li, Peiyu Yan and Rongwei Yang*

PRO 92: CAM 2014—2nd International Conference on advances in chemically-activated materials (2014) 392 pp., ISBN: 978-2-35158-141-4; e-ISBN: 978-2-35158-142-1; *Eds. Caijun Shi and Xiadong Shen*

PRO 93: SCC 2014—3rd International Symposium on Design, Performance and Use of Self-Consolidating Concrete (2014) 438 pp., ISBN: 978-2-35158-143-8; e-ISBN: 978-2-35158-144-5; *Eds. Caijun Shi, Zhihua Ou and Kamal H. Khayat*

PRO 94 (online version): HPFRCC-7—7th RILEM conference on High performance fiber reinforced cement composites (2015), e-ISBN: 978-2-35158-146-9; *Eds. H.W. Reinhardt, G.J. Parra-Montesinos and H. Garrecht*

PRO 95: International RILEM Conference on Application of superabsorbent polymers and other new admixtures in concrete construction (2014), ISBN: 978-2-35158-147-6; e-ISBN: 978-2-35158-148-3; *Eds. Viktor Mechtcherine and Christof Schroefl*

PRO 96 (online version): XIII DBMC: XIII International Conference on Durability of Building Materials and Components (2015), e-ISBN: 978-2-35158-149-0; *Eds. M. Quattrone and V.M. John*

PRO 97: SHCC3—3rd International RILEM Conference on Strain Hardening Cementitious Composites (2014), ISBN: 978-2-35158-150-6; e-ISBN: 978-2-35158-151-3; *Eds. E. Schlangen, M.G. Sierra Beltran, M. Lukovic and G. Ye*

PRO 98: FERRO-11—11th International Symposium on Ferrocement and 3rd ICTRC—International Conference on Textile Reinforced Concrete (2015), ISBN: 978-2-35158-152-0; e-ISBN: 978-2-35158-153-7; *Ed. W. Brameshuber*

PRO 99 (online version): ICBBM 2015—1st International Conference on Bio-Based Building Materials (2015), e-ISBN: 978-2-35158-154-4; *Eds. S. Amziane and M. Sonebi*

PRO 100: SCC16—RILEM Self-Consolidating Concrete Conference (2016), ISBN: 978-2-35158-156-8; e-ISBN: 978-2-35158-157-5; *Ed. Kamal H. Kayat*

PRO 101 (online version): III Progress of Recycling in the Built Environment (2015), e-ISBN: 978-2-35158-158-2; *Eds I. Martins, C. Ulsen and S. C. Angulo*

PRO 102 (online version): RILEM Conference on Microorganisms-Cementitious Materials Interactions (2016), e-ISBN: 978-2-35158-160-5; *Eds. Alexandra Bertron, Henk Jonkers and Virginie Wiktor*

PRO 103 (online version): ACESC'16—Advances in Civil Engineering and Sustainable Construction (2016), e-ISBN: 978-2-35158-161-2; *Eds. T.Ch. Madhavi, G. Prabhakar, Santhosh Ram and P.M. Rameshwaran*

PRO 104 (online version): SSCS'2015—Numerical Modeling—Strategies for Sustainable Concrete Structures (2015), e-ISBN: 978-2-35158-162-9

PRO 105: 1st International Conference on UHPC Materials and Structures (2016), ISBN: 978-2-35158-164-3; e-ISBN: 978-2-35158-165-0

PRO 106: AFGC-ACI-fib-RILEM International Conference on Ultra-High-Performance Fibre-Reinforced Concrete—UHPFRC 2017 (2017), ISBN: 978-2-35158-166-7; e-ISBN: 978-2-35158-167-4; *Eds. François Toutlemonde and Jacques Resplendino*

PRO 107 (online version): XIV DBMC—14th International Conference on Durability of Building Materials and Components (2017), e-ISBN: 978-2-35158-159-9; *Eds. Geert De Schutter, Nele De Belie, Arnold Janssens and Nathan Van Den Bossche*

PRO 108: MSSCE 2016—Innovation of Teaching in Materials and Structures (2016), ISBN: 978-2-35158-178-0; e-ISBN: 978-2-35158-179-7; *Ed. Per Goltermann*

PRO 109 (2 volumes): MSSCE 2016—Service Life of Cement-Based Materials and Structures (2016), ISBN Vol. 1: 978-2-35158-170-4; Vol. 2: 978-2-35158-171-4; Set Vol. 1&2: 978-2-35158-172-8; e-ISBN : 978-2-35158-173-5; *Eds. Miguel Azenha, Ivan Gabrijel, Dirk Schlicke, Terje Kanstad and Ole Mejlhede Jensen*

PRO 110: MSSCE 2016—Historical Masonry (2016), ISBN: 978-2-35158-178-0; e-ISBN: 978-2-35158-179-7; *Eds. Inge Rörig-Dalgaard and Ioannis Ioannou*

PRO 111: MSSCE 2016—Electrochemistry in Civil Engineering (2016); ISBN: 978-2-35158-176-6; e-ISBN: 978-2-35158-177-3; *Ed. Lisbeth M. Ottosen*

PRO 112: MSSCE 2016—Moisture in Materials and Structures (2016), ISBN: 978-2-35158-178-0; e-ISBN: 978-2-35158-179-7; *Eds. Kurt Kielsgaard Hansen, Carsten Rode and Lars-Olof Nilsson*

PRO 113: MSSCE 2016—Concrete with Supplementary Cementitious Materials (2016), ISBN: 978-2-35158-178-0; e-ISBN: 978-2-35158-179-7; *Eds. Ole Mejlhede Jensen, Konstantin Kovler and Nele De Belie*

PRO 114: MSSCE 2016—Frost Action in Concrete (2016), ISBN: 978-2-35158-182-7; e-ISBN: 978-2-35158-183-4; *Eds. Marianne Tange Hasholt, Katja Fridh and R. Doug Hooton*

PRO 115: MSSCE 2016—Fresh Concrete (2016), ISBN: 978-2-35158-184-1; e-ISBN: 978-2-35158-185-8; *Eds. Lars N. Thrane, Claus Pade, Oldrich Svec and Nicolas Roussel*

PRO 116: BEFIB 2016—9th RILEM International Symposium on Fiber Reinforced Concrete (2016), ISBN: 978-2-35158-187-2; e-ISBN: 978-2-35158-186-5; *Eds. N. Banthia, M. di Prisco and S. Soleimani-Dashtaki*

PRO 117: 3rd International RILEM Conference on Microstructure Related Durability of Cementitious Composites (2016), ISBN: 978-2-35158-188-9; e-ISBN: 978-2-35158-189-6; *Eds. Changwen Miao, Wei Sun, Jiaping Liu, Huisu Chen, Guang Ye and Klaas van Breugel*

PRO 118 (4 volumes): International Conference on Advances in Construction Materials and Systems (2017), ISBN Set: 978-2-35158-190-2; Vol. 1: 978-2-35158-193-3; Vol. 2: 978-2-35158-194-0; Vol. 3: ISBN:978-2-35158-195-7; Vol. 4: ISBN:978-2-35158-196-4; e-ISBN: 978-2-35158-191-9; *Ed. Manu Santhanam*

PRO 119 (online version): ICBBM 2017—Second International RILEM Conference on Bio-based Building Materials, (2017), e-ISBN: 978-2-35158-192-6; *Ed. Sofiane Amziane*

PRO 120 (2 volumes): EAC-02—2nd International RILEM/COST Conference on Early Age Cracking and Serviceability in Cement-based Materials and Structures, (2017), Vol. 1: 978-2-35158-199-5, Vol. 2: 978-2-35158-200-8, Set: 978-2-35158-197-1, e-ISBN: 978-2-35158-198-8; *Eds. Stéphanie Staquet and Dimitrios Aggelis*

PRO 121 (2 volumes): SynerCrete18: Interdisciplinary Approaches for Cement-based Materials and Structural Concrete: Synergizing Expertise and Bridging Scales of Space and Time, (2018), Set: 978-2-35158-202-2, Vol.1: 978-2-35158-211-4, Vol.2: 978-2-35158-212-1, e-ISBN: 978-2-35158-203-9; *Eds. Miguel Azenha, Dirk Schlicke, Farid Benboudjema, Agnieszka Knoppik*

PRO 122: SCC'2018 China—Fourth International Symposium on Design, Performance and Use of Self-Consolidating Concrete, (2018), ISBN: 978-2-35158-204-6, e-ISBN: 978-2-35158-205-3; *Eds. C. Shi, Z. Zhang, K. H. Khayat*

PRO 123: Final Conference of RILEM TC 253-MCI: Microorganisms-Cementitious Materials Interactions (2018), Set: 978-2-35158-207-7, Vol.1: 978-2-35158-209-1, Vol.2: 978-2-35158-210-7, e-ISBN: 978-2-35158-206-0; *Ed. Alexandra Bertron*

PRO 124 (online version): Fourth International Conference Progress of Recycling in the Built Environment (2018), e-ISBN: 978-2-35158-208-4; *Eds. Isabel M. Martins, Carina Ulsen, Yury Villagran*

PRO 125 (online version): SLD4—4th International Conference on Service Life Design for Infrastructures (2018), e-ISBN: 978-2-35158-213-8; *Eds. Guang Ye, Yong Yuan, Claudia Romero Rodriguez, Hongzhi Zhang, Branko Savija*

PRO 126: Workshop on Concrete Modelling and Material Behaviour in honor of Professor Klaas van Breugel (2018), ISBN: 978-2-35158-214-5, e-ISBN: 978-2-35158-215-2; *Ed. Guang Ye*

PRO 127 (online version): CONMOD2018—Symposium on Concrete Modelling (2018), e-ISBN: 978-2-35158-216-9; *Eds. Erik Schlangen, Geert de Schutter, Branko Savija, Hongzhi Zhang, Claudia Romero Rodriguez*

PRO 128: SMSS2019—International Conference on Sustainable Materials, Systems and Structures (2019), ISBN: 978-2-35158-217-6, e-ISBN: 978-2-35158-218-3

PRO 129: 2nd International Conference on UHPC Materials and Structures (UHPC2018-China), ISBN: 978-2-35158-219-0, e-ISBN: 978-2-35158-220-6

PRO 130: 5th Historic Mortars Conference (2019), ISBN: 978-2-35158-221-3, e-ISBN: 978-2-35158-222-0; *Eds. José Ignacio Álvarez, José María Fernández, Íñigo Navarro, Adrián Durán, Rafael Sirera*

PRO 131 (online version): 3rd International Conference on Bio-Based Building Materials (ICBBM2019), e-ISBN: 978-2-35158-229-9; *Eds. Mohammed Sonebi, Sofiane Amziane, Jonathan Page*

PRO 132: IRWRMC'18—International RILEM Workshop on Rheological Measurements of Cement-based Materials (2018), ISBN: 978-2-35158-230-5, e-ISBN: 978-2-35158-231-2; *Eds. Chafika Djelal, Yannick Vanhove*

PRO 133 (online version): CO2STO2019—International Workshop CO2 Storage in Concrete (2019), e-ISBN: 978-2-35158-232-9; *Eds. Assia Djerbi, Othman Omikrine-Metalssi, Teddy Fen-Chong*

PRO 134: 3rd ACF/HNU International Conference on UHPC Materials and Structures - UHPC'2020, ISBN: 978-2-35158-233-6, e-ISBN: 978-2-35158-234-3; *Eds. Caijun Shi & Jiaping Liu*

RILEM Reports (REP)

Report 19: Considerations for Use in Managing the Aging of Nuclear Power Plant Concrete Structures (ISBN: 2-912143-07-1); *Ed. D. J. Naus*

Report 20: Engineering and Transport Properties of the Interfacial Transition Zone in Cementitious Composites (ISBN: 2-912143-08-X); *Eds. M. G. Alexander, G. Arliguie, G. Ballivy, A. Bentur and J. Marchand*

Report 21: Durability of Building Sealants (ISBN: 2-912143-12-8); *Ed. A. T. Wolf*

Report 22: Sustainable Raw Materials—Construction and Demolition Waste (ISBN: 2-912143-17-9); *Eds. C. F. Hendriks and H. S. Pietersen*

Report 23: Self-Compacting Concrete state-of-the-art report (ISBN: 2-912143-23-3); *Eds. Å. Skarendahl and Ö. Petersson*

Report 24: Workability and Rheology of Fresh Concrete: Compendium of Tests (ISBN: 2-912143-32-2); *Eds. P. J. M. Bartos, M. Sonebi and A. K. Tamimi*

Report 25: Early Age Cracking in Cementitious Systems (ISBN: 2-912143-33-0); *Ed. A. Bentur*

Report 26: Towards Sustainable Roofing (Joint Committee CIB/RILEM) (CD 07) (e-ISBN 978-2-912143-65-5); *Eds. Thomas W. Hutchinson and Keith Roberts*

Report 27: Condition Assessment of Roofs (Joint Committee CIB/RILEM) (CD 08) (e-ISBN 978-2-912143-66-2); *Ed. CIB W 83/RILEM TC166-RMS*

Report 28: Final report of RILEM TC 167-COM 'Characterisation of Old Mortars with Respect to Their Repair (ISBN: 978-2-912143-56-3); *Eds. C. Groot, G. Ashall and J. Hughes*

Report 29: Pavement Performance Prediction and Evaluation (PPPE): Interlaboratory Tests (e-ISBN: 2-912143-68-3); *Eds. M. Partl and H. Piber*

Report 30: Final Report of RILEM TC 198-URM 'Use of Recycled Materials' (ISBN: 2-912143-82-9; e-ISBN: 2-912143-69-1); *Eds. Ch. F. Hendriks, G. M. T. Janssen and E. Vázquez*

Report 31: Final Report of RILEM TC 185-ATC 'Advanced testing of cement-based materials during setting and hardening' (ISBN: 2-912143-81-0; e-ISBN: 2-912143-70-5); *Eds. H. W. Reinhardt and C. U. Grosse*

Report 32: Probabilistic Assessment of Existing Structures. A JCSS publication (ISBN 2-912143-24-1); *Ed. D. Diamantidis*

Report 33: State-of-the-Art Report of RILEM Technical Committee TC 184-IFE 'Industrial Floors' (ISBN 2-35158-006-0); *Ed. P. Seidler*

Report 34: Report of RILEM Technical Committee TC 147-FMB 'Fracture mechanics applications to anchorage and bond' Tension of Reinforced Concrete Prisms—Round Robin Analysis and Tests on Bond (e-ISBN 2-912143-91-8); *Eds. L. Elfgren and K. Noghabai*

Report 35: Final Report of RILEM Technical Committee TC 188-CSC 'Casting of Self Compacting Concrete' (ISBN 2-35158-001-X; e-ISBN: 2-912143-98-5); *Eds. Å. Skarendahl and P. Billberg*

Report 36: State-of-the-Art Report of RILEM Technical Committee TC 201-TRC 'Textile Reinforced Concrete' (ISBN 2-912143-99-3); *Ed. W. Brameshuber*

Report 37: State-of-the-Art Report of RILEM Technical Committee TC 192-ECM 'Environment-conscious construction materials and systems' (ISBN: 978-2-35158-053-0); *Eds. N. Kashino, D. Van Gemert and K. Imamoto*

Report 38: State-of-the-Art Report of RILEM Technical Committee TC 205-DSC 'Durability of Self-Compacting Concrete' (ISBN: 978-2-35158-048-6); *Eds. G. De Schutter and K. Audenaert*

Report 39: Final Report of RILEM Technical Committee TC 187-SOC 'Experimental determination of the stress-crack opening curve for concrete in tension' (ISBN 978-2-35158-049-3); *Ed. J. Planas*

Report 40: State-of-the-Art Report of RILEM Technical Committee TC 189-NEC 'Non-Destructive Evaluation of the Penetrability and Thickness of the Concrete Cover' (ISBN 978-2-35158-054-7); *Eds. R. Torrent and L. Fernández Luco*

Report 41: State-of-the-Art Report of RILEM Technical Committee TC 196-ICC 'Internal Curing of Concrete' (ISBN 978-2-35158-009-7); *Eds. K. Kovler and O. M. Jensen*

Report 42: 'Acoustic Emission and Related Non-destructive Evaluation Techniques for Crack Detection and Damage Evaluation in Concrete'—Final Report of RILEM Technical Committee 212-ACD (e-ISBN: 978-2-35158-100-1); *Ed. M. Ohtsu*

Report 45: Repair Mortars for Historic Masonry—State-of-the-Art Report of RILEM Technical Committee TC 203-RHM (e-ISBN: 978-2-35158-163-6); *Eds. Paul Maurenbrecher and Caspar Groot*

Report 46: Surface delamination of concrete industrial ffioors and other durability related aspects guide—Report of RILEM Technical Committee TC 268-SIF (e-ISBN: 978-2-35158-201-5); *Ed. Valerie Pollet*

Contents

1 **General Introduction** ... 1
Erwan Hamard, Antonin Fabbri, and Jean-Claude Morel

2 **Characterization of Earth Used in Earth Construction Materials** ... 17
Jean-Emmanuel Aubert, Paulina Faria, Pascal Maillard, Kouka Amed Jérémy Ouedraogo, Claudiane Ouellet-Plamondon, and Elodie Prud'homme

3 **Hygrothermal and Acoustic Assessment of Earthen Materials** ... 83
Antonin Fabbri, Jean-Emmanuel Aubert, Ana Armanda Bras, Paulina Faria, Domenico Gallipoli, Jeanne Goffart, Fionn McGregor, Céline Perlot-Bascoules, and Lucile Soudani

4 **Mechanical Behaviour of Earth Building Materials** ... 127
H. N. Abhilash, Erwan Hamard, C. T. S. Beckett, Jean-Claude Morel, Humberto Varum, Dora Silveira, I. Ioannou, and R. Illampas

5 **Seismic Assessment of Earthen Structures** ... 181
Quoc-Bao Bui, Ranime El-Nabouch, Lorenzo Miccoli, Jean-Claude Morel, Daniel V. Oliveira, Rui A. Silva, Dora Silveira, Humberto Varum, and Florent Vieux-Champagne

6 **Durability of Earth Materials: Weathering Agents, Testing Procedures and Stabilisation Methods** ... 211
Domenico Gallipoli, Agostino W. Bruno, Quoc-Bao Bui, Antonin Fabbri, Paulina Faria, Daniel V. Oliveira, Claudiane Ouellet-Plamondon, and Rui A. Silva

7 **Codes and Standards on Earth Construction** ... 243
B. V. Venkatarama Reddy, Jean-Claude Morel, Paulina Faria, P. Fontana, Daniel V. Oliveira, I. Serclerat, P. Walker, and Pascal Maillard

8 Environmental Potential of Earth-Based Building Materials: Key Facts and Issues from a Life Cycle Assessment Perspective 261
Anne Ventura, Claudiane Ouellet-Plamondon, Martin Röck, Torben Hecht, Vincent Roy, Paula Higuera, Thibaut Lecompte, Paulina Faria, Erwan Hamard, Jean-Claude Morel, and Guillaume Habert

Contributors

H. N. Abhilash Coventry University, Coventry, UK

Jean-Emmanuel Aubert UPS, INSA, LMDC (Laboratoire Matériaux et Durabilité des Constructions), Université de Toulouse, Toulouse, France

C. T. S. Beckett The University of Edinburgh, Edinburgh, UK

Ana Armanda Bras Liverpool John Moores University, Liverpool, UK

Agostino W. Bruno School of Engineering, University of Newcastle, Newcastle, UK

Quoc-Bao Bui Faculty of Civil Engineering, Ton Duc Thang University, Ho Chi Minh City, Vietnam

Ranime El-Nabouch LOCIE, CNRS, University Savoie Mont Blanc, Chambéry, France

Antonin Fabbri LTDS-ENTPE, CNRS, University of Lyon, Vaulx-en-Velin, France

Paulina Faria CERIS and NOVA School of Science and Technology, NOVA University of Lisbon, Caparica, Portugal

P. Fontana BAM, Stuttgart, Germany

Domenico Gallipoli Laboratoire SIAME, Fédération IPRA, Université de Pau Et Des Pays de L'Adour, Anglet, France;
Dipartimento di Ingegneria Civile, Chimica e Ambientale, Università degli Studi di Genova, Genoa, Italy

Jeanne Goffart Université de Savoie Mont Blanc, LOCIE, Chambéry, France

Guillaume Habert Chair of Sustainable Construction, ETH Zurich, Zürich, Switzerland

Erwan Hamard University Gustave Eiffel MAST/GPEM, Bouguenais, France

Torben Hecht Graz University of Technology, Working Group Sustainable Construction, Graz, Austria

Paula Higuera University Gustave Eiffel MAST/GPEM, Bouguenais, France

R. Illampas University of Cyprus, Nicosia, Cyprus

I. Ioannou University of Cyprus, Nicosia, Cyprus

Thibaut Lecompte University Bretagne Sud, Lorient, France

Pascal Maillard CTMNC, Research and Development Department of Ceramic, Ester Technopole, Limoges, France

Fionn McGregor LTDS-ENTPE, CNRS, University of Lyon, Vaulx-en-Velin, France

Lorenzo Miccoli Bundesanstalt für Materialforschung und –prüfung (BAM), Division Building Materials, Berlin, Germany

Jean-Claude Morel Coventry University, Coventry, UK;
LTDS-ENTPE, CNRS, University of Lyon, Vaulx-en-Velin, France;
Faculty of Engineering, Environment and Computing, Coventry University, Coventry, UK

Daniel V. Oliveira ISISE & IB-S, University of Minho, Guimarães, Portugal

Kouka Amed Jérémy Ouedraogo UPS, INSA, LMDC, Université de Toulouse, Toulouse, France

Claudiane Ouellet-Plamondon Département de Génie de La Construction, École de Technologie Supérieure, Université du Québec, Montreal, QC, Canada

Céline Perlot-Bascoules Laboratoire SIAME, Fédération IPRA, Université de Pau Et Des Pays de L'Adour, Anglet, France

Elodie Prud'homme Université de Lyon, MATEIS, Matériaux: Ingénierie et Science, Université de Lyon, Lyon, France

B. V. Venkatarama Reddy Indian Institute of Science, Bangalore, India

Vincent Roy Département de Génie de La Construction, École de technologie supérieure, Montréal, Qc, Canada

Martin Röck Faculty of Engineering Science, Architectural Engineering Unit, KU Leuven, Leuven, Belgium;
Graz University of Technology, Working Group Sustainable Construction, Graz, Austria

I. Serclerat Lafarge Centre de Recherche, Saint-Quentin-Fallavier, France

Rui A. Silva ISISE & IB-S, University of Minho, Guimarães, Portugal

Dora Silveira ADAI-LAETA, Itecons, Coimbra, Portugal

Lucile Soudani LTDS-ENTPE, CNRS, University of Lyon, Vaulx-en-Velin, France;
Université de Savoie Mont Blanc, LOCIE, Chambéry, France

Humberto Varum CONSTRUCT-LESE, Faculty of Engineering, University of Porto, Porto, Portugal

Anne Ventura University Gustave Eiffel MAST/GPEM, Bouguenais, France

Florent Vieux-Champagne Université Grenoble Alpes, CNRS, Grenoble, France

P. Walker Universtiy of Bath, Bath, UK

Chapter 1
General Introduction

Erwan Hamard, Antonin Fabbri, and Jean-Claude Morel

Abstract Earth is a building material excavated from the subsoil which is employed by Mankind since the Neolithic time all over the world. Various building techniques are used with earth, to build monolithic walls, to produce bricks, as infill of walls or as plasters or mortars. Earth architectures undergo a rebirth since the 2000s because they are a way to save natural resources and energy and to provide a good indoor comfort and a good social impact. Nonetheless, earth building sector faces many challenges to be considered as a contemporary material. This book, which is produced in the framework of the TC 274, will focus on the estimation of the parameters which are necessary to properly design earthen constructions. After a general introduction on earthen materials and constructions, the state of the art on the material characterisation techniques, the assessment of hygrothermal performance, the mechanical and seismic behaviors and the durability will be presented, each in a dedicated chapter. A critical review of the standards which are used for earthen material will be presented in the last chapter.

Keywords Earth material · Resource · History · Building techniques · Environmental impact · Social impact · Modern construction

1.1 The Origin of Earth

Most of earth materials are excavated from subsoil horizons and some of them are excavated from alterites or soft rock deposits. Topsoil is sensitive to shrinkage and decay and is therefore unsuitable for building [17, 28]. These materials are

E. Hamard (✉)
University Gustave Eiffel MAST/GPEM, Bouguenais, France
e-mail: erwan.hamard@univ-eiffel.fr

A. Fabbri
LTDS-ENTPE, CNRS, University of Lyon, UMR 5513, Vaulx-en-Velin, France

J.-C. Morel
Coventry University, Coventry, UK

A. Fabbri et al. (eds.), *Testing and Characterisation of Earth-based Building Materials and Elements*, RILEM State-of-the-Art Reports 35,
https://doi.org/10.1007/978-3-030-83297-1_1

produced by geological, weathering and pedological processes. These processes are summarized hereafter.

When exposed to the surface, the physicochemical equilibrium of rocks is in imbalance with their new environment, and rocks are exposed to weathering. The effect of weathering tends to decrease from land surface downward to the unweathered bedrock, creating a weathering profile. Weathering generates loose material that are easily mobilized by ablation processes, i.e. colluvial, alluvial, eolian and glacial processes, to form superficial deposits. Superficial deposits differ from soil since they are subjected to sedimentary processes (erosion, transport and deposition).

When superficial deposits remain stable over time, they are affected by weathering and pedogenic processes. These processes, over time, tend toward the complete transformation of a parent material into more stable components and structures. This transformation depends on (1) mineralogical composition and structure of parent material; (2) climate, governing chemical weathering conditions; (3) soil living organisms, affecting chemical weathering to their profit; (4) relief, controlling horizontal transfers in soils; and (5) time, since an old soil will be more mature than a young one. Soil is an accumulation of parent material weathering products and biota degradation products. Fractions of these products are colloids (clays and humus) and are responsible for swelling and shrinking of soil by a change in moisture content. The repeated volumetric changes induced by seasonal moisture variations create vertical soil structural units, known as peds. Peds facilitate the downward movement of water and colloids inside soil and the differentiation of soil in pedological horizons. Soil formation is a 3-dimensional balance among gains, losses, internal redistribution, and chemical and physical changes.

The study of these materials concerns geology, geomorphology, pedology, agronomy and geotechnics. The frontiers between these disciplines are not clear and the terms and definitions used are different. A vertical cross-section of weathered materials from ground surface to unweathered bedrock, according to several disciplines' definitions, is proposed in Fig. 1.1. The possible material sources for earth building is highlighted on this figure. Before the invention of excavators, excavation was carried out by hand. This is why, in the past, excavations concerned the first(s) soil horizons of the subsoil, just beneath the topsoil.

1.2 Historical Overview

Since, at least, the very beginning of the Neolithic revolution, human beings have employed earth associated with timber, fibres and stones to build their shelters and dwellings [37]. Earth as a building material is attested in most of Neolithic centres of origin, like, for example, the fertile crescent [1, 11, 37], Mesoamerica [10] and China [47]. It was first used as mortars, as plasters and as infills of timber frame structures (wattle and daub) (Fig. 1.2) [1, 37]. During the early Neolithic, at least in the Near East, two new techniques are attested on archaeological sites. The first one consists in piling wet clods of earth to build a wall (cob). When cob technique emerged, it

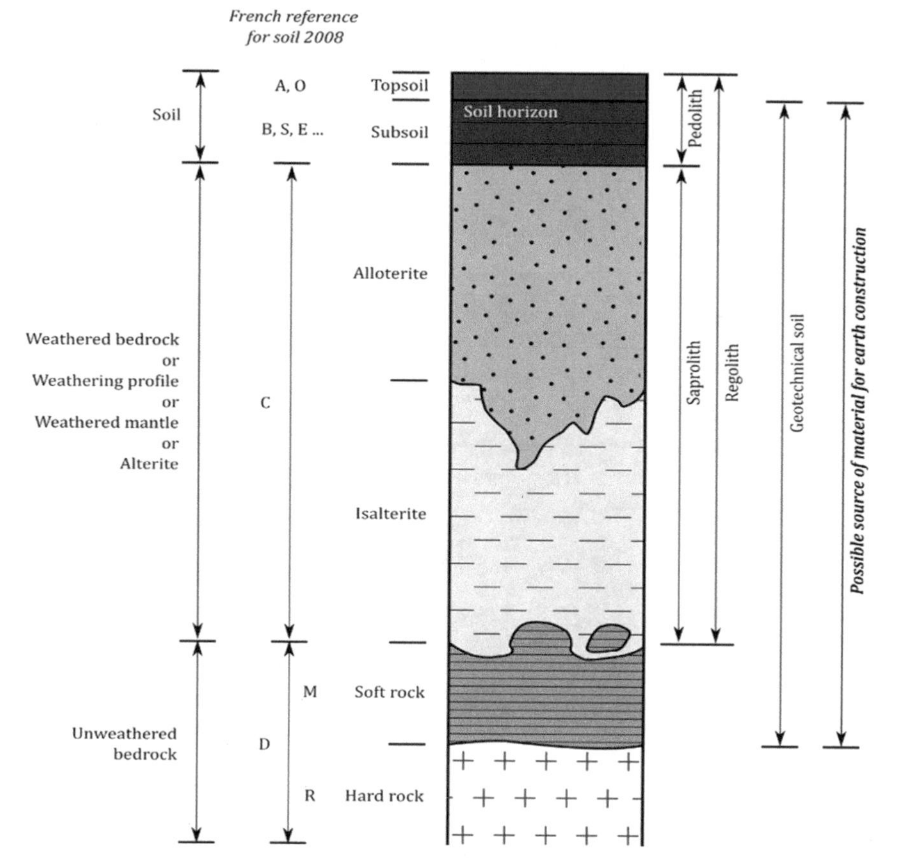

Fig. 1.1 Vertical cross-section of weathered materials from ground surface to unweathered bedrock, according to several disciplines definitions [19]

tended to replace wattle and daub [37]. The second building technique appears after cob and consist in laying sun-dried bricks shaped by hand (adobe). In western Near East, when adobe technique emerged, it largely replaced cob [1, 5, 37] (Fig. 1.2).

During the Iron Age, about six millennia after the invention of adobe, a new earth building technique, attributed to the Phoenicians, is attested in the West Mediterranean area [5, 22]. It consists in compacting earth, layer by layer, inside a formwork (rammed earth). Some attempts of compaction of earth in smaller moulds to make bricks (Compressed Earth Blocks, CEB) are recorded in the nineteenth century. But CEB were more commonly used after the development of powerful and functional presses, like the CINVA-RAM designed by Raul Ramirez at the CINVA center in Bogota in 1952 [33]. The last commonly used earth building technique arose in Germany after 1920 and was designed to improve the thermal insulation of walls by significantly increase the fiber content of wattle and daub mixtures, in order to reach

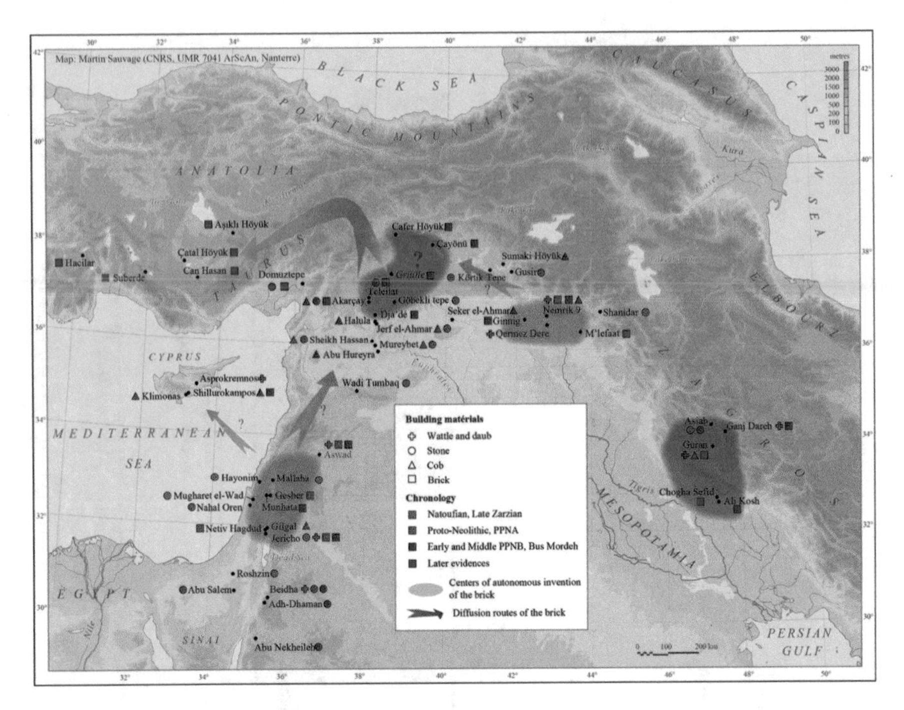

Fig. 1.2 Archaeological evidences of earth building techniques in the ancient Near East, adapted from [38], reproduced with the kind permission of the author (© Martin Sauvage, CNRS, UMR 7041, Nanterre)

densities lower than 1200 kg.m^{-3} [43]. It consists in binding bio-based aggregates or fibers with an earth slip (light earth).

These techniques have spread from their territories of invention and have been adopted by other groups of people. For example, adobe technique spread from the Near East across the Mediterranean sea from the Neolithic to the Iron Age period [6]. The adoption of building technique on a new territory requires favorable natural conditions, i.e. suitable soil, adapted water supply and climate conditions. Nonetheless, the succession of several techniques on a same territory, like for example in the Languedoc region (south of France), where cob was replaced by adobe in the Iron Age and adobe by rammed earth in the Middle Age [4], highlights that natural conditions are not enough to explain the propagation of a technique in new territories. The cultural and social acceptability is also a major factor. This acceptability is difficult to set, but the emergence of a new technique is favored if it meets a need, raises the social esteem, if it is affordable and not imposed by an authority [4, 5].

When adopted, people become more and more familiar with the technique and they adapt it to their natural environment and with their needs. This adaptation is illustrated, for example, by the cob walls in England that shift from self-standing walls in the fourteenth century to load-bearing walls in the seventeenth century [23]. It leads to the creation a wide variety of local construction cultures [9]. These

Fig. 1.3 Ksar of Aït-Ben-Haddou (Morocco) inscribed in the UNESCO World Heritage List (© Marc Marlier)

construction culture are not immutable, they appear, evolve, expand and disappear, depending on resource availability and social changes. Construction cultures are the result of vibrant processes and earth building, since its 10,000 years of existence, has experienced many golden ages, disuses and renaissances. Earth built heritage reflects the outcome of this long evolution. This heritage highlights the appropriation of earth as a building material by cultures all over the world and in all periods since the Neolithic time [7, 42] (Fig. 1.3). For example, 20% of cultural sites of the UNESCO World Heritage List are fully or partially made of earth and it is estimated that about one third of the word population lies in earthen houses [21].

In the middle of the twentieth century, a slow awakening of the impact of human activities on nature began. Since then, the consequences of the consumption of natural resources, fossil energies, greenhouse gas emissions and the artificialization of natural spaces, due to the lifestyles of Western societies, have revealed their unsustainability. Today, there is a strong societal desire to promote solutions in harmony with nature. It is in this context that earth building, like other natural and low-process building materials, is experiencing a new renaissance [24].

1.3 Classification and Definition of Earth Building Processes

A classification of earth building processes is proposed in Fig. 1.4. A first distinction is made to classify earth-building processes, regarding the hydric state of the mixture during their fabrication: plastic, solid and liquid. For plastic-state processes, earth mixture is employed in a plastic state and mechanical strength of the material is provided through drying shrinkage densification. For solid-state processes, earth mixture is employed at an optimum water content and mechanical strength is provided through compaction densification. For liquid-state process, an earth slip is used to bind plant particles. A second distinction is made regarding the type of implementation and the hydric state of the material during the implementation. It can be implemented right after the mixture fabrication, i.e. wet or moist, or after drying, i.e. using prefabricated elements. A final distinction is made regarding the

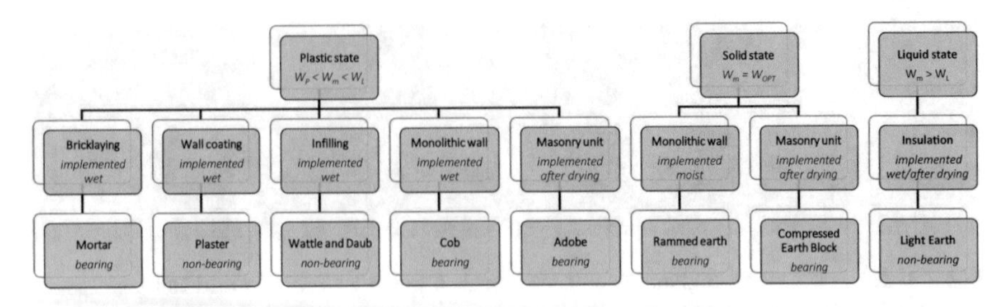

Fig. 1.4 Earth construction processes classification, adapted after [13, 19, 23, 25]. (W_m = water content of manufacturing stage, W_{OPT} = optimum water content; W_P = water content at plastic limit; W_L = water content at liquid limit)

structural role of the earth element that can be a load-bearing/freestanding wall or a non-load-bearing element.

Using this classification, it is possible to define earth construction processes as follow (cf. Fig. 1.5):

- ***Mortar***: earth mixture carried out at plastic state, implemented wet, in order to lay bricks or stones.
- ***Plaster and render***: earth mixture carried out at plastic state, implemented wet, to coat indoor or outdoor surfaces, respectively.
- ***Wattle and Daub***: earth elements mixed with fibres in a plastic state, implemented wet, in order to fill a timber frame load-bearing structure.
- ***Cob***: earth elements mixed in a plastic state, stacked wet, in order to build a monolithic and load-bearing or freestanding wall.
- ***Adobe***: masonry unit moulded at plastic state, dried and laid in order to build a load-bearing or freestanding wall.
- ***Rammed Earth***: earth compacted at optimum water content layers by layers inside a formwork in order to build a monolithic and load-bearing or freestanding wall.
- ***Compressed Earth Block*** (CEB): masonry unit compacted at an optimum water content, dried and laid in order to build a load-bearing or freestanding wall.
- ***Light earth***: earth slip at liquid state mixed with a large volume of plant particles (dry density less than 1200 $kg.m^{-3}$) in order to provide an insulation material, carried out on-site or prefabricated.

These simple definitions permit to distinguish one technique from one other, but they do not reflect the large diversity of these processes. For example, it can be estimated that hundreds of variations exist for cob process [18]. Every technique should be more considered as a family of processes with large variations.

Among these variations, some can be regarded as "missing link" between these techniques. Adobe and the cob process variations that consist of stacking cut or modelled plastic elements are quite similar, but adobes are implemented dry and require to be grouted with a mortar, whereas cob elements are implemented damp without mortar. Shuttered cob has great similarities with rammed earth, but, for rammed earth, shuttering is employed to make the ramming process efficient,

Fig. 1.5 Earth building techniques (© Univ. Eiffel, ENTPE, NOVA Univ. Lisbon)

whereas, for shuttered cob, shuttering is employed to avoid the trimming of the faces of the wall and therefore accelerate the wall faces rectification stage [18]. This proximity is the result of a shared history, every technique arising from a pre-existing one. There is thus a continuity between all earth construction techniques and they should be considered as a whole, and not as separated processes.

1.4 Why Building with Earth?

1.4.1 Saving Natural Resources

In the Western countries, the construction sector consumes a large volume of natural resources and is responsible for about 50% of wastes production in the European Union [26]. These wastes have a negative environmental impact and it is increasingly difficult to find suitable landfill areas. Among these wastes, about 75% are soils and stones [2]. These materials could be reused for earth building. For example, in Brittany (France) it was estimated that 23% of landfilled earth, i.e. 0.6 million tons every year, are suitable for earth building. The reuse of this high-quality building material would enable the construction of 52% of individual housing of Brittany [20]. This resource is widely available and is produced during earthworks, usually located near construction sites, limiting transportation. The reuse of those wastes for building might save natural resources required for conventional building material production and avoid landfilling.

Nonetheless, soil is a non-renewable material on the human time scale and it provides various ecosystem services concerning provisioning, regulating, cultural and supporting services [46]. Extraction of earth for construction might affect multi-functional roles of soil. Management of the consumption of this resource should therefore be carefully considered. If unstabilized, reversible clay binding allows a complete and low-energy reuse of earth at end of life. The unstabilized earth construction allows an almost infinite reuse of the material for construction or its return to agricultural land.

Considering that materials for earth construction are wastes of the construction sector and that unstabilized earth material is endlessly reusable, earth can be regarded as one of the load-bearing construction material that best meet the challenges of circular economy.

1.4.2 Energy

Embodied energy together with operational energy of the building sector represent approximately 40% of global energy use [27, 36]. As a consequence, the building sector is a major producer of greenhouse gases that contribute to climate change.

Until the 2000s, only operational energy was considered because of its dominant share in the total life cycle. Since then, the use of more efficient equipment and insulations modified the balance between embodied energy and operational energy so that the proportion of embodied energy increased [27, 32, 35]. In order to pursue energy saving effort, the next challenge for the building sector will be the reduction of embodied energy for new buildings and the development of low-impact insulation solutions to retrofit existing buildings. This involves good maintenance of existing buildings and the use of construction materials with low embodied energy [15, 16].

Historical builders mainly had animal energy and unprocessed local materials for construction purpose. As a consequence, embodied energy of earth built heritage is almost zero. Nowadays, in Western countries, excavations are carried out using mechanical diggers. On-site, earth dug from building foundations or from landscaping is used for construction. When not enough earth is available on-site or if on-site earth is unsuitable for construction, the material can be supplied from earthwork sites near the building site. Afterwards, implementation is conducted using manual and/or mechanized means. The recourse to mechanized means and possible transportation of earth increases the embodied energy of buildings. However, the embodied energy of modern unstabilized earth construction remains very low in comparison to other materials conventionally used in construction. For example, embodied energy of a wall made of earth is about 20 times less than this of a hollow cinder block wall [14, 23, 34]. As a consequence, earth construction is considered as a low greenhouse gas emitter.

This is not the case of stabilized earth construction. Indeed, even if stabilization could increase the durability of buildings, the stabilisation of thick earthen walls, even at low percentage, consume large amount of energy and prevent the reuse of the material at end of life [8, 31].

1.4.3 Indoor Comfort

Thanks to their high thermal mass, and their high hygroscopicity enabling water phase changes [39, 40], earthen walls buffer outdoor temperature variations. They are able to accumulate solar energy during the day and restore this energy during the night. These features provide to inhabitants of earthen buildings a good thermal comfort and more specifically during summer period.

Thanks to their high hygroscopicity, earthen materials are able either to adsorb rapidly or release a significant amount of water vapour in building indoor air. Indoor air quality is closely linked to relative humidity levels and therefore moisture buffering of earthen materials is beneficial for health and well-being of the occupants [3, 29].

Some authors mention several other beneficial properties of earthen buildings such as: good acoustic properties, fireproof properties, non-toxic and non-allergic properties and even a capacity to adsorb pollutant from the indoor air. These features have, however, yet to be clearly demonstrated but are explored further in subsequent chapters.

1.4.4 Social Impact

Earth meets human construction needs for more than 10,000 years and left us an important and rich architectural heritage worldwide. This heritage has a high historical and cultural value and should be properly maintained. This implies the preservation of a vibrant vernacular know-how. The vernacular construction strategies developed by past builders can be regarded as an optimum use of locally available resources under local natural conditions. With the loss of the vernacular know-how, the built heritage is the last witness of these strategies. Beyond its historical and cultural value, heritage should be also considered as a source of inspiration for modern sustainable building.

On earth building construction sites, implementation is carried out by skilled craftsmen whose expertise is recognized by other actors of the construction [12]. This increases their responsibility and thus contributes to the limitation of building defect risks. This also increases the esteem of mason's corporation and makes this profession more attractive to new mason generations. Since vernacular construction techniques depend on local conditions, the required skills to build with earth vary from a region to another. Earth construction thus creates jobs that cannot be relocated.

In emerging countries, there is a strong demand for affordable houses in high urbanization rate areas as well as in remote areas. Conventional construction requires the importation and the transportation of materials whereas the use of local materials, like earth, significantly reduces construction costs [14, 41]. In high-income countries earth material can be considered as free. The cost of earth construction is almost entirely due to salaries and social taxes. As a consequence, earth construction sector profits the local and social economy and has, therefore, a positive social impact.

1.5 The Challenge for Modern Earth Building

Earth construction will play an important role in the modern sustainable building of the twenty-first century if the actors of the sector adopt earth construction processes able to meet social demand, with low environmental impact and at an affordable cost. The study of earth heritage demonstrated the ability of historical earth builders to innovate in order to comply with social demand variations and technical developments. Earth construction benefits of an old and rich past and it would be a nonsense to leave this past behind. The analysis of earth heritage and the rediscovering of vernacular construction techniques is a valuable source of inspiration for modern earth construction. The valorisation of vernacular knowledge will save time, energy

Fig. 1.6 Examples of modern earthen constructions

and avoid repeating past mistakes. The future of earth construction should be a continuation of past vernacular earth construction.

Nonetheless, past vernacular processes are slow, time-consuming and require a large workforce, which is inappropriate in current Western modern economies [24, 44, 45]. In order to comply with this economic constraint, two options can be identified for earth: the recourse to self-build houses or the recourse to mechanisation and/or prefabrication. Self-builders have little site equipment and usually use the vernacular, low-impact, process. This solution may, however, satisfy only a small part of housing needs. The other solution is to go on with the development of mechanized/prefabricated earth process. Since the mechanical strength of earth is quite limited, for load-bearing walls, walls are quite thick and prefabricated wall elements heavy. Their transportation has therefore a high environmental and economic cost. To reduce the economic and environmental costs, the on-site prefabrication seems the more adapted. However, in specific context, like dense urban areas, external prefabrication processes, especially for non-bearing elements of smaller size, can be considered.

The earth material source is another issue since earth is a natural material and varies from a site to another. To overcome these variations, two different approaches are observed: (1) adapt the material to the process, thanks to a granular correction, forcing its particle size distribution into a grading envelope predetermined in the laboratory and/or addition of hydraulic binder, this solutions reduce the environmental benefits of earth; (2) adapt the process to the material [30], this solution optimizes the consumption of natural resources and relies on the expertise of skilled craftsmen, architects and on performance based tests. It, therefore, requires the education of specialist of earth construction (Fig. 1.6).

In addition, the development of this ancestral building technique notably suffers from a lack of appropriate standards. In consequence, they are disadvantaged compared to conventional construction techniques. The lack of knowledge of the material behavior can lead to apply common procedures and solutions, which are suitable for other building materials but which may be not adapted or even harmful when they are applied to earth buildings. There is a strong need for highlighting the particularities of earthen material and providing tools and methods to properly assess their performance.

If appropriately employed, earth construction offers many advantages in terms of resource management, environmental impact, indoor comfort and social impact. Earth has the potential to be one of the most sustainable building materials. However, inappropriate architectural design, long distance transportation of material, steel bar reinforcement, high impact admixture addition or significant grading correction can deeply alter its sustainability. These alterations usually come from economic and regulation constraints of the building sector imposing to speed up the construction process and to strengthen the material. A balance has to be found between a zero-emission vernacular material and a fast implemented and strengthens the material. The future of earth construction will be the result of an optimization of the economic and environmental sustainability of construction processes. The use of earth for construction should be justified by its beneficial effects. This is why earth construction goes hand in hand with sustainability assessment. To this aim, ecodesign and Life Cycle Assessment methods should be considered.

1.6 Conclusion

The bibliographic study of this chapter underlines that a good understanding of the earthen constructions requires taking into account their large variability. A first reason of this variability is that the local soils are used as building materials. The local soils are variable depending on the geology and local conditions of the site. Each construction can potentially be built with a different material and cannot be totally included in an industrial process. Then partly to adapt the building technique to the different soils, several construction techniques have been invented, which is the second reason of the variability. This book, which is produced in the framework of the TC 274, will focus on the estimation of the parameters which are necessary to properly design earthen constructions. After a general introduction on earthen materials and constructions, the state of the art on the material characterisation techniques, the assessment of hygrothermal performance, the mechanical and seismic behaviors and the durability will be presented, each in a dedicated chapter. A critical review of the standards which are used for earthen material will be presented in the last chapter.

References

1. Bicakci E (2003) Observations on the Early Pre-Pottery Neolithic architecture in the Near East: 1. New building materials and construction techniques. In: Ozdogan M, Hauptmann H, Basegelen N (eds) From villages to towns. Arkeoloji ve Sanat Yayinlari, Istanbul, pp 385–414
2. Cabello Eras JJ et al (2013) Improving the environmental performance of an earthwork project using cleaner production strategies. J Clean Prod 47:368–376. https://doi.org/10.1016/j.jclepro.2012.11.026

3. Cagnon H et al (2014) Hygrothermal properties of earth bricks. Energy Build 80:208–217. https://doi.org/10.1016/j.enbuild.2014.05.024
4. De Chazelles C-A (2009) Stabilité, disparition et fluctuation des traditions constructives en terre dans les pays méditerranéens, pp 139–150. Available at: http://halshs.archives-ouvertes.fr/halshs-00548051/en/
5. De Chazelles C-A (2010) Terre modelée et terre moulée, deux conceptions de la construction en terre. In: Arvais R et al (eds) Edifice et artifice, Histoire constructive. Picard. Paris, France, pp 411–419. Available at: http://halshs.archives-ouvertes.fr/halshs-00548027/en/
6. De Chazelles C-A (2011) La construction en brique crue moulée dans les pays de la Méditerranée, du Néolithique à l'époque romaine. Réflexions sur la question du moulage de la terre. In: de Chazelles C, Klein A, Pousthoumis N (eds) Les cultures constructives DE LA BRIQUE CRUE. 3 èmes échanges transdisciplinaires sur les constructions en terre crue. Table—ronde de Toulouse. Editions de l'Esperou, Montpellier, France, pp 153–164
7. Correia M (2010) A synopsis review of earthen archaeological. In: 6° Seminario Arquitectura de terra em portugal : 9° Seminário Ibero-americano de arquitectura e construçao com Terra. Coimbra, pp 29–33
8. Van Damme H, Houben H (2017) Earth concrete. Stabilization revisited. Cement Concr Res. https://doi.org/10.1016/j.cemconres.2017.02.035
9. Ferrigni F et al (2005) Ancient buildings and earthquakes, the local seismic culture approach: principles, methods, potentialities. Edipuglia. Centro Universitario Europeo per i beni Culturali, Bari, Italy
10. Flannery KV (2002) The origins of the village revisited: from nuclear to extended households. Am Antiq 67(3):417–433. https://doi.org/10.2307/1593820
11. Flohr P et al (2015) Building WF16: construction of a Pre-Pottery Neolithic A (PPNA) pisé structure in Southern Jordan. Levant 47(2):143–163. https://doi.org/10.1179/0075891415Z.00000000063
12. Floissac L et al (2009) How to assess the sustainability of building construction processes. In: Fifth urban research symposium, pp 1–17
13. Geological Society (2006) Earthen architecture. Engineering Geology Special Publications, v.21. London, UK, pp 387–400. https://doi.org/10.1144/GSL.ENG.2006.021.01.13
14. Ghavami K (2016) Introduction to nonconventional materials and an historic retrospective of the field. In: Harries K, Bhavna S (eds) Nonconventional and vernacular construction materials. Woodhead P. Elsevier, pp 37–61. https://doi.org/10.1016/B978-0-08-100038-0.00002-0
15. Giesekam J et al (2014) The greenhouse gas emissions and mitigation options for materials used in UK construction. Energy Build 78:202–214. https://doi.org/10.1016/j.enbuild.2014.04.035
16. Godwin PJ (2011) Building conservation and sustainability in the United Kingdom. Procedia Eng 20:12–21. https://doi.org/10.1016/j.proeng.2011.11.135
17. Hall M, Djerbib Y (2004) Rammed earth sample production: context, recommendations and consistency. Constr Build Mater 18(4):281–286. https://doi.org/10.1016/j.conbuildmat.2003.11.001
18. Hamard E et al (2016) Cob, a vernacular earth construction process in the context of modern sustainable building. Build Environ 106:103–119. https://doi.org/10.1016/j.buildenv.2016.06.009
19. Hamard E (2017) Rediscovering of vernacular adaptive construction strategies for sustainable modern building—application to cob and rammed earth. ENTPE. Available at: http://www.theses.fr/2017LYSET011
20. Hamard E et al (2018) A new methodology to identify and quantify material resource at a large scale for earth construction—application to cob in Brittany. Constr Build Mater 170:485–497. https://doi.org/10.1016/j.conbuildmat.2018.03.097
21. Houben H, Guillaud H (1994) Earth construction: a comprehensive guide. IT Publications, London
22. Jaquin PA, Augarde CE, Gerrard CM (2008) Chronological description of the spatial development of rammed earth techniques. Int J Archit Herit 2(4):377–400. https://doi.org/10.1080/15583050801958826

23. Keefe L (2005) Earth building—methods and materials, repair and conservation. Taylor & Francis Group, Abingdon, UK
24. King B (2010) The renaissance of earthen architecture—a fresh and updated look at clay-based construction. In: Buildwell symposium, pp 1–23
25. Kouakou CH, Morel J-C (2009) Strength and elasto-plastic properties of non-industrial building materials manufactured with clay as a natural binder. Appl Clay Sci 44(1–2):27–34. https://doi.org/10.1016/j.clay.2008.12.019
26. Llatas C (2011) A model for quantifying construction waste in projects according to the European waste list. Waste Manag 31(6):1261–1276. https://doi.org/10.1016/j.wasman.2011.01.023
27. Mandley S, Harmsen R, Worrell E (2015) Identifying the potential for resource and embodied energy savings within the UK building sector. Energy Build 86:841–851. https://doi.org/10.1016/j.enbuild.2014.10.044
28. Maniatidis V, Walker P (2003) A review of rammed earth construction. Report for DTI Partners in Innovation Project Developing Rammed Earth for UK housing. National Building Technology Group, University of Bath. DTi Partners in Innovation Project, Bath, UK, p 109
29. McGregor F et al (2016) A review on the buffering capacity of earth building materials. Proc Inst Civ Eng Constr Mater 1–11. https://doi.org/10.1680/jcoma.15.00035
30. Morel J-C et al (2001) Building houses with local materials: means to drastically reduce the environmental impact of construction. Build Environ 36(10):1119–1126. https://doi.org/10.1016/S0360-1323(00)00054-8
31. Morel JC et al (2013) Some observations about the paper "earth construction: lessons from the past for future eco-efficient construction" by F. Pacheco-Torgal and S. Jalali. Constr Build Mater 44:419–421. https://doi.org/10.1016/j.conbuildmat.2013.02.054
32. Passer A et al (2016) The impact of future scenarios on building refurbishment strategies towards plus energy buildings. Energy Build 124:153–163. https://doi.org/10.1016/j.enbuild.2016.04.008
33. Rigassi V (1995) Compressed earth blocks: Volume I: Manual of production. CRATerre-E. Eschborn (Germany): Deutsches Zentrum für Entwicklungstechnologien
34. Röhlen U, Ziegert C (2013) In Lefèvre C (ed) Construire en terre crue—Construction—Rénovation—Finition. Le Moniteur, Paris
35. Rosselló-Batle B et al (2015) An assessment of the relationship between embodied and thermal energy demands in dwellings in a Mediterranean climate. Energy Build 109:230–244. https://doi.org/10.1016/j.enbuild.2015.10.007
36. Sameh SH (2014) Promoting earth architecture as a sustainable construction technique in Egypt. J Clean Prod 65:362–373. https://doi.org/10.1016/j.jclepro.2013.08.046
37. Sauvage M (2009) Les débuts de l'architecture de terre au Proche-Orient. In: Achenza M, Correia M, Guillaud H (eds) Mediterra 2009, 1st Mediterranean conference on earth architecture. EdicomEdizioni, Cagliari, Italy, pp 189–198
38. Sauvage M (2020) Atlas historique du Proche-Orient ancien. Les Belles Lettres de Beyrouth, IFPO, Paris, France
39. Soudani L et al (2016) Assessment of the validity of some common assumptions in hygrothermal modeling of earth based materials. Energy Build 116:498–511. https://doi.org/10.1016/j.enbuild.2016.01.025
40. Soudani L (2016) Modelling and experimental validation of the hygrothermal performances of earth as a building material
41. Ugochukwu IB, Chioma MIB (2015) Local building materials: affordable strategy for housing the urban poor in Nigeria. Procedia Eng 118:42–49. https://doi.org/10.1016/j.proeng.2015.08.402
42. UNESCO (2013) Earthen Architecture in today's world. In: Eloundou L, Joffroy T (eds) Proceedings of the UNESCO international colloquium on the conservation of world heritage Earthen Architecture, 17–18 Dec 2012. United Nations Educational, Scientific and Cultural Organization, Paris, France, p 271. Available at: https://whc.unesco.org/document/126549

43. Volhard F (2016) Light earth building, a handbook for building with wood and earth. Birkhäuser, Basel, Switzerland
44. Watson L, McCabe K (2011) The cob building technique. Past, present and future. Inf Constr 63(523):59–70. https://doi.org/10.3989/ic.10.018
45. Williams C et al (2010) The feasibility of earth block masonry for building sustainable walling in the United Kingdom. J Build Apprais 6(2):99–108. https://doi.org/10.1057/jba.2010.15
46. Zakri AH, Watson R (2005) Ecosystems and human well-being, a framework for assessment, Island Pre, Millennium Ecosystem Assessment, Washington D.C. (USA). http://www.millenniumassessment.org/en/Framework.html#download
47. Zhuang Y, Kidder TR (2014) Archaeology of the Anthropocene in the Yellow River region, China, 8000–2000 cal. BP. Holocene 24(11):1602–1623. https://doi.org/10.1177/0959683614544058

Chapter 2
Characterization of Earth Used in Earth Construction Materials

Jean-Emmanuel Aubert, Paulina Faria, Pascal Maillard, Kouka Amed Jérémy Ouedraogo, Claudiane Ouellet-Plamondon, and Elodie Prud'homme

Abstract The objective of this chapter is to present the physical, geotechnical, chemical and mineralogical characterization techniques used to characterize the raw material (earth and mineral addition, such as sand and gravel) contained in the earth materials manufactured with different techniques: earth bricks, rammed earth or cob. This chapter will be divided into 6 sections. The first will present the method used to find the references considered in this state of the art and we will carry out a general qualitative analysis of these references. The other sections will deal respectively with granular, geotechnical, chemical and mineralogical characteristics and, finally, the last part will be dedicated to field tests.

Keywords Particle size distribution · Physical and geotechnical characterization · Chemical and mineralogical properties · Field tests

---

J.-E. Aubert (✉)
UPS, INSA, LMDC (Laboratoire Matériaux et Durabilité des Constructions), Université de Toulouse, Toulouse, France
e-mail: jean-emmanuel.aubert@univ-tlse3.fr

P. Faria
CERIS and NOVA School of Science and Technology, NOVA University of Lisbon, Caparica, Portugal

P. Maillard
CTMNC, Research and Development Department of Ceramic, Ester Technopole, Limoges, France

K. A. J. Ouedraogo
UPS, INSA, LMDC, Université de Toulouse, Toulouse, France

C. Ouellet-Plamondon
Département de Génie de La Construction, École de technologie supérieure, 1100, rue Notre-Dame Ouest, Montréal, Qc 3C 1K3, Canada

E. Prud'homme
Université de Lyon, MATEIS, Matériaux: Ingénierie et Science, Université de Lyon, Lyon, France

A. Fabbri et al. (eds.), *Testing and Characterisation of Earth-based Building Materials and Elements*, RILEM State-of-the-Art Reports 35,
https://doi.org/10.1007/978-3-030-83297-1_2

2.1 Introduction

Earth has always been used by man to build his habitat. Until the middle of the twentieth century, it was the most used building material with stone and wood. After the Second World War, concrete came to replace these materials, especially in western countries, thanks to its properties: speed of curing coupled with ease of mechanization of implementation, low cost, high mechanical performances and good durability. Concrete made from aggregates (mostly from natural origin), cement and admixture is being used at a fast rate worldwide. If cement did not pose any environmental problems (in particular because of its carbon footprint due to the significant CO_2 emission during its production), it is unlikely that researchers will once again be interested in natural materials, such as earth, as an alternative to cementitious composites. But the dramatic ecological situation in which the world is at the beginning of the third millennium forces humanity to reconsider how they consume, and in particular how they build. Furthermore, cementitious traditional components, especially sand scarcity, more and more highlights the economic effectiveness of using alternative available construction materials. Thus, materials that have been neglected for several decades, such as earth or biobased materials as the case of vegetable fibres, are finding renewed interest. This is due to their low environmental impact (abundance, renewability in the case of biobased materials, recyclability and low embodied energy) but also to their own properties, especially from the point of view of the comfort of inhabitants (see the chapter on the hygrothermal properties of earth construction materials). Today, it is estimated that over two billion people are still living in earthen buildings and that 10% of the architectural world heritage properties include earth structures [1]. Earth construction is part of the solution to ensure that humanity lives within the resources and climate planetary boundaries.

Historically, the formulation and the manufacture of earth construction materials have been done in an empirical way based on local constructive cultures, often orally passed down from generation to generation by the builders who were also mostly farmers who exploited earth for agriculture. Those builders learned to adapt to the constraints of local materials and, in particular, to the properties of the local soils. In fact, unlike today, it was not possible to transport materials over long distances and it was not possible to improve an inadequate earth by adding chemical stabilizers. This has led in particular to a regionalization of techniques which France can be used as an example as illustrated in Fig. 2.1.

Indeed, before it became cultural, humans adapted to the properties of the soil from their region, which oriented their choices on the technique used. For example, earth from the Rhone Valley contains significant amounts of coarse grains and if the builder of that period had wanted to manufacture adobe with this earth, they would have needed to sift (or to grind) his material which was not possible at that time. Thus, the best adapted technique to this granularity of earth was rammed earth, which explains why almost all the earth buildings in the Rhône Valley were built with rammed earth technology. There is a similar analysis for the cob in Normandy but this time, one of the reasons for the choice of this technique compared to others is not necessarily

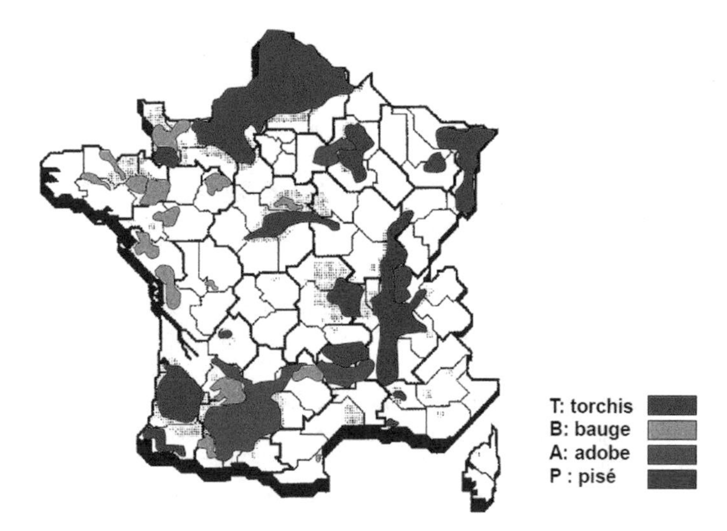

Fig. 2.1 Geographical distribution of earthen heritage in France ("pisé" = "rammed earth", "bauge" = "cob", "torchis" = "wattle and daub") based on [2]

related to the characteristics of the earth but more to the meteorological conditions. Even if the earth from Normandy has a granularity usable to manufacture adobe, the state of humidity in place and the difficulties of drying have oriented the builders towards "adobe put directly in place without drying", namely the cob technology. In cases where the earth used was too clayey, the high-water content of cob caused significant cracking of the materials during drying. The masons of the past had found a solution to this problem by adding vegetable fibres to the earthen mixture, which allowed both to limit cracking during drying but also to better structure the fresh pieces of cob. In the south-west, hot and dry summers coupled with Garonne Valley soil with excellent properties have led to the development of adobe in this region. Thus, the vernacular heritage shows that the builders, by experience, have learned to adapt to local materials. The same goes for the formulation of the materials: they used no weighing, no particle size distribution and even less chemistry or mineralogy! The formulas used were the result of the experience gained on the construction site by the builders.

The world of building materials and products has changed and today we talk about standards, control, performance and modelling. In addition, we are able to transport materials over greater distances (even if from an environmental point of view this is not desirable) or to transform soils not adapted to earth construction by using chemical additions especially hydraulic binders. In this logic, researchers are trying to rationalize the formulation of earth building materials or to understand and predict the behaviour of these materials from a thorough characterization of raw materials. The objective of this chapter is to present the physical, geotechnical, chemical and mineralogical characterization techniques used to characterize the raw material (earth and mineral addition, such as sand and gravel) contained in the earth materials manufactured with different techniques: earth bricks, rammed earth or cob.

2.2 Global Analysis of References Used for This Review—Methodology

The number of scientific studies on earth construction materials has increased rapidly in the last ten years, as shown in Fig. 2.2. This curve was obtained in 2017 after a search using the key following keywords: adobe, cob, wattle and daub, compressed earth bricks, rammed earth, earthen material and earth bricks. Then a sorting was done, in particular to eliminate the duplicates. We find a total of 422 references whose distribution by years is shown in Fig. 2.2.

The research was restricted to the Science Direct engine with the following combinations of keywords:

1. Earth bricks (in abstract, title, keywords): 146 results
2. Adobe alone does not work (too many references); so adobe (in abstract, title, keywords) and earth (in all fields): 143
3. Compressed earth blocks (in abstract, title, keywords): 58
4. Rammed earth (in abstract, title, keywords): 176
5. Cob alone does not work (too many references); so cob (in abstract, title, keywords) and earth (in all fields): 138
6. Wattle (in abstract, title, keywords) and earth (in all fields): 24
7. Daub (in abstract, title, keywords) and earth (in all fields): 12.

Search by keywords gives many results and sometimes some are not relevant. It was therefore necessary to perform an important manual sorting of these results by

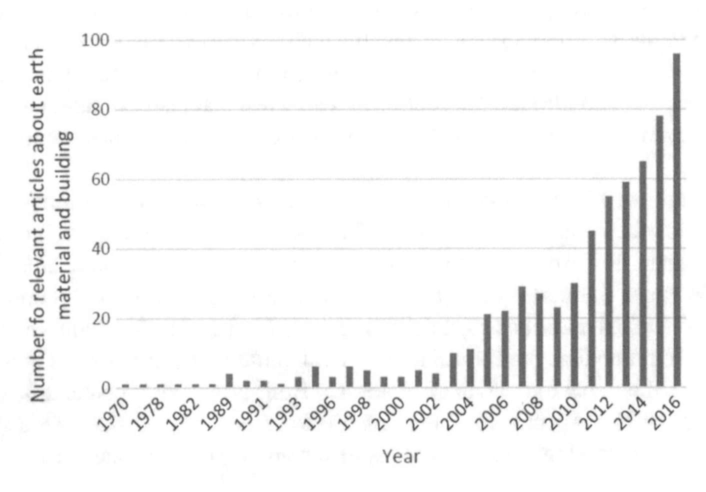

Fig. 2.2 Number of relevant articles about earth materials and buildings found in 2017

Table 2.1 Number of articles dealing with earth construction materials where raw materials are characterised

Technology	Unstabilized	Stabilized	Total
Extruded earth bricks	5	1	6
Compressed earth bricks	4	12	16
Adobe	18	7	25
Cob	5	0	4
Rammed earth	8	11	19
Total	40	31	71

removing irrelevant references. Then, we kept only the references in which the raw material was effectively characterized. To facilitate the reading and the comparison of the results, we separated these articles by technique (Extruded Earth Bricks (EEB), Compressed Earth Bricks (CEB), adobe, cob and rammed earth) and according to whether they were chemically stabilized or not. The results are shown in Table 2.1.

As shown in Table 2.1, the reference number for some materials is very low but this corresponds to a reality. Moreover, in the case of the cob, we had to add references that are not papers from international journals to supplement the data that were too scarce. The most studied materials in the literature are earth bricks (adobe and CEB) and rammed earth. It is important to remember that this research is not exhaustive and that the number of articles on earth construction materials in which raw material data are found is more important. But, with the partial research that we have done, we reach a relatively high reference number (71 articles) which therefore begins to be representative. In addition, some authors have written several articles using the same earth: in these cases, either we keep only the article in which the characterization is presented, or we keep the most recent reference if the data presented are the same.

In recent years, it has been usual to add chemical stabilizers to the earth. The reasons given are the improvement of the mechanical performance of the construction products (see Chap. 3) as well as the improvement of its resistance to liquid water (see Chap. 5). It is interesting to study the differences in the use or not of stabilization depending on the techniques used. For example, we note that the CEB are almost systematically stabilized. This can be explained by the fact that CEB are modern materials that emerged recently (mainly after the 1980s) in the history of earth construction. The objective is to accelerate the drying of the bricks by producing them with less water than in the traditional adobes: the consistency thus obtained no longer allows the material to be applied by moulding and must then be pressed (or compacted) in a mould. Since mould release is immediate, the CEB manufacturers quickly decided to add a hydraulic binder (lime with hydraulic properties classified based on EN 459-1 [3] or cement) in their material to improve its performances in a very short term to facilitate handling and storage of these bricks. Conversely, the earth of all the articles on cob is unstabilised. A large majority of articles are about unstabilised adobes probably because the materials from the vernacular heritage often studied in these articles are rarely stabilized. However, when the local earth had low clay content, air lime could be added for vernacular adobe production [4].

EEB are singular because, like CEB, they are modern materials, yet few studies focus on stabilization (only 1 out of 6). This can partly be explained by the method of manufacture of these bricks (extrusion after evacuation of the air under vacuum) which gives the bricks at the extruder outlet exceptional holding that allows them to be handled and stored easily while waiting for them drying. In fact, EEB are often similar to the ones that will be fired, but do not embody the firing energy of the latter. Finally, there is no strong trend for rammed earth where there are generally as many studies on stabilized or unstabilised materials.

Data was collected from these articles and separated into 4 types of material characterization:

- particle size distribution;
- physical and geotechnical characterization;
- chemical characterization;
- mineralogical characterization.

The accuracy of earth characterization in the papers dealing with earth building materials is highly variable. In almost every paper, the particle size distribution of the earth is presented: this characteristic can be considered as the basic characteristic of earth. Nevertheless, the particle size distribution is not always complete: sometimes, the fine fraction is missing, and in some cases it is not performed by wet method and, therefore, the fine fraction can be considered agglutinated in clods. However, the measurement of other characteristics is not systematic. To quantify this, we used three levels as follows:

- the number "3" corresponds to the most basic characterization, that means that only the particle size distribution of the earth is given;
- the number "2" is used when, in addition to particle size distribution, at least one other characteristic is given (physical and geotechnical characterization or chemical or mineralogical characterization);
- finally, the number "1" qualifies the articles in which we find: particle size distribution + physical and geotechnical characterization + chemical and/or mineralogical composition: this corresponds to the most thorough characterization.

Table 2.2 presents the distribution of these levels of deepening of the characteri-

Table 2.2 Deepening levels of the characterization of raw earth

Technology	"1" Deepest	"2" Average	"3" The most basic	Total
Extruded earth bricks	3	2	1	6
Compressed earth bricks	4	11	1	16
Adobe	8	12	5	25
Cob	1	1	3	5
Rammed earth	3	10	6	19
Total	19	36	16	71

Table 2.3 Proportions of the main characteristics of raw earth found in the papers dealing with earth construction materials

Technology	Particle size distribution	Physical and geotechnical characterization	Chemical characterization	Mineralogical characterization
Extruded earth bricks	6 (100%)	2 (33%)	4 (67%)	4 (67%)
Compressed earth bricks	16 (100%)	14 (88%)	5 (31%)	4 (25%)
Adobe	24 (96%)	16 (64%)	7 (28%)	9 (36%)
Cob	5 (100%)	2 (50%)	0 (0%)	2 (50%)
Rammed earth	19 (100%)	8 (42%)	3 (16%)	7 (37%)
Average value	70 (99%)	42 (71%)	19 (27%)	26 (37%)

zation according to the type of materials.

Generally, the majority of the items we analysed are in the middle level "2". It is also interesting to note that it depends a lot on the type of material studied. For example, in the case of the cob, the majority of references are level "3" whereas in the case of the EEB, it is the opposite: the studies present thorough characterization of the earth. However, for both technologies, the total of references is very low. This could be explained by the fact that the cob is a more traditional material whereas the EEB are more modern and more technological materials. For these latter, a better knowledge of the characteristics of the components is needed to efficiently optimize them.

Table 2.3 presents these results in a different and more precise way. In this table, the proportions of the different characteristics found in the articles are presented.

As previously noted, when the level was established, all selected items (except one) present the particle size distribution of the earth. For the other characteristics, the proportions are much more variable. Physical and geotechnical characteristics come in second place with an average of 71% appearance. As a general rule, most of the paper dealing with earth bricks contains the Atterberg limits and the particle size distribution of earth. For rammed earth articles, the Atterberg limits are often replaced by the Proctor tests, which is consistent because of the differences in the use of these materials, namely the use of humid compacted earth for rammed earth instead of plastic moulded earth for adobe. The mineralogical characterization of the earth appears in about a third of the papers, which is still relatively high. However, we will analyse these data in Sect. 2.6.2 of this chapter and we will see that most of the data presented are qualitative and that many questions raise about the procedures used to obtain some results and about the accuracy of these results. Finally, the least studied characteristics are the chemical characteristics. No data are available for the cob and, for rammed earth, there is very little data: there are measurements of earth pH in two articles and measurements of organic matter content in two others.

Considering the 19 articles dealing with rammed earth, no basic chemical analysis of earth is given what may seem surprising. One can try to explain this by a different cultural approach of the researchers working on rammed earth and those working on earth bricks and namely on EEB. Indeed, for rammed earth, one often uses the earth which is directly available on the building site and which showed by experiment that is suitable. The researchers working on these materials are essentially specialists in mechanics or more recently in hygrothermal behaviour of materials, and are probably less sensitive to the chemical composition of earth. In addition, the studies of the rammed earth are often made on a macroscopic scale because of the specific implementation of these materials (a wall is directly built and not elements of a wall as in the case of bricks), and the heterogeneous granular material, with presence of coarse aggregates. The presence of those coarse aggregates turns low the clay content of the earth. Therefore, rammed earth is not as much influenced by the characteristics of the clay as other more clayey materials used for brick production. In the field of research on bricks, the analyses are often done on a finer scale and the researchers are often specialists in physico-chemistry of materials.

After this first chapter of global analysis of the elements of the bibliography that we collected for this state of the art, we will present different sections corresponding to the different families of characteristics presented previously. Each of these sections will have the same structure and will be divided into two parts. In the first part, we will present the test procedures found in the literature based on some international standards, when they exist. The aim here is not to make a comprehensive review of existing international standards but to provide examples of the most used standards. Thus, and also in relation to the origin of the co-authors of this chapter, we will use the following norms: American, Quebec, English and European. The second part of each section will be devoted to the analysis of the results extracted from the bibliographical references in order to see if trends exist according to the type of materials or the type of technique used. Something that is very important for earth characterization is the representativeness of the samples. A representative earth sample depends on the tests to perform and, therefore, the type of earth building technology that is intended. However, the earthen sample size is often not presented in scientific articles. Furthermore, in case of earth architectural heritage characterization, the sampling is frequently limited due to restrictions of the heritage property [5]. The first of the characteristic studied that is most present and certainly the most important for earth construction materials is the particle size distribution that is the focus of the next section.

In this review we have used the term earth for the building material. However, namely for geotechnical characterization, frequently the term soil is used not only for the non-extracted earth. Therefore in the following sections both terms are used to follow the original authors' terminology.

2.3 Particle Size Distribution

As commented before, the particle size distribution is one of the most important physical characteristics of soil. Classification of soils is mainly based on the particle size distribution. Many geotechnical and geohydrological properties of soil are related to the particle size distribution. The particle size distribution provides a description of soil based on a subdivision in discrete classes of particle sizes.

2.3.1 Procedures and Standards

Standards

Several standards for soil classification and particle size distribution exist. It is possible to separate them in two types: the wet sieving particle size for the coarser particles (> 80 μm) and the sedimentometry for the fine fraction (1–80 μm). It is important to specify that the laser granulometry is not suitable for the measurements of the granularity on clay soils, mainly because of the difficulties of dispersion of the particles. To be applied the previous dissolution of the soil in water and a wet method should be used.

North American standards [6] deal with wet sieving and [7] with sedimentometry. The British BS 1377 Part 2.9 [8] and Canada BNQ-2501-025 [9] standards include procedures for wet sieving and sedimentometry. It is the same for the European standard EN ISO 17892-4 [10]. Whether they are North American, British or European, they are very close or even similar in particular in characterization methods (sieving and sedimentation). The principle of these test procedures is described later in this section.

Procedures

Coarse soils are usually tested by sieving, but fine and mixed soils are usually tested by a combination of sieving and sedimentation, depending on the composition of the soil. The sieving method described is applicable to all non-cemented soils with particle sizes less than 125 mm. Two sedimentation methods are described: the hydrometer method and the pipette method.

The test method or combination of methods should be specified prior to testing or be selected on the following basis. If a sample has less than about 10% of particles smaller than 0.063 mm, sedimentation test is not normally required. If all particles of the sample are smaller than 2 mm and the sample has less than about 10% of particles larger than 0.063 mm, a full-sieve test is not normally required. For all other samples, a combination of a sieve test and a sedimentation should be performed in order to determine the full-particle size distribution.

Sieving method: The test consists of separating the agglomerated grains from a known mass of soil by fractionating it under water with a series of sieves and weighing the cumulative and dried rejection on each sieve (dried usually at 105 °C). The mass of the cumulative rejection for each sieve is related to the total dry mass of the soil sample submitted for analysis. Either a moist or a dry sample may be tested. The sieve test consists in the determination of the masses of material retained on the various sieves with decreasing diameter sizes. The number of sieves used and their aperture sizes shall be sufficient to ensure that any discontinuities in the grading curve are detected. In the standard EN ISO 17892-4 [10], it is recommended (but not imposed) to use the sieves of 63, 20, 6.3, 2.0, 0.63, 0.20, 0.0063 mm because these values represent the size limits for coarse materials as defined in EN ISO 14688-1 [11].

Dry sieving is not appropriate particularly for clayey earths/soils because grains that result from the agglomeration of particles are sieved without separation.

Sedimentation: Based on the Stokes' law, the method is based on the measurement of the sedimentation time of solid particles in suspension in a solution of water mixed with sodium hexametaphosphate as a deflocculating agent. The sedimentation analysis is an analysis completing the sieving analysis for particles usually with a diameter of less than 80 μm. The test is based on the fact that in a liquid in which a deflocculating agent has been added (sodium hexametaphosphate), the decantation rate of the fine particles depends on their size. The principle follows Stokes' law linking the diameter of the grains and their sedimentation rate. By convention, this law is applied to the elements of a soil to determine the equivalent diameters of the particles. The test can be carried out using two different methods:

- Hydrometer method: A part of the soil is dried then mixed with water containing the dispersing agent, and then the hydrometer is introduced into the graduated cylinder. The density of the mixture is measured with the hydrometer at various time intervals (e.g.: 30 s, 1 min, 2 min, 4 min, 8 min, 30 min, 1 h, 2 h and 24 h). From the density measured at a given time, the size of the suspended particles can be determined. The hydrometer shall be torpedo-shaped, made of glass, as free as possible from visible defects and preferably manufactured to a national standard. The hydrometer stem and bulb shall be circular in cross section and symmetrical around the main axis, without abrupt change in cross section.
- Pipette method: Based on the same principle and theory, the pipette method consists of taking a fraction of the mixture (soil dispersed in water containing a dispersant) at different times and depths, and then drying and weighing the residue. It is also possible to initially define the particle sizes in order to know their quantity, and then calculate the corresponding sampling times. The pipette shall have a nominal volume of 2% of the volume of the soil suspension and shall be mounted in a pipette configuration.

This sedimentation measurement method has also been automated and modernized with the use of a sedigraph. An X-ray beam measures the concentration of suspended particles at a sedimentation height that decreases with time. The particle diameters are obtained instantly corresponding to the elapsed time and sedimentation height.

A source of error in these different procedures could be linked to the incomplete dispersion of soil clays. If clay particles are not separated correctly, they form aggregates with a larger size. It results in low values for clay and high values for silt and sand. The rate of sedimentation is also affected by temperature, the density of the dispersing solution and by a too abrupt introduction of the hygrometer or of the pipette.

Table 2.4 Particle size fractions according to the EN ISO 14688-1 [11]

Soil group	Particle size fractions	Range of particle sizes (mm)
Very coarse soil	Large boulder	> 630
	Boulder	> 200 to ≤ 630
	Cobble	> 63 to ≤ 200
Coarse soil	Gravel	> 2.0 to ≤ 63
	Coarse gravel	> 20 to ≤ 63
	Medium gravel	> 6.3 to ≤ 20
	Fine gravel	> 2.0 to ≤ 6.3
	Sand	> 0.063 to ≤ 2.0
	Coarse sand	> 0.63 to ≤ 2.0
	Medium sand	> 0.20 to ≤ 0.63
	Fine sand	> 0.063 to ≤ 0.20
Fine soil	Silt	> 0.002 to ≤ 0.063
	Coarse silt	> 0.02 to ≤ 0.063
	Medium silt	> 0.0063 to ≤ 0.02
	Fine silt	> 0.002 to ≤ 0.0063
	Clay	≤ 0.002

Soil classification

As defined in the standard EN ISO 14688-1 [11], Table 2.4 shows the terms to be used for each size fraction, together with the corresponding range of particle sizes. Clay can be defined from a granular point of view (particle size) and also from a geological point of view (mineral composition). But, in most publications, clay is defined as a particle with a diameter of less than 2 μm. According to the standards and their origin, the limits between the particle size and their names can vary, especially the limit silt–sand. In the standards EN ISO 14688-1 [11], USDA [12] and ASTM-D2487 [13], this limit is fixed respectively to 0.063 mm, 0.05 mm and 0.075 mm.

Study of data from literature

The particle size distributions of the soils studied in the literature are presented in Tables 2.18, 2.19, 2.20, 2.21 and 2.22 of Appendix 1. Some cells were deliberately grayed out: this corresponds to the data that we were unable to use in our study because the granular classes used do not correspond to the conventional classes.

2.3.1.1 Earth Bricks

Figures 2.3 and 2.4 show the particle size distributions of the earth bricks studied in different papers, all techniques being considered (moulding, compression and

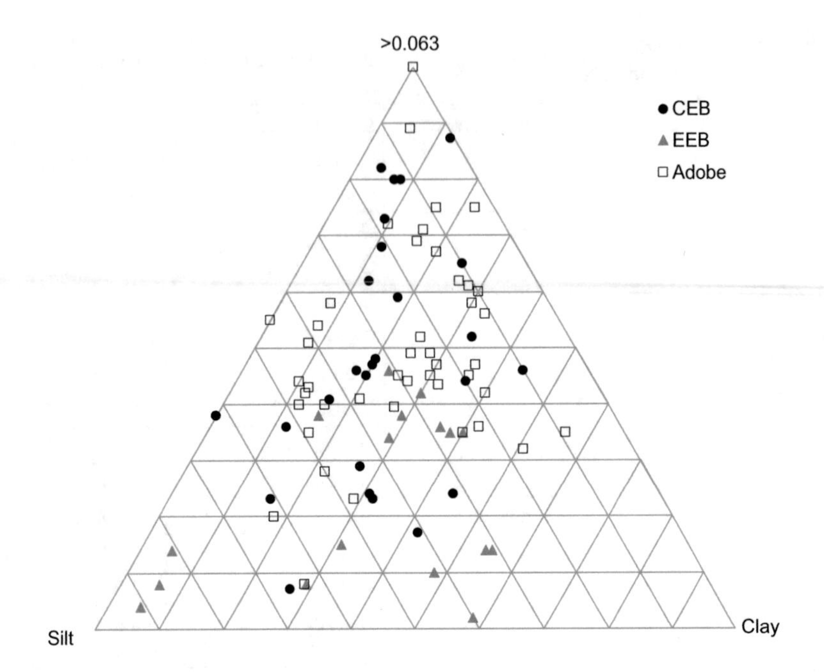

Fig. 2.3 Particle size distribution or raw earth used for earth bricks in a ternary diagram

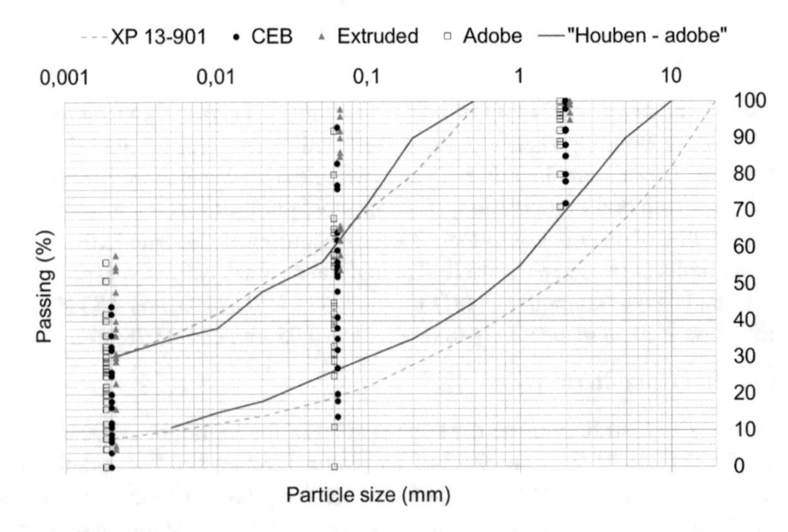

Fig. 2.4 Particle size distribution or raw earth used for earth bricks

extrusion) (see details in Tables 2.18, 2.19, 2.20, 2.21 and 2.22 given in Appendix). Although the ternary diagram is not the most used in the literature, it allows here to represent the different sizes. No clear trend appears but earth used for EEB seems to be thinner than those used for adobe and CEB, and the silt and clay quantities are generally higher, which makes sense because of the particular manufacturing process of EEB. For adobes and CEB, there is no significant difference observed. However, one technique moulds a plastic earth mixture while the latter compresses a humid mixture (with much lower water content), and the composition (mineralogical but also in terms of particle size distribution) could be different. Moreover, the ternary diagram shows specific earth as for example earth containing little or no clay. That seems to be strange because clay, like cement for concrete, is the binder of earth building materials. As mentioned above, certain values may be due to handling errors (insufficient dispersion, etc.). Moreover, these "special" values also highlight the role of a binder (lime or cement) added to the soil to ensure better cohesion. Although stabilization can improve the mechanical performance of bricks, it is important that the earth alone already has good cohesion as indicated in French standard XP P13-901 [14].

On Fig. 2.4, the limits recommended by XP 13–901 for CEB are also presented. Many soils studied in the literature do not respect these limits. Moreover, Houben and Guillau [15] gave other limits for adobes (clay: 10–30%, silt: 15–33%, sand + gravel: 37–75%) while specifying that the recommended area is only approximate. Other specific cases of study have been observed particularly in the French heritage: an example in south-western France with adobes of the nineteenth century which contain pebbles of several centimetres (Fig. 2.5) questions the relevance of these limits.

Fig. 2.5 Adobes from southern France (Montricoux) with coarse aggregates (XIXth century) ((c) Pays Midi-Quercy; Conseil départemental de Tarn-et-Garonne; Inventaire général Région Occitanie)

2.3.1.2 Cob

A significant review was done by Hamard et al. [16, 17]. The authors indicate that the cob is less studied, unlike other techniques, the number of results on the soils used for this technique is therefore rather low and it is difficult to generalize the few data found in the literature. In contrast to other techniques, it is not necessary to differentiate between stabilized and non-stabilized materials, since all references only deal on earthen materials that are not chemically stabilized. Moreover, the straw content is only very seldom specified: only one reference [18] provides the straw content of the cob (1% wt.). Fibres that are generally largely present in this type of material make particle size analysis difficult. Indeed, for these types of materials including plant fibres, it is necessary to separate the mineral phase from the plant phase to characterize the particle size distribution the more precisely as possible. This separation is often difficult.

Table 2.21 of Appendix A presents the particle size distribution of soils found in the literature. Some data are special as that of [19] because the soil used for the renovation of the studied buildings is a soil rich in clay intended for the manufacture of clay bricks. Other older data indicate only two lots of grains (clay + silt and sand + gravel). In [18], two of the five soils studied have very few fine particles (clay and silt contents of between 5 and 7%) whereas for one earth, the content of fine particles is very high (clay + silt equal to 94%). The researcher specifies that the five soils tested in compression (cylindrical sample) have good mechanical properties, which shows that it is possible to produce cob having good mechanical performance with very different particle size distributions of soil. Nevertheless, Saxton [20] recommends an ideal particle size distribution (clay + silt = 35%, sand = 35% and gravel = 30%) with a wider acceptable zone (clay + silt = 25–45%, sand = 25–45% and gravel = 20–40%). Same limits (1/3 clay + silt, 1/3 sand, 1/3 gravel) are indicated by Harries et al. [21]. It can be seen that none of the nine soils used for cobs in the literature and presented in Table 2.21 meet these recommendations.

Finally, Hamard et al. [22] have developed a new methodology to identify and quantify material resources at a large scale for earth construction especially applied to the cob in Brittany. This methodology is based on the cross-referencing of spatialized pedological and heritage data. The methodology applied at the regional scale in France (for a given area of 27,200 km^2 in Brittany) enabled to specify five new texture classes (balance between clay, silt, sand and gravel content) of suitability for cob soils. For a further discussion on the identification and quantification of soils for construction, the researchers propose to apply the same methodology to other regions with different earth construction techniques.

2.3.1.3 Rammed Earth

Figure 2.6 presents the particle size distributions of 19 bibliographic references dealing with rammed earth. Compared to the bricks described in Fig. 2.3, a trend is clearly apparent: the earth generally contains more than 60% of grains with a

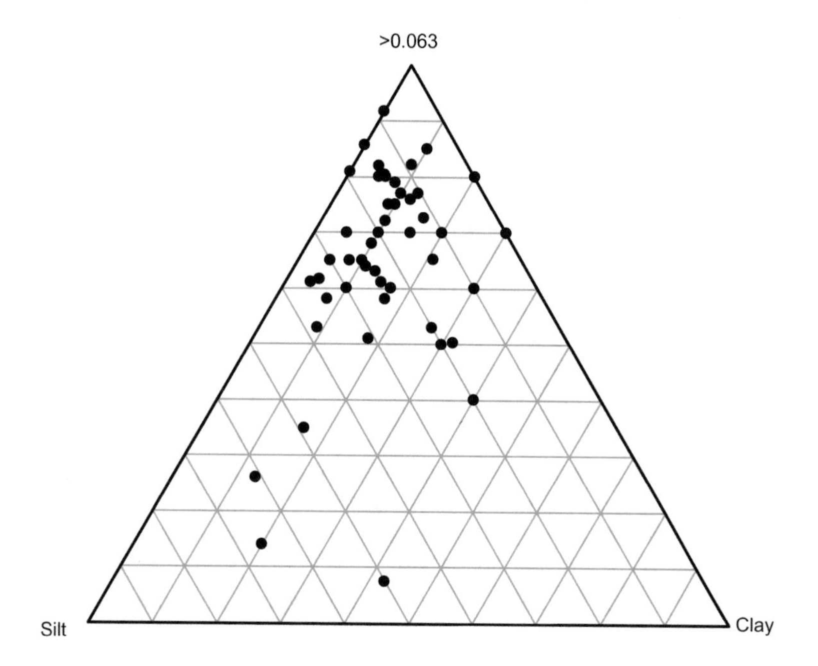

Fig. 2.6 Particle size distribution or raw earth used for rammed earth in a ternary diagram

diameter greater than 0.063 mm and less than 20% of clay (Ø < 2 μm). These values follow the recommendations of the Walker and Australian standard HP195 (2001) which indicates the following limits: clay up to 20%, silt between 10 and 30%, sand and gravel between 45 and 75%. Some points highlight clay-rich soils, up to 40% of clay, that do not respect the recommended limits [23, 24].

Nevertheless, no soil is likely to be ideal; therefore researchers usually published in the past upper and lower limits for each of the main soil elements (Table 2.5) [25, 26]. As it was the case for bricks, the particle size distribution of earth materials often comes out of these limits, which was found by Gomes et al. [25] on the six earth materials from existent constructions they have studied.

Table 2.5 Recommendations concerning the particle size distribution of soils for rammed earth construction

References	Clay (%)	Silt (%)	Sand + gravel (%)
[27]	5–15	15–30	50–70
[28]	10–25	15–30	45–75
[29]	5–15	15–30	50–70
[30]	7–15	10–18	75
[31]	5–20	10–30	45–80
[15]	0–20	10–30	45–75

Adapted from [25]

2.4 Physical and Geotechnical Characterization

2.4.1 Procedures and Standards

Atterberg limits

British (UK) and North American standards on how to measure the Atterberg limits are [32], BNQ 2501-090 [33], BS 1377-2 [8] and [34]. In Europe, the test method is defined by EN ISO 17892-12 [35].

The liquid limit (LL) and the plastic limit (PL) classify fine-grained earth and the fine fraction of mixed soil. They are commonly required for geotechnical engineering tests, both in industry and in academic research [36]. In our literature review, it is the most frequent geotechnical test performed to characterize the earth for construction. According to the standard BS 1377-2 [8], LL is the empirical moisture content at which a soil passes from a liquid state to a plastic state. The LL is measured either with the cone penetrometer or the Casagrande method. The definitions of PL slightly differ depending of the standard considered. In BS 1377-2 [8], the PL is the empirical moisture content at which a soil is too dry to be plastic, which is the transition from a ductile to a brittle behaviour [36]. In [32], the PL is the percentage of water content of a soil at the boundary between the plastic and the semi-solid state.

The PL is measured by rolling a thread of soil. The plasticity index (PI) is calculated as the numerical difference between the LL and the PL. The graphical representation of the PI allows to classify cohesive soils [37] and to determine boundaries between consistency states of plastic soil, which means the relative ease in which soil can be deformed [32]. These tests are not much precise and the standard BS 1377-2 [8] specifies that the results remain variable with the judgment of the operator. These tests originate from the work of Atterberg, which was then standardized. The PL and LL are often collectively referred as the Atterberg limits. Actually Atterberg defined 7 limits [38]. Many countries have their own version of the standard, which means that the variability according to the testing method has been a subject of discussion. As a start point, the [32, 34] specify to measure the plasticity on the fraction finer than 0.425 mm, while finer than 0.400 mm in the standards from Quebec [33] and in Europe [35] or Internationally [39]. Increasing the sand content decreases plasticity, while the fine organic content increases plasticity.

To determine the LL limit, the fall cone penetrometer is preferred in the UK [34] and by the International [39] and European [35] standards because it is a static test [36]. The Casagrande test can introduce a dynamic effect and is susceptible to variability between operators [8]. The cone penetrometer gives results that are more reproducible than the Casagrande method [40]. The fall-cone test assesses the soil shear strength by relating the soil's undrained shear strength to the fall-cone weight divided by the square of the penetration depth [36]. The LL is measured with the Casagrande percussion cup in ASTM and Canadian standards. A portion of the earth specimen is spread in a brass cup, divided in two by a grooving tool, and subsequently allowed to flow together from the shocks caused by dropping the cup in a standard mechanical device [32]. According to the number of drops, the test follows a one-point method or a multipoint method. The number of drops to decide on the method may vary according to standards. The multipoint method is generally more precise. The water content is determined on the soil in the cup at the end of the test.

The plastic limit is measured internationally by pressing and manually rolling a thread of plastic soil on a glass plate until the water content is reduced to the point the thread crumbles and can no longer be pressed again and rerolled [32, 36]. The soil is rolled to a thread diameter of 3.0 mm in the UK and Quebec standard, while it is 3.2 mm in the ASTM standard. The soil water content is determined at the breaking point. The repeatability of the thread rolling tests varies with the number of operators. For example, the standard deviation was less than 1% when one operator was considered and up to 3% when considering 41 operators from different laboratories [41].

Test methods for laboratory compaction characteristics of soil using standard effort (Proctor tests)

The standard test methods for laboratory compaction characteristics of soil using standard effort is commonly known as the Proctor test in reference to the equipment and procedure proposed by R. R. Proctor in 1933 [42, 43], BS 1377, 1990 [44]. This test determines the relationship between moulding water content and dry unit weight of soils compacted in mould with a 24.5 N rammer dropped in a free fall from a height of 305 mm producing a compacting volumetric energy of 600 kJ/m^3. The diameter of the cylindrical mould is 101.6 mm or 152.4 mm and the height is 116.4 mm. In the original Proctor test, the rammer blows were applied as firm strokes, producing variable compaction effort with the operator. Compactability is the ability of earth to be compacted by static pressure or dynamic compaction so that its volume is reduced. To attain maximum compaction, the earth must have a specific water content, the so-called "optimum water content," which allows particles to be moved into a denser configuration.

Moisture content test

The moisture content is determined on soil samples and the mass of the sample depends on the maximum grain size. In BNQ 2501-170 [45], it varies from 10 g for soil with maximum particle size of 400 μm to 1000 g for soil with maximum particle size of 56 mm. The sample is heated to 110 °C ± 5 °C until a constant mass is obtained. These parameters vary between various international standards [46], BS 1377, 1990; [39]. For example, for the ISO standard the drying temperature is equal to 105 °C ± 5 °C and the mass of the sample according to the particle size is different [39]. The standards specify that this method does not suit for soils containing gypsum, hydrated minerals and organic matter. The amount of water can also vary for soil with significant content in salt and other dissolved materials.

Specific gravity test

The specific density is a relative density, the ratio of mass of an aggregate to the mass of a volume of water equal to the volume of the aggregate particles, also known as the absolute volume of the aggregates. This relative density is needed to calculate the volume occupied by the components of earth mixes. The specific gravity can be determined with a pycnometer for the gravimetric procedure. A Le Chatelier flask is used in the volumetric procedure. The sample is dried at 110 °C until the mass is constant, or the natural conditions of the aggregates. Several international standards exist for the measurement of this characteristic: ASTM C128-15 [47], BS 1377 (1990) and EN ISO 17892-3 [48].

Methylene blue value and the activity of clay minerals

The methylene blue test aims to detect clay minerals in aggregates fines. It is based on an ion exchange phenomenon between methylene blue cations and clay ions that is possible thanks to the large surface area and negative charge of clays. The amount of absorbed methylene solution varies according to the amount of clay minerals and clay type, cation exchange capacity and specific surface area. Based on this test the specific surface area of soils can be determined. Generally, the sieve at 400 μm is used for this test. There are two main test methods for the methylene blue test: titration method and "spot-test" method.

The first method is described in [49]. The test specimen in a methylene blue solution is shaken twice for 60 s, with a rest of 180 s between the two shakings. The mixture is filtered and a sample from the filtered solution is diluted for the measurement with a colorimeter. The concentration adsorbed is calculated from the initial concentration and the final concentration.

The "spot-test" is described in NF P 94-068 [50], EN 933-9+A1 [51] and ASTM C837 [52]. Depending on the soil, a mass of 30–60 g is taken for high clayey soil and 60 to 120 g for less clayey soil. The soil sample is then dissolved in 500 ml of distilled water, each time 5 ml of methylene blue solution (10 g/L) is added to the soil solution. One drop of the mixture is placed onto a paper filter after 1 min. The test ends when the dye forms a second lighter coloured blue halo around the aggregate dye spot and stays stable over five consecutive spots without addition of methylene blue to the soil solution. NF EN 933-9+A1 [51] follows the same procedure but with a soil sample mass higher than 200 g depending on the sample moisture. ASTM C837 [52] follows the same procedure but involves the use of 2 g of soil samples and of acidic solutions (pH value from 2.5 and 3.8). However due to the small amount of investigated soil, ATSM C837 [52] is more suitable for homogeneous and fine materials.

The methylene blue value (MBV or V_{BS}) is reported in mg/g. A high methylene blue value indicates the presence of clays and allows the definition of six categories of soil as described in Table 2.6 [53].

Table 2.6 Definition of soil categories according to methylene blue value (V_{BS}) [53]

V_{BS}	Soil categories
$0.1 \leq V_{BS} < 0.2$	Water insensitive
$0.2 \leq V_{BS} < 1.5$	Sandy and silty
$1.5 \leq V_{BS} < 2.5$	Sandy-clay
$2.5 \leq V_{BS} < 6$	Silty moderately plastic
$6 \leq V_{BS} < 8$	Clayey
$8 \leq V_{BS}$	Heavy clayey

2.4.2 *Study of Data from Literature*

For physical and geotechnical properties, the most numerous results in the literature concern the Atterberg limits. Tables 2.23, 2.24, 2.25, 2.26 and 2.27 in Appendix 2 show the Atterberg limit values for various techniques of earth construction. The most numerous results concern CEB (16 in 10 papers) and adobes (55 in 14 papers). For the other techniques, the number of values available in the literature is too low to be meaningful. We will therefore focus our analysis of the Atterberg limits on soils used for CEB and adobes. Figure 2.7 shows the frequencies of occurrence of these limits for these two techniques.

Figure 2.7 also shows that the liquid limits are relatively dispersed. For CEB, 75% of the soils studied have liquid limits ranging between 25 and 45% that is slightly higher than for adobes. For adobes, 67% of soils have liquid limits ranging between 20 and 40%. However, if we compare the averages obtained for the values of the two techniques, they are exactly the same for the two materials (37%).

For the plastic limit, the results are much less dispersed and quite similar for the two techniques: the majority of the plastic limit values ranges between 15 and 29% (94% and 76% of the values for the CEB and the adobes respectively).

Finally, the liquid limits being different, it is therefore normal to observe marked differences for the plasticity index between the two techniques. For CEB, the frequency distribution is very centred: 50% of the values range between 30 and 34%. For adobes, the results are much more dispersed between 20 and 39%. A hypothesis to explain this difference could be the existence of a standard (French but used in many other countries) on the CEB which indicates a zone recommended for the Atterberg limits [14]. For the manufacture of adobes, such normative recommendation does not exist and researchers working on the subject often use the soil on the site directly without seeking to modify their granularity or their Atterberg limits, what leads to a greater variability of characteristics. Furthermore, the majority of the characterization studies on adobe are on vernacular constructions, built when normative requirements were not established.

The results of Fig. 2.7 are placed in the Casagrande plasticity chart on Fig. 2.8. According to the standards [13, 13], the signification of the acronyms is:

- CL = Lean Clay: inorganic clays, sandy clays, silty clays, lean clays, low to medium plasticity, no or slow dilatancy;

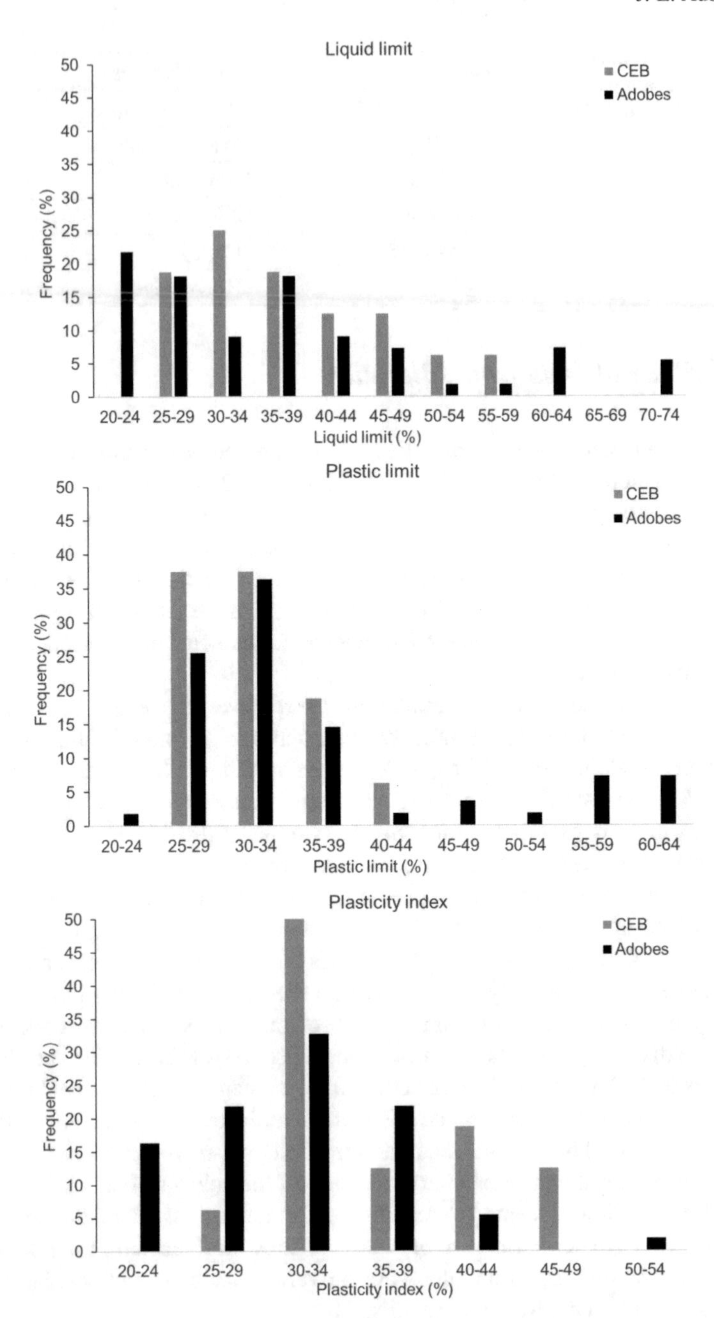

Fig. 2.7 Frequency of occurrence of Atterberg limits for CEB and adobes

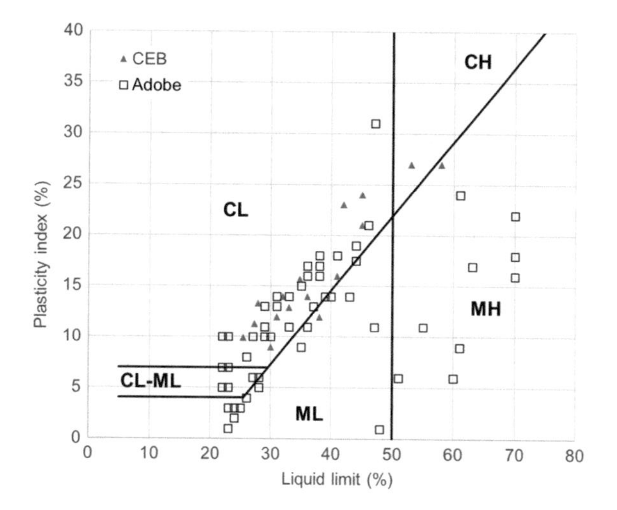

Fig. 2.8 Casagrande plasticity chart of CEB and adobes

- ML = Silt: inorganic silts and very fine sands, rock flour, silty or clayey fine sands, slight plasticity to non-plastic, slow to rapid dilatancy;
- CH = Fat Clay: inorganic clays, fat clays, high plasticity, no dilatancy;
- MH = Elastic Silt: micaceous or diatomaceous fine sandy and silty soils, elastic silts, low to medium plasticity, no to slow dilatancy;
- CL-ML = Silty Clay: mixed zone where both CL and ML soils plot.

The results of Fig. 2.8 show that the vast majority of results are in the CL category corresponding to lean clay. For adobes, few analyses are in the MH category corresponding to elastic silt but this concerns a minority of the values (around 16%). These analyses correspond to the higher values of plastic limits (higher than 40%). The zones for the Atterberg limits of CEB and adobes recommended by Houben and Guillaud [15] are possible to add on this chart. It is interesting to note that for the CEB, these are the same limits that have been included in the French standard [14]. The results with the recommended limits are shown in Fig. 2.9.

Figure 2.9 shows that the majority of the results obtained for the CEB are within the recommended limits. But, for adobes, this is not really the case and the Atterberg limits of adobes are more within the limits recommended for CEB with the exception of the few analyses in the MH category with high plastic limits. As we have seen for particle size distribution, these recommended areas are essentially indicative and many soils used with success in earth construction materials do not respect them.

For the other geotechnical characteristics, the studies on adobes give some values of MBV (13 in 4 articles). These values are between 0.16 and 0.60 mg/g with an average value equal to 0.34 mg/g, which would correspond to the soil category "sandy and silty" in comparison with the values in Table 2.6. However, the number of Methylene Blue values (V_{BS}) available is too limited to enable draw generalizable conclusions. For the other techniques, the number of results is much lower (3 in 2

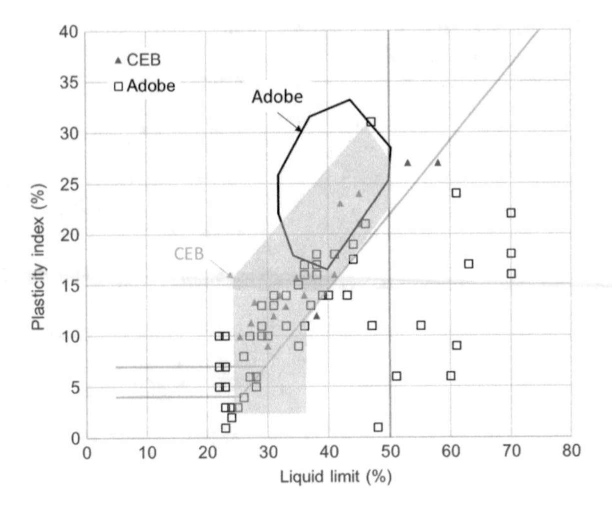

Fig. 2.9 Recommendations on Atterberg limits for CEB and adobes

articles for CEB, 0 for rammed earth and cob and 3 in one article for extruded earth bricks).

Concerning the optimal compaction characteristics of soil using the Proctor test, the trend is a little opposite because these data are more available for CEB (11 in 6 articles) and rammed earth (19 in 7 articles) in comparison to adobes (1 in 1 article), extruded bricks (0) and cob (4 in an article). This could be easily explained by the methods used to manufacture CEB and rammed earth, which is based on optimal compaction of the earth and which therefore uses the results of Proctor tests as formulation parameters. The values of optimum moisture content and dry density of earth used for CEB and rammed earth are given in Tables 2.28 and 2.29. The Proctor test results are reported for rammed earth, CEB and cob in Fig. 2.10.

The earth optimum water content is lower for rammed earth, while it is higher for the cob. The higher densities are found at lower water content. For rammed earth, the optimum water content varies between 8 and 21.5% and most values are below 12%. For CEB, the optimum water content is found between 9.8 and 18%. For cob,

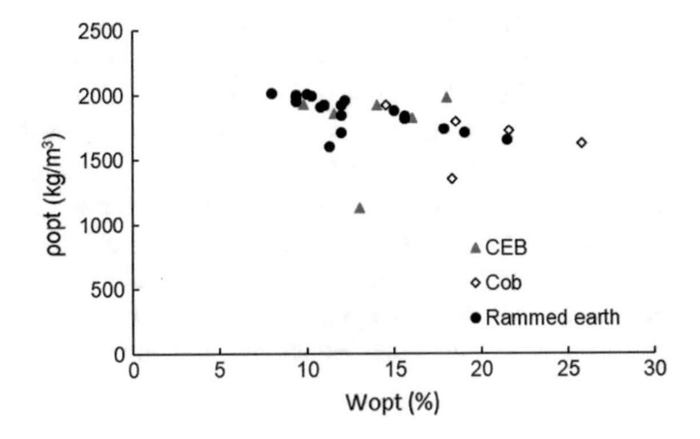

Fig. 2.10 Results from the Proctor tests (optimum density (ρ_{opt}) and optimum water content (w_{opt}) for CEB, cob and rammed earth

the optimum water content is between 14.5 and 25.8%. More characterisation is needed on the earth properties that allow engineering design, such as compressive strength, soil friction angle, California bearing ratio (CBR) and vane strength. The German standard on earth blocks and earth masonry mortar specify strength classes [54]. More data is needed to correlate the geotechnical properties of earth used in construction to engineering design properties.

2.5 Chemical Characterisation

The chemical characterization aims at determining the chemical properties of the soils used for earth construction material. Many chemical properties could be measured but the most important is then the elemental chemical composition since it permits, in combination with the mineralogical qualitative analysis, to calculate the mineral composition of the material (see Sect. 1.4). For physico-chemical analysis, the samples must be prepared based on [55]. Other chemical parameters, such as the amount of organic matters, the soluble salts or the pH, are important especially in the case of stabilization using mineral binders because they could affect the reaction with the binders.

2.5.1 *Procedures and Standards*

Chemical composition (major elements)

Two main techniques are used in the literature: the Inductively Coupled Plasma—Atomic Emission Spectrometry (ICP-AES) (or Inductively Coupled Plasma—Optical Emission Spectrometry (ICP-OES)) and the X-ray fluorescence spectroscopy.

Analysis by the ICP-AES (with the exception of laser ablation systems), requires samples to be completely dissolved (digested) into a solution. Hence, a sample of the material in powder ($\leq$ 80 μm) is melted in combination with lithium metaborate ($LiBO_2$) and/or lithium tetraborate ($Li_2B_4O_7$) to form beads. These beads are dissolved in one or more acids as hydrofluoric (HF), nitric (HNO_3) or hydrochloric (HCl) acids. The obtained solution is then used for the ICP analysis.

The X-ray fluorescence spectroscopy test is carried out either on beads prepared with the same procedure as for the ICP test, or on pressed tablets of the material in powder ($\leq$ 80 μm), aggregated with an organic waxy binder or not. The principle of X-ray fluorescence is the analysis of the X-ray emitted by the matter excited by an incident X-ray source.

Two other techniques could be used too and are based on the same principle: it consists in the analysis of the X-ray emitted by a beam of electrons. The interactions between the electron beam and the matter are used in two techniques: the Electron

Table 2.7 Requirements for organic matter

References	Requirements for organic matter
[27]	"Soil shall be free of organic materials" (p. 18)
[57]	"Soil shall not be used if contains organic matter prone to rot or breakdown within the wall" (p. 15)
[29]	"Soil should be free from organic material" (p. 6)
[58]	"A musty aroma indicates an unacceptable quantity of organic matter and the soil should, therefore, be rejected" (p. 131)
[31]	< 2% by mass (p. 37)
[15]	< 2 to 4% by mass (p. 34)
[59]	"The soil shall be free of all organic matter" (p. 5)
[60]	"The smell test is sufficient for rejection of a soil: organic soil is identifiable by its strong smell of humus. The smell test should be performed immediately after extraction of the soil" (p. 8)

Adapted from [25]

Dispersive Spectrometer (EDS) analysis coupled with Scanning Electron Microscope (SEM) or microprobe analysis. The main inconvenient of these techniques is that the result may not be representative of the entire material since the analysis is carried out on a small zone of the sample but they allow having an image of this zone.

Organic matter content

The organic matter is rarely a problem because earth used for construction is often extracted from the geotechnical soil (cf. Fig. 1.1) and in this case, the amount of organic matter is negligible. But, in some specific cases, e.g. some Canadian soils rich in organic matter, it could be necessary to measure the organic matter content of the soils [56]. Van Damme and Houben [1] also reviewed the soil classification systems and, preventing organic soil for construction, refer that subsoil is preferred. Gomes et al. [25] have extracted the requirements for organic matter from the literature and their results are presented in Table 2.7.

This content can be determined either by the calcination method [61] or the chemical method [62]. For the calcination method, several dried samples are respectively weighed before and after a 3 h (or more if necessary to achieve a constant mass) heat at 450–550 °C. The mass loss is assumed to represent the organic matter. The materials' organic matter content is then computed as the mean value of the mass loss percentages of the respective samples. The chemical method consists in determining the carbon content of a soil sample by mixing it with an oxidizing solution (potassium dichromate with sulphuric acid). Once the oxidation has been completed, the quantity of products, which has reacted with the carbon of the soil, is measured.

Calcite content

Free carbonates in soils, such as calcite, aragonite and dolomite, affect their physical and chemical properties. The determination of calcite content or equivalent CO_2

is an important point. The evaluation of calcite content can be realised by various techniques: titration [37], gravimetric [63] or volumetric measurement [64, 65]. In all cases, the measurement principle is based on the determination of the volume of CO_2 released by the soil sample under the action of excess hydrochloric acid. A quantity of soil (10 g for soils estimated to be low in carbonates and 0.25 g for chalky soils) is mixed with a solution of hydrochloric acid. Carbonates being unstable at a pH value lower than 7, dissociate leading to the formation of CO_2.

According to EN ISO 10693 [64], ASTM D4373 [66] and NF P94-048 [65], the CO_2 is recovered by the intermediary of a calcimeter. The volume of gas produced allow the evaluation of CO_2 content in the sample.

According to [37], the excess hydrochloric acid is dosed from a NaOH solution and an indicator solution. Thus, evaluating the amount of acid consumed by the reaction, it is possible to calculate the amount of CO_2 initially present in the sample.

Loss on ignition

The loss on ignition (LOI) is the mass fraction lost by a dried sample by ignition at a specified temperature. The LOI is related to the organic content of certain soils (sandy, clayey, chalky). The procedure for soil is specified in [37] and XP P94-047 [61]. A dried soil sample passing the 2 mm test sieve is heated to a constant mass at 550 °C, not less than 3 h. The LOI is determined by the ratio between the mass loss of samples during heat treatment and the initial dried mass of the sample.

When the purpose is to determine all volatile species (hydroxyls, carbonates), the test is then conducted on samples at the temperature of 1000 °C during two hours. The change in mass after 1000 °C heating represents the total mass loss of the sample including organic matter, hydroxyl for clay minerals and carbonate. Usually the loss on ignition is also determined during the chemical composition test by ICP.

pH measurement

The soils' pH measurement is useful to know its minerals solubility and its ion mobility. The test is performed on air dried samples of the soil without coarse grains ($\leq$ 2 mm). A given amount of the material is mixed with either a pure water, a 0.01 mol/L solution of chloride calcium ($CaCl_2$) or 1 mol/L solution of potassium chloride. The suspension is stirred for a few minutes, covered with a cover glass and allowed to stand for a couple of hours before pH value measurement. The suspension needs to be stirred right before the pH value measurement. The European standard [67] and International standard [64] recommend a volumetric ratio of 1:5 and a rest at most 3 h, while the British standard [37] recommend a volumetric ratio of 1:2.5 and a rest of at least 8 h, and the American one (ASTM D 4972 [68] states a mass concentration (10 g of air-dried soil for 10 mL of solution) and a rest of 1 h. This last procedure can induce a bias in the case of lightweight soil. Generally, standards recommend performing a test with pure water and one with chloride solution for the determination of the soils' pH value. Both preparations are required to fully assess the soils' pH. A pH meter or a pH paper (low accuracy) can be used for the measurement.

Cation exchange capacity test

The clay minerals in fine soils have a negative surface charge that is balanced by bound cations at the mineral surface. These bound cations can be exchanged by other cations in the pore water. Cation-exchange capacity (CEC) is defined as the amount of positive charge, generally calcium (Ca), sodium (Na), magnesium (Mg) and potassium (K), that can be exchanged per mass of soil. This test makes it possible to estimate the behaviour of the soil during stabilisation with inorganic binder and, therefore, the cationic exchanges between inorganic binder and clays. It has two origins: the isomorphic substitution in the tetrahedral and/or octahedral sheet of clays, which is not dependent of the system pH value, and the dissociation of aluminium groups on the edge of the sheet of clays, which is pH value dependant. CEC is usually measured in centimoles of positive electric charge ($cmol_c/kg$). Numerous techniques were developed for the CEC measurement of soil. Commonly CEC is measured by displacing all the bound cation with a concentrated solution of another cation, and then measuring either the displaced cations or the mount of added cations that is retrained.

Different solutions are used based on salt (ammonium acetate [69], sodium acetate [70], barium chloride [71], cationic surfactants [72], metal–organic complex (cobalt hexamine [73–75], silver-thiourea [76], copper bisethylenediamine [77] or copper triethylenetetramine [78]. Standards and literature focused mostly on three procedures using ammonium acetate [79, 80], cobaltihexammine trichloride [80, 81] and barium chloride [82].

Cobalt hexamine chloride method [81]—The exchange is carried out by simply shaking the test portion in the reagent. For a given volume of reagent (100 mL), the quantity of samples weighed (2.5, 5 or 10 g) is such that a sufficient concentration of cobalt hexamine ions remains in solution. This concentration is determined by spectro-colorimetry without chemical pretreatment of the solution. The loss of cobalt hexamine from solution gives the CEC of the sample. Exchangeable cation contents are measured either by flame atomic emission spectrometry for K or by flame atomic absorption spectrometry for Ca and Mg. This procedure is recommended for soil with a natural pH value lower than 6.5.

Barium chloride method [82]—A soil test portion of 2.5 g (< 2 mm) is shaken for 1 h with 30 mL of 0.1 mol L^{-1} $BaCl_2$ solution. The solid and liquid phases are separated by centrifugation. This operation is repeated twice and the three supernatants are collected for the determination of exchanged cations. After equilibrating under shaken overnight the soil with 30 mL of 0.0025 mol L^{-1} $BaCl_2$, the solid phase is shaken once again, but this time with 30 mL of 0.02 mol L^{-1} magnesium sulphate ($MgSO_4$) solution overnight. The adsorbed barium exchanges with magnesium and precipitates in the form of $BaSO_4$. The residual content of magnesium in leaching solution is measured by flame atomic absorption spectrometry and subtracted from the initial content. The difference gives the CEC value.

Ammonium acetate method [80]—Widely used throughout the world, the ammonium acetate method was proposed by Metson [83]. The saturation of the exchange sites by ammonium is carried out by percolating a 1 mol L^{-1} ammonium acetate solution (75 mL) through a test portion of 2.5 g of soil. The excess reagent is eliminated with several rinses with ethanol (75 mL). After drying in air, the solid phase is agitated in 50 mL of a 1 mol L^{-1} solution of sodium chloride. The exchanged ammonium is measured by spectrocolorimetry, which permits the measurement of CEC. However, the obtained solution must be used with caution. In fact, the use of ammonium acetate induces a measurement carried out at a pH value of 7, due to the large excess of sodium acetate, and the dissolution of a part of carbonate species.

Ideally, the CEC measurement should be performed at the natural soil pH value in order to avoid the modification of electrical charges [84]. This induces a dissolution of carbonate and a modification of CEC value. Most of the literature agree that the use of cobalt hexamine trichloride procedure gives reliable and accurate value of the effective CEC, that is to say the CEC value at soil natural pH value [85].

Soluble salt content (nitrate, sulfate, chloride)

The most common soluble salts in soils are the cation calcium (Ca^{+2}), magnesium (Mg^{+2}) and sodium (Na^+), and the anion chloride (Cl^-), sulphate ($SO_4{}^{2-}$) and bicarbonate ($HCO_3{}^-$). Potassium (K^+), ammonium ($NH_4{}^+$), nitrate ($NO_3{}^-$) and carbonate ($CO_3{}^{2-}$) can also be found in most soils in a lower quantity. The soluble salt content of the soil is an important element that determines the quality of the soil for its use in construction (Table 2.8).

The determination of the soluble salt content can be approached in different ways. It is indeed possible to assess this content from a qualitative point of view. In this case, standard [86] is used. The soil sample is extracted with water with an extraction ratio of 1:5 (m/V). The specific electrical conductivity of the extract is then measured. The higher the concentration of salt in a solution, the higher will be the electrical conductance (the reciprocal of resistance) (Table 2.9).

Another possibility is to try to quantify the amount of each salt, especially sulphate, nitrate and chloride. Sulphate quantification is described in [37, 88]. The sulphate is extracted from the dried soil samples using dilute hydrochloric acid or water in a soil/added water ratio of 1:2 or 1:5 (m/V). The sulphate content of these extracts is determined by a gravimetric method according to which barium chloride is added to the aqueous or acid extract. The precipitate of barium sulphate is dried and weighed

Table 2.8 Requirements for salt content

References	Requirements for salt content
[57]	"Shall not be used soils containing water soluble salts to an extent which will impair the strength or durability of the wall" (p. 15)
[29]	"Soil should be free from salts such as sulphates" (p. 6)
[31]	< 2% (p. 37)
[15]	"Sulphates of sodium, magnesium and calcium are dangerous to soils used in earth construction, since they crystallize, making it easily broken" (p. 23)
[59]	< 2% (p. 5)

Adapted from [25]

Table 2.9 Classification of salinity according to soil electrical conductivity (ECe in μS/cm at 25 °C) [87]

Grade	CEe (μS/cm)	Soil quality
I	0–500	Unsalty
II	500–1000	Slightly salty
III	1000–2000	Salty
IV	2000–4000	Very salty
V	> 4000	Extremely salty

and the sulphate content is then calculated from the mass of the soil used in the analysis and the mass of precipitated barium sulphate.

Chloride quantification is described in [37] based on Charpentier-Volhard's method. The chloride is extracted from the dried soil samples water. Silver nitrate ($AgNO_3$) is added to the aqueous soil extract. The solution is then diluted and then titrated using a solution of potassium thiocyantate (KSCN) in the presence of ammonium ferric sulphate as a coloured indicator. The soluble chloride is calculated based on the volume of silver nitrate added and on the mass of the soil used in the analysis.

Nitrate, nitrite and ammonium quantification are described in ISO 14256-2 [89] using automatic measurement by spectrophotometry. The homogenized soil samples are extracted using a potassium chloride solution (1 mol/L). The concentrations of mineral nitrogen compounds, namely nitrate, nitrite and ammonium, are determined in the extracts by automated spectrophotometric methods.

2.5.2 *Study of Data from Literature*

Table 2.10 shows the chemical composition of the soils used in the studies of the literature for Extruded Earth Bricks (EEB), Compressed Earth Bricks (CEB) and adobes.

The study of this table shows that these results are difficult to exploit and that it is difficult to compare materials with each other. Nevertheless, we can extract some information from Table 2.10. First of all, we note that the following elements can be considered, in all the soils analysed, as "minor", that means that their concentration is always lower than 2% (expressed as oxides): Na_2O, TiO_2, P_2O_5 and MnO. If we consider the average concentration, we can classify the other oxides in ascending order: K_2O, MgO, Fe_2O_3, Al_2O_3, CaO and SiO_2. The cases of iron and calcium are interesting because depending on the type of soils, the contents could be very low or very high. For iron, the concentrations range from 1.7 to 15.0%. This element will play a relatively small role in the behaviour of earth materials. The iron oxides, being often in the form of goethite (FeO(OH)), play the role of inert raw earth materials. However, the nature and the content of iron oxides will have a very important influence on the colour of the earth, especially in the field of fired clay bricks. In the case of calcium, some soils may not contain it at all (0.03% for soil B studied by Ammari et al. [92]) and others contain very large amounts: 31.8% for soil M studied by Ammari et al.

Table 2.10 Chemical composition of the soils used in the studies of the literature

References	Name	Type	Major elements (%)												
			SiO_2	Al_2O_3	CaO	Fe_2O_3	K_2O	MgO	Na_2O	TiO_2	P_2O_5	MnO	SO_3	Cl	LOI
[90]	A	EEB	64.22	14.59	1.04	5.66	3.08	1.51	0.35	0.9	n.m	n.m	n.m	n.m	8.75
	B		50.72	12.83	1.61	15.03	2.63	2.27	0.28	0.8	n.m	n.m	n.m	n.m	14.11
	C		73.97	8.82	4.18	3.22	2.0	0.73	1.01	0.77	n.m	n.m	n.m	n.m	6.03
[91]			64.7	16.6	1.1	4.8	4.0	1.1	0.2	0.6	0.1	0.1	n.m	n.m	5.9
[92]	M	CEB	22.3	6.0	31.8	1.7	0.5	2.1	0.3	0.2	0.8	0.1	1.29	n.m	32.9
	B		79.4	11.9	0.03	1.8	0.3	tr	0.0	1.3	–	0.0	0.0	n.m	6.3
[93]			54.7	19.7	0.9	8.6	3.9	3.6	1.8	1.0	0.2	0.1	n.m	n.m	5.0
[94]			18.73	7.47	35.3	3.39	0.9	1.27	0.09	0.39	0.09	0.03	n.m	n.m	31.92
[95]		Adobe	77.81	10.06	0.33	3.23	1.6	0.39	0.24	1.08	0.08	0.04	n.m	n.m	
[96]	PrA1		29.7	6.38	41.0	11.9	2.32	6.3	0.48	1.11	–	0.12	0.12	–	–
	PrA2		10.7	2.4	81.0	2.95	0.72	1.39	–	0.39	–	–	–	–	–
	PrA3		10.8	1.72	48.3	1.66	0.47	1.72	1.31	–	0.35	–	33.2	–	–
	PrA4		22.6	4.26	46.6	5.43	1.44	2.55	0.75	0.69	0.21	–	14.9	–	–
	PrA5		13.9	2.59	50.1	3.05	0.82	1.81	1.02	0.29	0.32	–	25.6	–	–
	PrA6		22.5	3.6	54.4	3.35	1.97	2.51	1.09	0.74	0.71	0.11	1.99	0.46	–
	PrA8		18.8	2.72	59.5	2.81	1.59	1.69	0.71	0.66	0.97	0.10	2.82	0.30	–
	OA1		41.6	7.53	28.4	11.4	1.74	5.26	1.83	1.3	–	–	0.32	0.24	–
	OA2		28.8	5.45	46.8	9.01	1.74	4.2	1.2	0.91	–	–	0.18	1.21	–
	OA3		24.1	4.46	61.0	4.99	1.32	2.42	0.17	0.72	–	–	0.12	0.26	–
	OA4		44.6	8.96	21.0	13.4	2.21	6.87	1.0	1.15	–	–	0.12	0.18	–

(continued)

Table 2.10 (continued)

References	Name	Type	Major elements (%)												
			SiO_2	Al_2O_3	CaO	Fe_2O_3	K_2O	MgO	Na_2O	TiO_2	P_2O_5	MnO	SO_3	Cl	LOI
	OA5		30.4	6.04	48.2	7.84	1.85	3.42	0.43	0.97	–	–	0.22	0.25	–
	OA6		40.4	8.18	27.3	12.8	1.72	6.05	1.46	1.32	–	–	0.20	0.10	–
	LyB		40.7	8.1	26.9	12.3	1.56	6.02	1.85	1.26	–	–	0.35	0.59	–
	AthB		26.1	4.5	55.6	5.84	1.6	3.29	0.61	0.73	0.43	–	0.70	0.23	–
[97]			50.6	23.44	2.48	5.81	1.69	0.02	0.05	0.65	0.17	0.02	n.m	n.m	14.91
[98]			49.88	25.95	0.05	9.69	0.39	0.17	0.03	1.12	0.04	0.05	n.m	n.m	11.78
[99]			66.13	14.38	0.41	6.68	1.0	0.45	0.24	1.09	0.06	0.15	n.m	n.m	8.93
Maximum (%)			79.40	25.95	81.00	15.03	4.00	6.87	1.85	1.32	0.97	0.15	33.20	1.21	32.90
Average (%)			39.96	9.21	28.71	6.61	1.67	2.56	0.71	0.85	0.32	0.08	5.13	0.38	12.61
Minimum (%)			10.70	1.72	0.03	1.66	0.30	0.00	0.00	0.20	0.04	0.00	0.00	0.10	4.77

LOI: loss on ignition

[92] or 35.3% for the soil studied by Laborel-Préneron et al. [94]. In these two cases, the loss on ignition is also very high (32.9% and 31.9% respectively), which leads to the conclusion that calcite ($CaCO_3$) is present (approximately 60% in both cases), which shows that these soils are strongly calcareous.

If we do not consider the calcium present in calcareous soils, the two major elements of the soils are silicon and aluminium, which seems to be logical for clayey soils: clays being phyllosilicates are rich in these two elements. The results of Castrillo et al. [96] were obtained by X-ray fluorescence spectroscopy and do not take into account the loss on ignition (LOI). The researcher normalized the concentrations of the oxides to 100%, which does not allow a direct comparison with the results of the other studies that integrate the loss on ignition. Finally, the measurement of total sulphates and chlorides contained in soils is relatively rare (only two out of nine studies). In some studies, the contents of soluble chlorides and sulphates are measured but this is very rare too. We can cite the study of Galan-Marin et al. [93] who measured soluble chloride contents equal to 0.03% or the study of [100] with content of sulphates and soluble chlorides equal to 0.64 and 0.07% respectively. Generally, these soluble salt contents are extremely low and rarely measured by the researcher.

In addition to the chemical composition, the two most frequently encountered chemical characteristics in the literature are the measurements of pH value and organic content. The values found in the literature are given in Table 2.11. It is important to note that in all the references studied almost none give CEC values in exception to [22, 101].

Although the measurement of the $CaCO_3$ content is relatively simple, this content is rarely given in the studies. This is no doubt explained by the fact that calcite could be considered as an inert in earth construction materials and the knowledge of its content in the soil is therefore not essential. The measurement of calcite content could be important in some specific cases of study as for instance for lime stabilised earth constructions. There are in Portugal examples of "military" rammed earth that was used since the XII century mainly for fortresses and that was stabilised with air lime [5]. For these specific cases of study, the knowledge of calcite content is useful because it permits to determine the amount of lime used for the stabilisation of these materials.

The pH values presented in Table 2.11 on soils reported in the literature vary from 4.8 to 9. It is difficult to compare these values directly because we have seen during the presentation of the procedures that the methods used for the measurement of pH value strongly vary (in particular the volumetric ratio), which has consequences on the pH value. Despite this, except the soil "U" studied by Dove et al. [101] which has a pH of 4.8, the pH values measured on the different soils are relatively close (between 6.6 and 9).

Finally, the organic contents of the soils studied in the literature are often less than 2% as recommended in Table 2.5 but in some studies, the contents may be higher. This is the case in the study of [25] who worked on unstabilised rammed earth collected from old constructions in south Portugal. In this study, some samples

Table 2.11 pH values, organic and $CaCO_3$ contents of the soils used in the studies of the literature

References	Name	Type	pH	Organic content (%)	$CaCO_3$ (%)
[101]	U	CEB	4.8		
	V		6.8		
	W		6.9		
[93]			8	0	12.4
[100]	1			0.78	16
[102]	BCS		8.95	0.67	
	TB1		8.05	2.32	
	TB2		7.22	1.40	
	TB3		6.58	1.26	
[94]					60
[103]	CLG-1S			0.73	
	CTL-2S			1.89	
	CTL-2A			1.58	
	CHG-3SS			1.44	
	CHG-3SP			0.78	
	CHG-3AR			1.25	
	CHG-3AA			1.48	
	CGR-4S			0.58	
	CHJJ-1S			0.44	
	CHJJ-5CA			1.47	
	LMD-6S			1.42	
	MNG-7A			2.66	
	NDD-8A			1.27	
	NPK-9S			0.76	
[104]				2.1	
[105]			7.4	1.7	22
[25]	Av	Rammed earth		0.9	
	PD			4.5	
	VC			3.5	
	CZ			1.8	
	Cl			3.6	
	Ar			5.4	
[106]	H2		5.8		
[107]	S1		7.7	0.9	
	S2		7.8	0.6	

(continued)

Table 2.11 (continued)

References	Name	Type	pH	Organic content (%)	$CaCO_3$ (%)
	S3		8	0.5	
	S4		8.1	0.4	
	S5		8.3	0.3	

(Ar, Cl and particularly PD) contained large sized organic matter particles that could explain the high organic content of these materials.

2.6 Mineralogical Characterisation

Chemical analysis provides important data on the chemical compounds contained in soils but they do not permit to know under which forms these elements are in the material. The behaviour of clay materials will essentially depend on the mineral form in which these elements are. For example, silicon will not react at all if it is under the form of quartz (SiO_2), clay, feldspar or mica. The mineralogical characterization therefore permits to complete the chemical analysis by determining the nature and, under certain conditions, the quantity of the minerals contained in a sample. Many techniques exist but the most used for the characterization of clay materials are X-Ray Diffraction (XRD), thermal analyses, infrared spectroscopy and microscopic observations.

2.6.1 *Procedures and Standards*

X-ray diffraction

XRD is an analytical technique used for phase identification of a crystalline material and can provide information on unit cell dimensions. The principle consists on placing the crystallized material in an intense beam of X-rays, usually of a single wavelength (monochromatic X-rays). This beam of X-rays is diffracted by the materials: the angles and intensities of diffracted X-rays are measured, with each compound having a unique diffraction pattern. By comparison with standards obtained on reference minerals, it is possible to determine the nature of the crystallized phases contained in the sample.

Conventionally, the tests are carried out on the sample crushed $< 80\ \mu m$. This technique is sufficient for materials without clay minerals or materials containing illite and kaolinite. But, if the sample contains clay minerals with a basal reflection (001) at 14 Å (typically chlorite, vermiculite or smectite), it is necessary to complete this first test by another one carried out on oriented aggregates using three preparations: air-dried or natural, after glycolation and after heat treatment at 500 °C [108, 109].

Thermal Analysis

Three types of thermal analysis exist and could be used to complete the XRD analysis: Differential Thermal Analysis (DTA), Differential Scanning Calorimetry (DSC) and Thermal Gravimetric Analysis (TGA). The principle of TGA consists in weighing the sample over time as the temperature increases. DTA and DSC are relatively similar techniques. In DTA, the material under study and an inert reference are submitted to identical thermal cycles while recording any temperature difference between the two samples. Changes in the sample, either exothermic or endothermic, are thus detected by comparison to the inert reference. In DSC, the difference in the amount of heat required to increase the temperature of a sample and a reference is measured as a function of temperature (generally, the temperature increases linearly as a function of time).

Concerning the mineralogical characterization of earth construction materials, DTA could be used in addition to the XRD to determine the nature of the minerals contained in soil. However, it is relatively rare that it brings new results and it often confirms the qualitative characterization performed by XRD. The results of DSC are less used for the characterization of soil but could be used for the determination of the thermal properties of earth construction materials such as the heat capacity. TGA is very useful because it permits to calculate the content of some minerals contained in soil such as goethite (FeO(OH)) or gibbsite ($Al(OH)_3$) (loss of weight around 300 °C for both minerals), clay minerals (dehydroxylation around 500 °C) and calcite (decarbonation around 700 °C).

Infrared spectroscopy

Infrared spectroscopy is based on the absorption phenomenon occurring when infrared radiation passes through a material. When a molecule is excited to its own energy of vibration, it absorbs the incident energy, thus allowing the study of the various bonds present in the material. The soils are mainly composed of silicon, aluminium, calcium, iron, alkaline elements and metals. The presence of these elements will induce a large number of possible atomic bonds (Si–O–Si, Si–O–Al, etc.), each one having different vibrational modes. The signal processing by Fourier transform allows to highlight vibration bands linked to the covalent bonds of the materials. Certain characteristic bands allow the identification of materials. Analysis can be done in transmission or Attenuated Total Reflection (ATR). In all cases, the range of analysis is in the mid-infrared between 4000 and 600 cm^{-1} (or if possible up to 250 cm^{-1}). Clay minerals are identified from the signal of chemical groups in specific regions [110, 111].

Transmission analysis requires the use of potassium bromide (KBr) for the production of pellets. The preparation is performed by mixing 0.5–1.5 mg of prepared soil with 100 mg of KBr in a mortar. The mixture is very finely ground. The mixture is then put into a 13 mm diameter steel pressing die and pressed at 10 t for 2 min. At the end of this time, the pellet is removed from the press and placed in a support suitable for the spectrometer.

The analysis in attenuated total reflection (ATR) allows the study of the soil, without addition of chemical products. The sample is placed on a crystal, usually diamond type, during the analysis. The infrared beam penetrates into the material over a thin thickness of the order of 5 μm. This rapid type of analysis leads to the identification of crystalline or amorphous mineralogical phases, but not to their quantitative evaluation.

Microscopy

Some researchers use SEM (Scanning Electron Microscopy) to complete the mineralogical characterization of soil. That permits to show some pictures of the microstructure of the soil. These observations could be completed by a very useful isolated chemical analysis using Energy Dispersive X-ray Spectroscopy (EDS).

2.6.2 *Study of Data from Literature*

Table 2.12 presents the numbers of qualitative and quantitative mineralogical compositions found in the literature. The number of articles where these compositions are presented is also given.

Table 2.12 shows that qualitative mineralogical analyses are relatively numerous even if they are found only in 24 articles out of 71 (Table 2.1). All these analysis use X-ray diffractions. Figures 2.10 and 2.11 illustrate a complex case of mineralogical characterization by XRD of clay soil from the study by Ouedraogo et al. [112]. This study focuses on the effects of stabilization by cement or lime of two soils of different mineralogy. XRD patterns of the two soils measured on crushed powder are presented on Fig. 2.11.

The XRD pattern on crushed powder leads to the identification of the main constituents of the two soils: quartz (SiO_2), calcite ($CaCO_3$), feldspar (albite ($NaSi_3AlO_8$) and orthoclase (KSi_3AlO_8) and goethite (FeO(OH). The most interesting result of this analysis is the differences in the nature of clay minerals. Soil B contained illite/muscovite (it is not possible to distinguish these two phases using XRD) and kaolinite. For soil N, it is not possible to determine the nature of clays with

Table 2.12 Numbers of mineralogical studies in the literature (with the number of articles where this qualitative or quantitative mineralogical composition is found)

Technology	"Qualitative" *(articles)*	"Quantitative" *(articles)*
EEB	9 *(4)*	1 *(1)*
Adobes	46 *(10)*	6 *(6)*
CEB	10 *(4)*	1 *(1)*
Rammed earth	13 *(5)*	1 *(1)*
Cob	4 *(1)*	0
Total	82 *(24)*	9 *(9)*

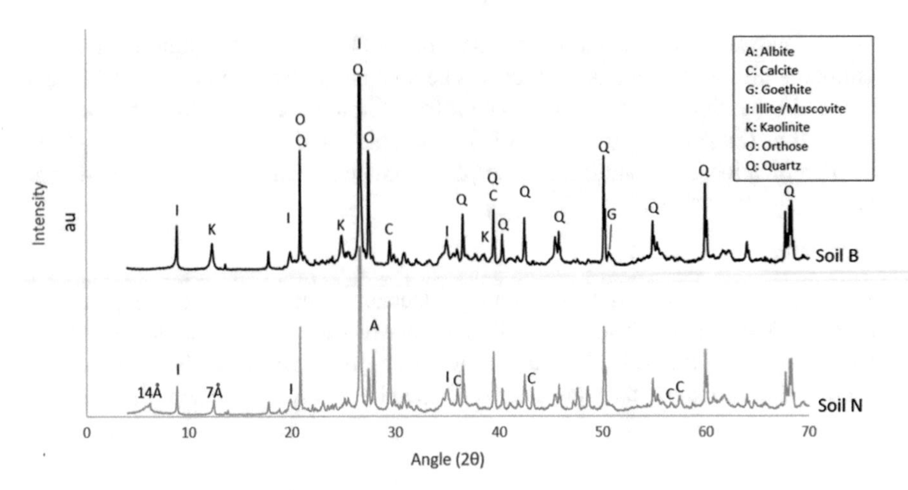

Fig. 2.11 X-ray diffraction patterns of two soils studied by Ouedraogo et al. [112]

the diagram of Fig. 2.11 because the peaks at 14 Å can correspond to various types of clay. It is necessary in this case of study to use the oriented aggregate technique, the XRD patterns for which are presented on Fig. 2.12.

The analysis of the evolution of the first four peaks of the pattern during the various preparations permits to conclude that the soil N contains three types of clay minerals: illite, chlorite and montmorillonite.

Quantitative mineralogical studies are much rarer (9 in 9 papers). In some papers, the authors provide semi-quantitative studies based on the intensity of the diffraction peaks of X-ray powder diagrams. In these papers, some tables present the minerals present with semi-quantitative criteria, as for example in the study by Gomes et al.

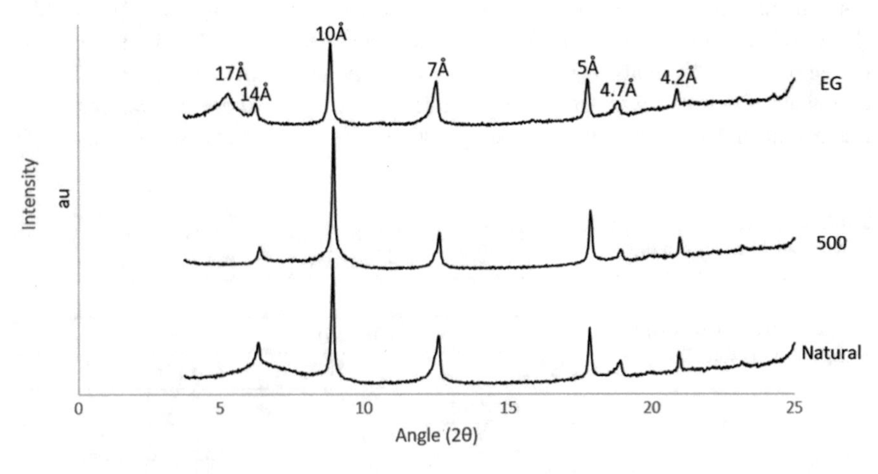

Fig. 2.12 X-ray diffractograms of oriented aggregates (EG: ethylene glycol, 500: heated at 50 °C and natural) [112]

[25]: "+++" corresponds to high proportion, "++" to intermediate proportion and "+" to low proportion. These semi-quantitative analyses are only indicative because all minerals do not diffract X-rays with the same intensity. These semi-quantitative studies are not considered in the quantitative studies listed in Table 2.12.

In several papers presenting quantitative mineralogical compositions, the methodologies used are questionable. For example, Duarte et al. [103] have determined the semi-quantitative abundance of the minerals using the Reference Intensity Ratio (RIR) method by XRD. In their article, no detail on this technique was given and the chemical composition of the studied samples was not presented too. It is then impossible to verify the correctness of the compositions given by this method by comparing the calculated mineralogical compositions with the measured chemical compositions. In other articles, Maskel et al. [113] or Wouatong et al. [114] give accurate mineralogical compositions but both groups of researchers do not present the method they used to measure these compositions and they do not give any chemical composition that could allow to check their results. The most serious example is undoubtedly that of the use of the Rietveld method for the calculation of the mineralogical compositions of earth building materials. Rietveld's method consists in calculating a X-ray diagram from crystallographic characteristics of reference minerals and in deducing the proportions of these phases by comparison to the real X-ray diagram of the studied sample. This quantification method is both very powerful and very robust, but also very dangerous for inexperienced users. Indeed, there are a lot of quantification software using Rietveld's method, which is very easy to use, automated and which will always give a final result to the user even if this result is perfectly aberrant. This quantification method is based on many hypotheses, which are not always verified and can lead to aberrant results. To illustrate this, it is possible to refer to the results obtained by Costi de Castrillo et al. [96] but other studies have certainly used this method as the previously mentioned studies in which the quantification method was not specified. Costi de Castrillo et al. [96] have compared adobes from pre-history to date. They characterized in depth 15 samples assessing in particular the chemical composition using XRF (Table 2 in the article) and the mineralogical composition with very high precision using the Rietveld method (Table 3 in the article). A quick comparison of the results shows that there is no consistency between the values of these two tables. It is likely that the chemical composition was correctly carried out, which means that the results of the quantification using Rietveld's method are false. To illustrate this, we will take a very simple example: the adobe "OA4" contains 1.00% of Na_2O. If we consider that all the sodium is in the form of albite (which is probably not true because these soils contain clay minerals and it is possible that a part of the sodium is present in some of these clays), this amount of sodium would correspond to 8.00% of albite (Si_3O_8AlNa) while the computer using the Rietveld's software finds 41.53% of albite! The same observations could be done with all the results of these two tables. It is possible that the Rietveld's method can strongly help for the quantification of the mineral phases contained in an earth material but the use of this tool requires significant know-how and skills in crystallography and mineralogy and it is therefore advisable to be very careful about its use.

There are examples in other articles where the researchers use robust methods of quantification based on the qualitative characterization by XRD coupled with a calculation using the chemical composition of the soils studied [94, 97–99]. A simple example of this calculation is given in the study of [98] on earth blocks stabilized by cow dung. The XRD spectrum of the soil used in this study showed that it only contained kaolinite ($Al_2(Si_2O_5)(OH)_4$), quartz (SiO_2) and goethite ($FeO(OH)$). By using the chemical composition of this soil, it is very easy to calculate the amount of kaolinite (by using the content of Al_2O_3), of quartz (content of SiO_2 corrected by the amount of SiO_2 contained in kaolinite) and of goethite (amount of Fe_2O_3). To make this calculation, it is nevertheless necessary to know the chemical formula of the minerals present in the soils and in particular the clay minerals. For the quoted studies, the calculations are relatively simple because the studies relate to soils which mainly contain kaolinite whose chemical formula is simple. However, in the case of soils containing other clay minerals such as illite, montmorillonite or chlorite for example, the chemical compositions of these minerals are complex and variable, which extremely complicates the calculation of the mineralogical composition [115].

In such complex cases of study, thermal gravimetric analysis (TGA) is a complementary tool that can help to improve the accuracy of the calculation or to check the correctness of the results obtained by the calculation. As an example, the differential thermal gravimetric analyses (DTGA) of the two soils studied by Ouedraogo et al. [112] are presented on Fig. 2.13.

The DTGA presented in Fig. 2.13 shows the presence of the same four main peaks for the two soils even if their qualitative mineralogical compositions are different (especially the nature of clay minerals):

- 100–200 °C—loss of hygroscopic water (water strongly linked to the material);
- 300 °C—dehydroxylation of goethite (FeO(OH));

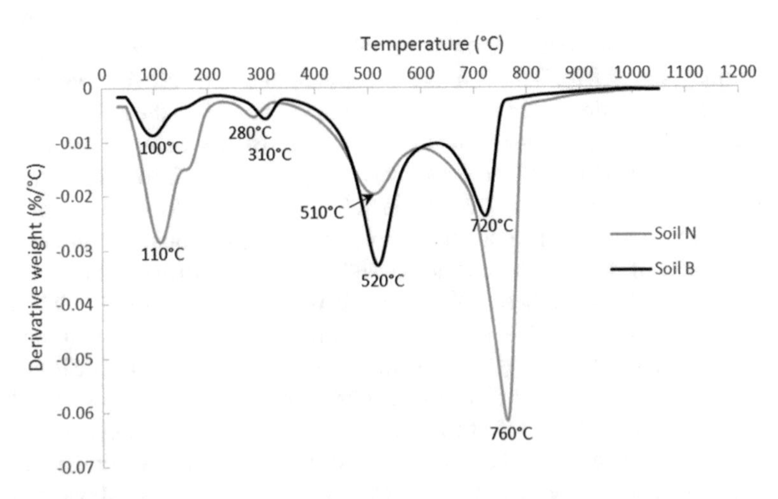

Fig. 2.13 Differential thermal gravimetric analyses of the two soils studied by Ouedraogo et al. [112]

- 500–550 °C—dehydroxylation of clay minerals;
- 700–800 °C—decarbonation of calcite ($CaCO_3$).

Some differences of intensity between the peaks of the two soils exist and especially for the first peak, corresponding to hygroscopic water. This peak is much higher for soil N that is consistent with the nature of the clay minerals it contains (chlorite and essentially montmorillonite are able to "stock" a lot of water in opposition of kaolinite contained in soil B). Finally, the main interest of DTGA is the possibility to quantify the numbers of some minerals contained in the soils. In this case of study, it is possible to quantity calcite and goethite for the two soils: they are equal to 11.6% (respectively 1.3%) for soil N and 5.3% (respectively 2.2%) for soil B.

Finally, examples of the use of IR spectroscopy to characterize earth materials are rare. We can quote the studies of [99, 112]. In these studies, the analysis of the IR spectra carried out on the soils confirms the qualitative analyses obtained by XRD and TGA but it does not bring any new results. This technique is still underused for the moment but it is likely that it could be very useful to improve the understanding of the phenomena that occur during the chemical stabilization of earth materials.

2.7 Field Tests

2.7.1 Procedures and Standards

Several field-test procedures are used. They have been developed since long, based on the expertise of professionals. To allow comparison between qualitative tests, in some cases each sample is assigned a score based on the soils' performance. Some examples of score assignments for the qualitative tests are presented with the test itself. In other cases, a table is presented. They can provide information namely about particle size and shape, presence of organic compounds, clay, silt and sand content and adequability to a determined building technique.

Visual inspection of soils

Visual inspection of a soil can be carried out based in [116] and other simple test procedures, evaluating more objectively properties such as colour, texture (angularity, shape of particles) and odour of a soil.

The visual particle size test: a thin layer of soil is placed on a surface and particles are pressed with fingers to be sure that clay granules are scrapped. The grain particles that are visible are the sand and gravel; the rest is clay and silt. If the volume of clay and silt is bigger than the rest, the soil is considered as not sandy [117]. The observation of the sand and gravel allows to assess angularity.

The touch test: a small portion of dried soil is picked between two fingers. Sand is detected because the grains can be felt. If the touch is silky, there is clay or silt.

Table 2.13 Exudation test assessment [117]

Reaction	Number of bangs	Effect	Classification
Rapid	5–10	Water appear at the surface; it disappears with finger impression; higher impression disintegrates the sample	Low plasticity; fine inorganic sand or coarse silt; sandy or silty soil
Slow	20–30	Water appears and slowly disappears; finger pressure deforms the sample	Slightly plastic silt or clayey-silt soil
Very slow	> 30	No significative change	High plasticity clayey soil

The wash test: a portion of soil is placed in the hand and washed with water. A content of fines is washed and coarser particles (namely sand) can be observed, allowing to assess their particle size and shape.

The exudation test: the test assesses the soil plasticity function of water retention. It consists in adding water to a soil sample, moulding and placing it in the palm of the hand. With the other hand, the sample is banged so that water gets out and the sample presents a shiny bright surface. Qualitative classification can be assessed by Table 2.13.

The colour and odour test: a portion of dry soil is observed: light and bright colours are characteristics of inorganic soils while dark colours are characteristics of organic ones. A strong smell when the soil is moistened is also characteristic of organic soil [117].

Ball and stick soil tests

The ball test: a portion of soil is moistened and moulded to form a ball. The content of clay is directly related with the easiness to mould a ball. The ball is dried and if the form is maintained, the soil can be considered apt to construction without need of stabilization [117]. If it disintegrates, it has too low clay content and is not apt to earth construction.

The dropping ball test: a ball similar to the previous one is moulded with about 3 cm diameter and let fall from 1 m high. The disintegration of the ball is observed: for sandy soils, the ball disintegrates in small portions while for clayey soils the ball just deforms by the impact with the floor [57, 117]. This test is considered adequate to evaluate the optimum moisture content for non-plastic earthen techniques, such as rammed earth or compressed earth blocks:

- if the ball disintegrates in small portions, the soil is too dry;
- if it breaks in 4–5 pieces, the moisture content is optimized;
- if it just deforms and stays in one portion, it is too wet [118].

The stick test: a sample of non-sandy soil is added water to form a paste that is rolled into a compact ball by hand, and stabbed with a knife. The amount of soil that clings to the knife when it is removed is observed. Soil sticking to the knife indicates high clay content. In their study, because they used artificial neural networks, Sitton et al. [119] need values to compute statistics in the method they developed for rapid soil classification for use in constructing compressed earth blocks. The qualitative classification they proposed is the following:

- 1: soil is adhered to the blade when the knife is pulled out of the sample;
- 0: there are streaks of soil residues on the blade when the knife is pulled out;
- −1: there are no soil residues when the knife is pulled out of the sample.

The shine test: a sample of soil is added water to form a paste that is rolled into a compact ball hand sized, that is cut in half with a knife. The cross section of the ball is observed. A glossy/shine cross section indicates high clay content, a dull cross section indicates higher silt or sand content. As it was the case for the stick test, Sitton et al. [119] have proposed a qualitative classification:

- 1: the ball cross-section appears glossy and reflects light: it is a clayey soil,
- 0: the cross-section appears somewhat glossy but is not very glossy and does not reflect much light: it is a silty soil,
- −1: the cross-section is not glossy and does not reflect light: the soil is sandy.

Cord and ribbon soil tests

The cord test: the test assesses soil cohesion and plasticity with a determined moisture content to classify the type of soil. A soil sample is moistened so that, moulding by hand on a flat surface, it is possible to form a soil cord that brakes when it has about 3 mm diameter. A ball is moulded immediately with that 3 mm cord and is pressed between thumb and forefinger [117]. The force to achieve that and the classification obtained are presented in Table 2.14.

The ribbon or cigar test: This test is correlated to the plasticity of the soil [58, 120]. Water is mixed with a soil sample to form a paste, as for the cord test. The paste is rolled on a flat surface and moulded manually into a long cigarette shape (approximately 1 cm diameter and 8 cm long; if longer, it should be cut). The soil is classified with low plasticity if it breaks before it can reach the diameter of the cigarette. With the thumb and forefinger, the cylinder is pressed to form a stripe with 3–6 mm thickness and as long as possible [117]. For that, part of the soil stripe is in the vertical position (draped over the side of the hand) while the rest of the cylinder

Table 2.14 Assessment of the cord test and soil classification [117]

Qualitative ball rupture	Soil classification
Hard to press and does not disintegrate	High clay content; high plasticity
Low resistance; cracks and disintegrate easily	Clay-silty, sandy or sandy-silty soil; medium plasticity
Brakes when pressed and cannot be moulded again	High silt or sand content and low clay content; low plasticity
Resilient when pressed	Organic soil, not adequate for construction

is being pressed. If it can be rolled to the diameter of a cigarette and can support its own weight when draped over the side of the hand, the soil has high plasticity. The following qualitative classification is adapted from Neves et al. [117] and Sitton et al. [119]:

- 2: the ribbon does not break when draped over the side of the hand up to 25–30 cm long—high clay content and classified as high plasticity soil;
- 1: the ribbon does not break when draped over the side of the hand up to 5–10 cm long—clay-silty or sand-clayey soil, classified as medium plasticity soil;
- 0: the cylinder holds together and can be rolled out; however, it breaks before it reaches approximately the diameter of the cigar or the ribbon breaks when draped over the side of the hand—silty or sandy soil, with low clay content, classified as low plasticity soil;
- −1: the cylinder does not hold together or cannot be rolled out without crumbling.

The pen or roll test: this test can be performed to assess if a clayey sample of soil is adequate for ramming. A paste of soil sample, already kneaded with water, is moulded on a plane surface with one rounded section to provide a 20 cm long soil cord with 2.5 cm diameter, similar to a pen. The soil pen is placed perpendicular to the round section of the plane surface and is slowly pulled forward so that the first centimetres of the soil pen get in the vertical position, while the rest is still horizontal, maintained by the hand, until a rupture occurs [117]. The following qualitative classification is proposed by Neves et al. [117]:

- 1: too high clay content for ramming is obtained with a vertical segment with more than 12 cm;
- 0: a vertical segment between 8 and 12 cm indicate an ideal clay content for ramming;
- −1: the soil has not enough clay content for ramming if the vertical segment has less than 8 cm.

Sedimentation field tests

The tube particle gradation test: a dry soil sample is dispersed in water in a test tube marked "sand". The sample will start to slowly settle out of suspension at the bottom of the test tube. After different time increments, the portion of the sample remaining in suspension is transferred to a "silt" test tube and then to a "clay" test tube. Essentially, the test uses gravity to separate the soil sample into different groups based on particle size and can give an approximation of the soils' particle gradation and fine content [119].

The Jar test: based on [58], a soil sample is placed inside a glass jar with a flat base. Water is added up to 2/3 of the jar height and a low amount of sodium chloride can be added to act as deflocculant agent [117]. The jar is capped, vigorously agitated, rested during 1 h and agitated again. Each soil component type, with different loose bulk density, settles out with different velocities. Therefore, once all of the samples have settled out, different strata can be observed and measured. If there are organic compounds, they will float at the surface. Nevertheless, sometimes it is difficult to locate and measure the strata between finer sands and coarser silts, finer silts and clays using this method. Furthermore, Silva et al. [121] refers the need of considering that flocculated clays occupy a much larger volume than the same clay in deflocculated state.

Linear and volumetric shrinkage test

The shrinkage tests can be easily performed in situ and allow assessing the viability of an earth or/and earthen mix to a defined technique. An earthen material, mixed with a water content that is suitable for the building technique (only moistened and compressed if it is for rammed earth or compressed earth blocks, in plastic state and not as much compressed if it is for adobe or cob) fills a longitudinal mould and is let to dry.

The linear shrinkage test can be visually evaluated by the cracking and space between the sample and the moulds. It can be quantified by the perceptual metric difference of the sample to the mould. Some recommendations for the moulds' dimensions, the moisture content, sampling and drying were gathered by Gomes et al. [25] and presented in Table 2.15.

In some cases of earth construction, the volumetric shrinkage should also be considered. The same samples used to assess linear shrinkage can be used and the difference on samples high is also considered to evaluate volume change [25].

The dry resistance tests

A soil sample is moistened and mixed to produce a planar specimen with 1 cm thick. Cylindrical samples with 3 cm diameter are cut from the specimen and let to dry. Each dry sample is firstly pressed by hand between the thumb and forefinger and, afterwards, if it has strength enough, it is broken using thumbs and forefingers of both hands [15]. A qualitative classification was proposed by Neves et al. [117] as presented in Table 2.16.

2.7.2 Study of Data from Literature

Silva et al. [121] analysed four superficial soils collected in North Portugal. By visual field tests, the soils were characterized by colour: light tones such as grey and yellow. In terms of angularity, the soils present subangular particles and one also presents sub-rounded particles. The soil particles were not elongated nor flat in terms of shape, whereby researchers expected that destabilised rammed earth walls built with those soils were expected to have lower mechanical properties than those built with schist residual soils from southern Portugal. Odour was not identified which indicates absence of organic matter.

The same soils were characterized by the Jar test. Three of them present very low clay content and researchers suggest they are not suitable to destabilised earth construction, compared to recommended values from literature [15, 58]. Only one tested soil present clay content within recommended values for rammed earth. When

Table 2.15 Recommendations and requirements for linear shrinkage test

Reference	Box dimensions (cm)	Water content	Material	Drying	Maximum shrinkage (%)
[27]	60 × 4 × 4	Optimum moisture content	Same material than in the wall	3 days in the sun	2[a]
[57]	60 × 5 × 5	Same moisture content than in the wall	Same material than in the wall	7 days with the sample covered by a plastic sheet. 21 days in the air out of direct sun light	0.05
[58]	60 × 4 × 4	Optimum moisture content	Size fractions 6.00 mm sample with 2–2.5 kg	3–7 days in the sun	2.5[b]
[30]	60 × 5 × 5	Optimum moisture content	Same material than in the wall	Until complete drying	0.25
[60]	60 × 5 × 5	Not mentioned	Remove the coarse fraction (quantitative values are not specified)	Until complete drying	2

Adapted from [25]
[a]For higher shrinkage values, the reference recommends adding a certain percentage of cement or of low clay content soil (sand/aggregate)
[b]For stabilized rammed earth with 4–6% cement content; the document provides threshold values for cement contents from 4–6 to 10%; the threshold value increases with the cement content.

Table 2.16 Dry resistance test assessment and soil classification

Dry resistance	Rupture	Behaviour	Classification
High	Resistant	Does not turn to dust	Inorganic high plasticity earth; clayey soil
Medium	Medium resistant	Pieces can be turn to dust	Clay-silty soil, clayey-sand soil or clayey sand; do not use if an organic clay
Low	Does not resist	Easily disintegrates	Lack of cohesion; sandy soil, inorganic silt or other soil type with low clay content

Adapted from [117]

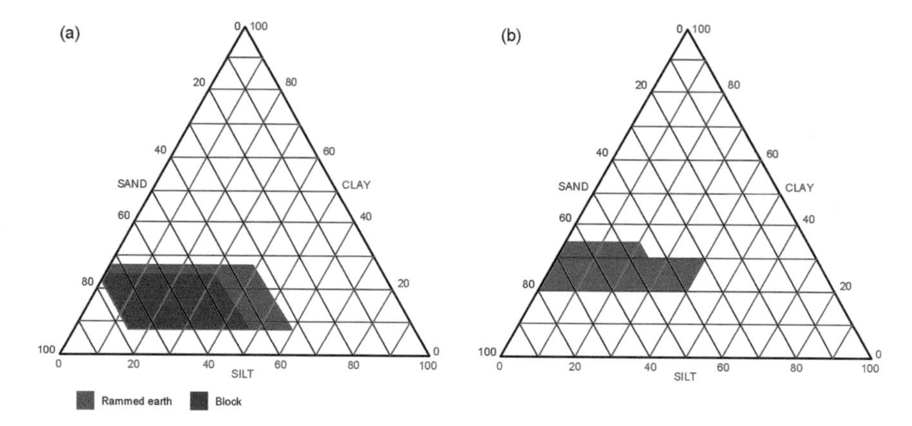

Fig. 2.14 Combined percentages of sand, silt and clay of earth material for blocks and rammed earth: **a** for stabilization; **b** without need of stabilization (adapted from [117])

the soils were tested by the ribbon test, results agreed with the jar test: the clay content of the three soils was very low because it was not possible to make a cylinder with them. The fourth soil had ribbons with an average of 45 mm. Therefore, in accordance with [58] it was suitable for rammed earth or stabilized compressed earth blocks. For the drop ball test, moulding the ball was difficult for the three soils, confirming the low clay content determined by previous tests. For the dry strength test, samples of 4 cm diameter and 1 cm thick were made. The three soils presented low strength when compared with the fourth, evidencing the higher clay content of the latter. Therefore, based on field-test results, this soil was the only considered adequate for ramming.

Neves et al. [117] recommend combined percentages of sand, silt and clay of soils to be used for earth blocks and rammed earth, based on the jar test results (Fig. 2.14). Similar combinations, mostly presented by a ternary diagram, are presented by other researchers [119].

Sitton et al. [119] considers that the Jar test has proven to be most valuable when used to assess the expansiveness of a soil, that is generally linked to the type of clay. Therefore, the test can be an easily indirect way to assess the existence in significant contents of expansive clays, such as montmorillonite. A soil that expands significantly after the addition of water in the jar is expected to cause problems with shrinkage cracks if used to produce for instance CEBs. Sitton et al. [119] consider that may be controlled by stabilization, although problematic.

Based on several field tests (cord, cigarette, exudation, dry resistance tests), Neves et al. [117] presents a classification of the type of soil and the earth construction techniques that are more adequate (Table 2.17).

Table 2.17 Field-test-based classification of soils and more adequate earth building techniques

Cord test	Cigarette and ribbon test	Exudation test	Dry resistance	Soil type	Building technique
Fragile cord and very low strength	Short cigarette and ribbon not produced	Fast reaction	Very low	Sandy, sand-silty, sand-clayey, silt-clayey	CEB, adobe, rammed earth
Fragile and soft cord	Short ribbon	Slow reaction	Low	Silty	Low content stabilized CEB, rammed earth, adobe
Soft cord	Medium ribbon	Very slow reaction	High	Clayey with gravel, clay-sandy, clay-silty	Stabilized rammed earth or CEB, adobe with fibres
Hard cord	Long ribbon	Without reaction	Very high	Clayey	Adobe with fibres

Adapted from [117]

2.8 Conclusion

This first chapter of this book on earth construction materials focused on the characterization of raw materials. We have seen that there are many characterization techniques in many fields: physics, geotechnics, chemistry and mineralogy. Deep knowledge of characteristics of raw materials is essential for optimizing the performance of earth materials but also for improving understanding of the phenomena. We have seen that the most measured characteristic is the particle size distribution of the earth materials. This test is the most common certainly because it is the simplest to perform and also because it plays a significant role in the behaviour of earth materials for building technologies but it is not essential. Indeed, for all the properties that we will study in the following chapters (hygrothermal and acoustic properties—Chap. 2, mechanical behaviour—Chaps. 3 and 4, and durability—Chap. 5), the most important is to know the nature and proportion of the active phase (namely the clay minerals) of earth materials. This is often known indirectly by the measurement of the particle size distribution (which notably caused a confusion in the use of the term "clay") or by the measurement of geotechnical and chemical properties such as the Atterberg limits, the methylene blue value or the cation-exchange capacity. The objective of these techniques is to assess the reactivity of the clayey active phase, which depends on the type of clay and its proportion in the material. A thorough mineralogical characterization would allow measuring these parameters but, as we have seen, this characterization is complex and still relatively rare in the studies of the literature. The chemical composition is a necessary tool to correctly carry out this mineralogical characterization. Finally, we have seen that there is also an even more global scale of analysis: the field tests. Field tests, although only qualitative, have the advantages of being easily performed, without significant costs. Therefore, results are obtained within a short period of time, turning them very useful in the working site. Nevertheless, they should be performed and results analysed by experienced professionals because they are mainly gathered by comparison.

Although characterization tests of soils are numerous, this literature review noted that only a handful of tests are commonly used, and few studies present a complete characterization of soils. This is due to the cost of the tests (in time and money) but also to the need. In fact, a complete characterization is not always necessary depending on the objective of the study.

Based on a set of tests, the local earth mix can be optimized for a defined building technique or the building technique can be adjusted to the local earth. The influence of sieving to decrease the content on a particle size fraction of the earth (and increase in others), the effect of additions such as plant fibres or a low binder content, can be assessed so that the material is optimized for the construction. We have just proposed that the understanding and the optimization of the characteristics of earth materials requires knowledge of the nature and the proportions of the clayey active phase. It is important to note that this comment is only valid for unstabilized earth materials. Indeed, the addition of plant aggregates or fibres or low binder content will completely modify the properties of earth materials. In these cases, the physico-chemical interactions will be numerous and complex and they will require further characterization.

Appendix 1: Particle Size Distribution of Earth Construction Materials

See Tables 2.18, 2.19, 2.20, 2.21 and 2.22.

Appendix 2: Atterberg Limits of Earth Construction Materials

See Tables 2.23, 2.24, 2.25, 2.26 and 2.27.

Appendix 3: Optimum Proctor Characteristics of Earth Construction Materials

See Tables 2.28 and 2.29.

Table 2.18 Particle size distribution of earth used for compressed earth bricks

References	Origin	Stabilizer	< 0.002 mm	0.063–0.002 mm	2–0.063 mm	> 2 mm
			Clay (%)	Silt (%)	Sand (%)	Gravel (%)
[92]	M (Morocco)	Cement	8	12	52	28
	B (Burkina Faso)	Cement	12	26	60	2
	A (Algeria)	Cement	0	62	30	8
[122]	France	None	16	43	40	0
[123]	USA	None	12	1	87	0
[101]	U (UK)	Alginate	31	45	24	
	V (UK)	Alginate	27	44	29	
	W (UK)	Alginate	16	61	23	
[124]	Cameroon	Cement	44	10	46	0
[93]	UK	Alginate, lignum	32	45	23	0
[100]	1 (France)	Cement	18	36	34	12
	2 (France)	Cement	20	32	33	15
[125]	TEO (Algeria)	Cement	7	13	60	20
	TMA (Algeria)	Cement	20	35	30	15
	TRM (Algeria)	Cement	25	10	65	0
[94]	France	None	27	66	7	0
[126]	Portugal	Alkaline activation	4	14	60	22
[127]	UK	Cement, lime, NaOH	18	23	59	0
[128]	France	None	12	52	36	0
[129]	Indonesia	Lime, RHA	20	33	47	0
[130]	India	Cement	9	18	73	0
[102]	BCS (India)	Lime	36	20	36	8
	TB1 (India)	Lime	44	32	24	0
	TB2 (India)	Lime	33	15	52	0
	TB3 (India)	Lime	42	41	17	0
[131]	Maroc	Cement	11	21	46	22

Table 2.19 Particle size distribution of earth used for adobe

References	Origin	Stabilizer	< 0.002 mm	0.063 mm-0.002 mm	2 mm-0.063 mm	> 2 mm
			Clay (%)	Silt (%)	Sand (%)	Gravel (%)
[132]	Italy	tomato, beetroot	0	45	47	8
[133]	Np	cement, gypsum	29	63	5	3
[95]	France	none	18	62	38	0
[134]	Has (Turkey)	gypsum, lime	0	0	95	5
	Tas (Turkey)	gypsum, lime	0	0	88	12
[135]	Turkey	ciment, pozzolana, gypse, chaux	32	25	43	0
[96]	PrA1 (Cyprus)	None	64		25	10
	PrA2 (Cyprus)	None	46		42	12
	PrA3 (Cyprus)	None	66		26	9
	PrA4 (Cyprus)	None	67		31	2
	PrA5 (Cyprus)	None	57		35	8
	PrA6 (Cyprus)	None	46		22	32
	PrA8 (Cyprus)	None	56		25	19
	OA1 (Cyprus)	None	74		26	0
	OA2 (Cyprus)	None	72		24	5
	OA3 (Cyprus)	None	79		20	1
	OA4 (Cyprus)	None	79		21	0
	OA5 (Cyprus)	None	57		30	13
	OA6 (Cyprus)	None	76		24	0

(continued)

Table 2.19 (continued)

References	Origin	Stabilizer	< 0.002 mm	0.063 mm-0.002 mm	2 mm-0.063 mm	> 2 mm
			Clay (%)	Silt (%)	Sand (%)	Gravel (%)
	LyB (Cyprus)	None	91		8	1
	AthB (Cyprus)	None	61		26	13
[99]	Burkina Faso	cement	30	23	42	5
[136]	EB1 (India)	None	49		51	0
	EB2 (India)	None	47		53	0
	EB6 (India)	None	84		16	0
	EB7 (India)	None	62		35	3
	EB8 (India)	None	68		32	0
	EB9 (India)	None	63		36	1
	EB13 (India)	None	46		54	0
[103]	CLG-1S (Angola)	None	30	25	16	29
	CTL-2S (Angola)	None	16	15	49	20
	CTL-2A (Angola)	None	16	13	67	4
	CHG-3SS (Angola)	None	30	10	60	0
	CHG-3SP (Angola)	None	56	9	33	2
	CHG-3AR (Angola)	None	33	11	56	0
	CHG-3AA (Angola)	None	26	12	62	0
	CGR-4S (Angola)	None	28	11	61	0
[103]	CHJJ-1S (Angola)	None	5	6	89	0
	CHJJ-5CA (Angola)	None	51	17	32	0
	LMD-6S (Angola)	None	30	12	47	11
	MNG-7A (Angola)	None	42	22	36	0
	NDD-8A (Angola)	None	16	9	75	0

(continued)

Table 2.19 (continued)

References	Origin	Stabilizer	< 0.002 mm	0.063 mm-0.002 mm	2 mm-0.063 mm	> 2 mm
			Clay (%)	Silt (%)	Sand (%)	Gravel (%)
	NPK-9S (Angola)	None	36	19	45	0
[137]	Brick 1 (Italy)	None	40	18	42	0
	Brick 2 (Italy)	None	16	13	71	0
	Brick 3 (Italy)	None	22	3	75	0
	Brick 4 (Italy)	None	20	13	67	0
	Brick 5 (Italy)	None	27	29	44	0
	Brick 7 (Italy)	None	25	23	52	0
	Local earth (Italy)	None	28	23	49	0
[138]	Ly (Cyprus)	None	78–91		8–18	1–4
	Ath (Cyprus)	None	61–86		11–26	3–13
[139]	np	None	12	45	43	0
[97]	France	None	25	30	45	0
[98]	Burkina Faso	Cow dung	36	17	43	4
[140]	Italy	None	10	18	64	8
[141]	Italy	None	22	50	25	3
[142]	Italy	None	29	48	13	10
[104]	np	fly ash	40	25	35	0
[105]	Turkey	None	27	33	37	2
[114]	MW4 (Cameroon)	None	16	49	33	2
	MW3 (Cameroon)	None	10	46	44	0
	MW2 (Cameroon)	None	8	41	49	2
	MW1 (Cameroon)	None	8	38	50	4
	HC3 (Cameroon)	None	21	38	36	5

(continued)

Table 2.19 (continued)

References	Origin	Stabilizer	< 0.002 mm	0.063 mm-0.002 mm	2 mm-0.063 mm	> 2 mm
			Clay (%)	Silt (%)	Sand (%)	Gravel (%)
	HC2 (Cameroon)	None	8	34	56	2
	HC1 (Cameroon)	None	16	44	38	2
	ME2 (Cameroon)	None	12	46	42	0
	BE1 (Cameroon)	None	12	48	40	0
	B (Cameroon)	None	25	26	28	21
[143]	China	None	89		11	0

Table 2.20 Particle size distribution of earth used for extruded earth bricks

References	Origin	Stabilizer	< 0.002 mm	0.063 mm-0.002 mm	2 mm-0.063 mm	> 2 mm
			Clay (%)	Silt (%)	Sand (%)	Gravel (%)
[91]	France	none	36	28	36	0
[144]	1 (France)	none	30	28	42	0
	2 (France)	none	29	37	33	1
	3 (France)	none	23	31	43	3
	4 (France)	none	29	33	37	1
	5 (France)	none	38	27	35	0
[90]	A (France)	none	6	86	8	0
	B (France)	none	5	91	4	0
	C (France)	none	5	81	14	0
[145]	France	none	29	63	8	0
	France	none	54	32	14	0
[146]	B1 (France)	none	40	25	35	0
	B2 (France)	none	55	31	14	0
	B3 (France)	lime	31	54	15	0
	B4 (France)	none	58	40	2	0
	B5 (France)	none	48	42	10	0
[147]	UK	cement or lime	16	46	33	5*

Table 2.21 Particle size distribution of earth used for cob

References	Origin	Stabilizer	< 0.002 mm	0.063 mm-0.002 mm	2 mm-0.063 mm	> 2 mm
			Clay (%)	Silt (%)	Sand (%)	Gravel (%)
[18]	Crediton (UK)	None	3	2	48	42
	Tedburn (UK)	None	26	25	13	36
	Halstow (UK)	None	49	45	4	1
	Bridgnorth (UK)	None	3	5	67	27
[21]	Carboniferous (UK)	None	28	34	16	22
	Permian (UK)	None	18	42	38	2
[139]	"Cob" (Germany)	None	21	61	18	
[19]	Original soil (Italy)	None	34	49	17	0
	Yellow Soil (Italy)	None	36	50.5	13.5	0

Table 2.22 Particle size distribution of earth used for rammed earth

References	Origin	Stabilizer	< 0.002 mm	0.063–0.002 mm	2–0.063 mm	> 2 mm
			Clay (%)	Silt (%)	Sand (%)	Gravel (%)
[106]	H2 (Japan)	$MgCl_2$	38	56	6	
	M (Japan)	CaO	28	70	2	
[148]	P (Australia)	None	20	66	14	0
	ELS (Australia)	Cement, fly ash or calcium carbide	20	9	60	10
[149]	FRE (France)	None	16	49	35	
	MRE (France)	None	8	27	49	16
	MRES (France)	Lime	8	27	49	16
[150]	A (France)	None	5	30	49	16
	B (France)	None	4	35	59	2
	C (France)	None	9	38	50	3
	D (France)	None	10	30	12	48
	E (France)	None	10	22	43	25
[151]	A (France)	None	10	25	18	47
	B (France)	None	5	30	49	16
	C (France)	None	8	34	8	50
[23]	1 (Australia)	None	5	25	50	20
	2 (Australia)	None	30	0	50	20
	3 (Australia)	None	15	15	50	20
	4 (Australia)	None	30	20	40	10
	5 (Australia)	None	40	20	20	20
	6 (Australia)	Cement	10	15	50	25
	7 (Australia)	Cement	10	5	40	45

(continued)

Table 2.22 (continued)

References	Origin	Stabilizer	< 0.002 mm	0.063–0.002 mm	2–0.063 mm	> 2 mm
			Clay (%)	Silt (%)	Sand (%)	Gravel (%)
	8 (Australia)	Cement, lime	20	0	60	20
	9 (Australia)	Cement, lime	30	10	20	40
	10 (Australia)	Cement	5	25	50	20
[152]	Portugal	Sodium silicate, sodium hydroxide	6	14	46	34
[153]	BRS (Brazil)	lime, fly ash	5	33	62	0
[154]	Belgium	None	13	64	26	
[25]	Av (Portugal)	None	9	16	67	8
	PD (Portugal)	None	27	20	23	30
	VC (Portugal)	None	18	31	17	34
	CZ (Portugal)	None	10	18	47	25
	Cl (Portugal)	None	13	24	29	34
	Ar (Portugal)	None	17	23	33	27
[155]	433 (UK)	Cement	10	20	31	39
	613 (UK)	Cement	10	20	46	24
	703 (UK)	Cement	10	20	55	15
[16, 17]	CRA (France)	None	12	33	53	2
[118]	S (Sri Lanka)	Cement	8	59	32	
	HL (Sri Lanka)	Cement	14	30	56	
	C (Sri Lanka)	Cement	19	30	51	
[139]	Germany	None	11	25	64	
[156]	USA	None	15	24	61	
[24]	Spain	Lime, alabaster	42	50	8	
[121]	S1 (Portugal)	None	6	14	45	35
	S2 (Portugal)	None	5	15	59	21
	S3 (Portugal)	None	4	14	60	22
	S4 (Portugal)	None	12	12	53	23
[157]	India	Cement	8	13	79	
[107]	S1 (India)	Cement	32	18	50	
	S2 (India)	Cement	21	14	65	
	S3 (India)	Cement	16	12	73	
	S4 (India)	Cement	13	10	77	
	S5 (India)	Cement	9	9	82	

Table 2.23 Atterberg limits of earth used for compressed earth bricks

References	Origin	Stabilizer	w_l (%)	w_p (%)	Ip
[122]	France	None	33.0	20.1	12.9
[101]	U (UK)	Alginate	27.3	16.0	11.3
	V (UK)	Alginate	27.9	14.6	13.3
	W (UK)	Alginate	25.4	15.3	9.9
[93]	UK	Alginate, lignum	34.8	19.1	15.7
[100]	1 (France)	Cement	38	26	12
	2 (France)	Cement	36	22	14
[94]	France	None	30	21	9
[128]	France	None	31	21	12
[129]	Indonesia	Lime, RHA	41	25	16
[130]	India	Cement	45	21	24
[102]	BCS (India)	Lime	53	26	27
	TB1 (India)	Lime	58	31	27
	TB2 (India)	Lime	42	19	23
	TB3 (India)	Lime	45	24	21
[131]	Maroc	Cement	32	18	14

Table 2.24 Atterberg limits of earth used for adobe

References	Origin	Stabilizer	w_l (%)	w_p (%)	Ip
[132]	Italy	Tomato, beetroot	24	21	3
[95]	France	None	23	20	3
[134]	Has (Turkey)	Gypsum, lime	23	13	10
	Tas (Turkey)	Gypsum, lime	33	22	11
[96]	PrA2 (Cyprus)	None	27	17	10
	PrA4 (Cyprus)	None	28	22	6
	PrA5 (Cyprus)	None	28	22	5
	PrA6 (Cyprus)	None	41	23	18
	PrA8 (Cyprus)	None	38	21	17
	OA1 (Cyprus)	None	35	25	9
	OA2 (Cyprus)	None	33	19	14
	OA3 (Cyprus)	None	38	22	16
	OA4 (Cyprus)	None	40	25	14
	OA5 (Cyprus)	None	30	20	10
	OA6 (Cyprus)	None	39	25	14
	LyB (Cyprus)	None	46	25	21
	AthB (Cyprus)	None	37	24	13
[99]	Burkina Faso	Cement	31	17	14
[136]	EB1 (India)	None	23	22	1
	EB2 (India)	None	23	22	1
	EB6 (India)	None	24	22	2
	EB7 (India)	None	25	22	3
	EB8 (India)	None	27	21	6
	EB9 (India)	None	26	22	4
	EB13 (India)	None	24	21	3
[103]	CLG-1S (Angola)	None	31	18	13
	CTL-2S (Angola)	None	22	17	5
	CTL-2A (Angola)	None	23	18	5
	CHG-3SS (Angola)	None	29	18	11
	CHG-3SP (Angola)	None	61	37	24
	CHG-3AR (Angola)	None	23	16	7
	CHG-3AA (Angola)	None	29	16	13
	CGR-4S (Angola)	None	36	20	16
	CHJJ-5CA (Angola)	None	43	29	14
	LMD-6S (Angola)	None	29	19	10
	MNG-7A (Angola)	None	35	20	15

(continued)

Table 2.24 (continued)

References	Origin	Stabilizer	w_l (%)	w_p (%)	Ip
	NDD-8A (Angola)	None	22	15	7
	NPK-9S (Angola)	None	44	25	19
[138]	Ly (Cyprus)	None	44	26.5	17.5
	Ath (Cyprus)	None	36	25	11
[97]	France	None	38	20	18
[141]	Italy	None	26	18	8
[104]	np	Fly ash	47	16	31
[105]	Turkey	None	22	32	10
[114]	MW4 (Cameroon)	None	61	52	9
	MW3 (Cameroon)	None	60	54	6
	MW2 (Cameroon)	None	48	47	1
	MW1 (Cameroon)	None	70	54	16
	HC3 (Cameroon)	None	63	46	17
	HC2 (Cameroon)	None	47	36	11
	HC1 (Cameroon)	None	70	48	22
	ME2 (Cameroon)	None	51	45	6
	BE1 (Cameroon)	None	55	44	11
	B (Cameroon)	None	70	52	18
[143]	China	None	36	19	17

Table 2.25 Atterberg limits of earth used for extruded earth bricks

References	Origin	Stabilizer	w_l (%)	w_p (%)	Ip
[90]	A (France)	None	100	28	72
	B (France)	None	60	29	31
	C (France)	None	24	21	3
[147]	UK	Cement or lime	24	16	8

Table 2.26 Atterberg limits of earth used for cob

References	Origin	Stabilizer	w_l (%)	w_p (%)	Ip
[18]	Crediton (UK)	None	36.6 ± 0.6	22.2 ± 1.5	17.6 ± 6.2
	Tedburn (UK)	None	48.1 ± 4.5	27.2 ± 0.9	20.9 ± 3.9
	Halstow (UK)	None	69.6 ± 5.5	34.1 ± 1.6	35.5 ± 4.1
	Bridgnorth (UK)	None	21.1 ± 1.7	3.5 ± 7.8	17.6 ± 6.2
[21]	Carboniferous (UK)	None	43	27	15
	Permian (UK)	None	28	20	8

Table 2.27 Atterberg limits of earth used for rammed earth

References	Origin	Stabilizer	w_l (%)	w_p (%)	Ip
[149]	FRE (France)	None	27	19	8
[23]	1 (Australia)	None	16	13	3
	2 (Australia)	None	26	12	14
	3 (Australia)	None	18	8	10
	4 (Australia)	None	25	11	13
	5 (Australia)	None	35	16	18
	6 (Australia)	Cement	15	10	5
	7 (Australia)	Cement	17	9	8
	8 (Australia)	Cement, lime	22	10	12
	9 (Australia)	Cement, lime	39	15	23
	10 (Australia)	Cement	15	11	4
[154]	Belgium	None	32.5	15	17.5
[25]	PD (Portugal)	None	41.2	25.1	16.1
	VC (Portugal)	None	46.1	26.7	19.4
	Cl (Portugal)	None	35.5	22	13.5
	Ar (Portugal)	None	26	20	6
[107]	S1 (India)	Cement	40	19	21
	S2 (India)	Cement	32	12	20
	S3 (India)	Cement	27	9	18

Table 2.28 Optimum moisture content and dry density of earth used for CEB

References	Origin	Stabilizer	w_{opt} (%)	ρ_{dopt} (kg/m^3)
[123]	USA	None	9	1785
[101]	U (UK)	Alginate	16	1820
	V (UK)	Alginate	18	1980
	W (UK)	Alginate	14	1920
[100]	1 (France)	Cement	9.8	1930
	2 (France)	Cement	11.6	1860
[125]	TEO (Algeria)	Cement	12.2	1880
	TMA (Algeria)	Cement	12.6	1760
	TRM (Algeria)	Cement	12.0	1750
[94]	France	None	14	1988
[126]	Portugal	Alkaline activation	12.1	1710

Table 2.29 Optimum moisture content and dry density of earth used for rammed earth

References	Origin	Stabilizer	w_{opt} (%)	ρ_{dopt} (kg/m^3)
[106]	H2 (Japan)	$MgCl_2$	10.3	1996
[152]	Portugal	Sodium silicate, sodium hydroxyde	12.2	1958
[154]	Belgium	None	15	1876
[25]	Av (Portugal)	None	8	2018
	PD (Portugal)	None	17.8	1733
	VC (Portugal)	None	21.5	1651
	CZ (Portugal)	None	11.3	1600
	Cl (Portugal)	None	15.6	1814
	Ar (Portugal)	None	8	2018
[121]	S1 (Portugal)	None	12	1920
	S2 (Portugal)	None	12	1840
	S3 (Portugal)	None	12	1710
	S4 (Portugal)	None	10	2010
[157]	India	Cement	19	1710
[107]	S1 (India)	Cement	15.6	1835
	S2 (India)	Cement	10.8	1910
	S3 (India)	Cement	9.4	2000
	S4 (India)	Cement	9.4	1980
	S5 (India)	Cement	9.4	1950

References

1. Van Damme H, Houben H (2018) Earth concrete Stabilization revisited. Cem Concr Res 114:90–102. https://doi.org/10.1016/j.cemconres.2017.02.035
2. Guillaud H (2018) CRATerre-ENSAG. In: Conférence sur le patrimoine architectural en terre de la France, post-master—DSA-Terre de l'ENSAG
3. EN 459-1 (2015) Building lime; Part 1: Definitions, specifications and conformity criteria. European Committee for Standardization
4. Parracha JL, Lima J, Freire MT, Ferreira M, Faria P (2019) Vernacular earthen buildings from Leiria, Portugal: material characterization. Int J Archit Herit. https://doi.org/10.1080/15583058.2019.1668986
5. Parracha JL, Santos Silva A, Cotrim M, Faria P (2019) Mineralogical and microstructural characterisation of rammed earth and earthen mortars from 12th century Paderne Castle. J Cult Herit. https://doi.org/10.1016/j.culher.2019.07.021
6. ASTM C136 (2014) Standard test method for sieve analysis of fine and coarse aggregates. ASTM International
7. ASTM D422 (2011) Standard test method for dispersive characteristics of clay soil by double hydrometer. ASTM International
8. BS 1377-2 (1990) Methods of test for soils for civil engineering purposes—Part 2: Classification tests. British Standards Institution
9. BNQ 2501-025 (2013) Sols-Analyse granulométrique des sols inorganiques. Bureau de Normalisation du Québec

10. EN ISO 17892-4 (2018) Geotechnical investigation and testing—Laboratory testing of soil—Part 4: Determination of particle size distribution. European Committee for Standardization
11. EN ISO 14688-1 (2018) Geotechnical investigation and testing—Identification and classification of soil—Part 1: Identification and description. European Committee for Standardization
12. USDA (1987) Soil mechanics level 1, USDA textural soil classification. United States Department of Agriculture
13. ASTM D2487 (2017) Standard practice for classification of soils for engineering purposes (unified soil classification system). ASTM International
14. XP P13-901 (2001) Blocs de terre comprimée pour murs et cloisons : Définitions—Spécifications—Méthodes d'essai—Conditions de réception. Association Française de Normalisation
15. Houben H, Guillaud H (2006) Traité de construction en terre. Editions Paranthèses
16. Hamard E, Cazacliu B, Razakamanantsoa A, Morel JC (2016) Cob, a vernacular earth construction process in the context of modern sustainable building. Build Environ 106:103–119. https://doi.org/10.1016/j.buildenv.2016.06.009
17. Hamard E, Cammas C, Fabbri A, Razakamanantsoa A, Cazacliu B, Morel JC (2016) Historical rammed earth process description thanks to micromorphological analysis. International journal of architectural heritage. https://doi.org/10.1080/15583058.2016.1222462
18. Coventry K (2004) Specification development for the use of Devon Cob in earth construction. University of Plymouth, UK
19. Quagliarini E, Stazi A, Pasqualini E, Fratalocchi E (2010) Cob construction in Italy: some lessons from the past. Sustainability 2:3291–3308. https://doi.org/10.3390/su2103291
20. Saxton RH (1995) Performance of cob as a building material. Struct Eng 73(7):111–115
21. Harries R, Saxton B, Coventry K (1995) The geological and geotechnical properties of earth material from central Devon in relation to its suitability for building in cob. In: Read at the annual conference of the Ussher Society, pp 441–444
22. Hamard E, Lemercier B, Cazacliu B, Razakamanantsoa A, Morel JC (2018) A new methodology to identify and quantify material resource at a large scale for earth construction—application to cob in Brittany. Constr Build Mater 170:485–497. https://doi.org/10.1016/j.conbuildmat.2018.03.097
23. Ciancio D, Jaquin P, Walker P (2013) Advances on the assessment of soil suitability for rammed earth. Constr Build Mater 42:40–47. https://doi.org/10.1016/j.conbuildmat.2012.12.049
24. Serrano S, Barreneche C, Rincón L, Boer D, Cabeza LF (2013) Optimization of three new compositions of stabilized rammed earth incorporating PCM: thermal properties characterization and LCA. Constr Build Mater 47:872–878. https://doi.org/10.1016/j.conbuildmat.2013.05.018
25. Gomes I, Gonçalves TD, Faria P (2014) Unstabilised rammed earth: characterization of material collected from old constructions in South Portugal and comparison to normative requirements. Int J Archit Herit 8(2):185–212. https://doi.org/10.1080/15583058.2012.683133
26. Maniatidis V, Walker P (2003) A review of rammed earth construction. University of Bath. Available at: http://staff.bath.ac.uk/abspw/rammedearth/review.pdf
27. Keable J (1996) Rammed earth structure—a code of practice. Intermediate Technology Publications, London, UK
28. Norton J (1997) Building with Earth: a handbook, 2nd edn. Intermediate Technology Publications, London, UK
29. SAZS 724 (2001) Standard code of practice for rammed earth structures. Standards Association of Zimbabwe
30. Keefe L (2005) Earth building—methods and materials, repair and conservation. Taylor & Francis Group. https://doi.org/10.4324/9780203342336
31. Walker P, Keable R, Martin J, Maniatidis V (2005) Rammed earth: design and construction guidelines. BRE Bookshop
32. ASTM D4318 (2017) Standard test methods for liquid limit, plastic limit, and plasticity index of soils. ASTM International

33. BNQ 2501-090 (2005) Sols-Détermination de la limite de liquidité à l'aide de l'appareil de Cassagrande et de la limite de plasticité. Bureau de Normalisation du Québec
34. BS 5930 (2015) Code of practice for ground investigations. British Standards Institution
35. EN ISO 17892-12 (2018) Geotechnical investigation and testing—Laboratory testing of soil—Part 12: Determination of Atterberg limits. European Committee for Standardization
36. O'Kelly BC, Vardanega PJ, Haigh SK (2018) Use of fall cones to determine Atterberg limits: a review. Géotechnique 68:843–856. https://doi.org/10.1680/jgeot.17.R.039
37. BS 1377-3 (1990) Methods of test for soils for civil engineering purposes—Part 3: Chemical and electro-chemical testing. British Standards Institution
38. Vardanega PJ, Haigh SK (2014) The undrained strength—liquidity index relationship. Can Geotech J 51:1073–1086. https://doi.org/10.1139/cgj-2013-0169
39. ISO 17892-1 (2005) Geotechnical investigation and testing—Laboratory testing of soil—Part 1: Determination of water content. International Organization for Standardization
40. Sherwood PT, Ryley MD (1970) An investigation of a cone-penetrometer method for the determination of the liquid limit. Géotechnique 20:203–208. https://doi.org/10.1680/geot.1970.20.2.203
41. Sherwood PT (1970) The reproducibility of the results of the soil classification and compaction tests. Road Research Laboratory, Crowthorne, Berkshire, UK, p 55
42. ASTM D698 (2012) Standard test methods for laboratory compaction characteristics of soil using standard effort (12,400 ft-lbf/ft^3 (600 kN-m/m^3)). ASTM International
43. BNQ 2501-250 (2005) Sols-Détermination de la relation teneur en eau-masse volumique-Essai avec energie de compactage normal (600 kNm/m^3). Bureau de Normalisation du Québec
44. NF P94-093 (2014) Sols : reconnaissance et essais—Détermination des références de compactage d'un matériau—Essai Proctor Normal—Essai Proctor modifié'. Association Française de Normalisation
45. BNQ 2501-170 (1986) Sols-Détermination de la teneur en eau. Bureau de Normalisation du Québec
46. ASTM D2216 (2010) Standard test methods for laboratory determination of water (moisture) content of soil and rock by mass. ASTM International
47. ASTM C128-15 (2015) Standard test method for relative density (specific gravity) and absorption of fine aggregate. ASTM International
48. EN ISO 17892-3 (2005) Geotechnical investigation and testing—Laboratory testing of soil—Part 3: Determination of particle density—Pycnometer method. European Committee for Standardization
49. ASTM C1777 (2015) Standard test method for rapid determination of the methylene blue value for fine aggregate or mineral filler using a colorimeter. ASTM International
50. NF P94-068 (1998) Sols : reconnaissance et essais—Mesure de la capacité d'adsorption de bleu de méthylène d'un sol ou d'un matériau rocheux—Détermination de la valeur de bleu de méthylène d'un sol ou d'un matériau rocheux par l'essai à la tâche. Association Française de Normalisation
51. EN 933-9+A1 (2013) Tests for geometrical properties of aggregates—Part 9: assessment of fines—Methylene blue test. European Committee for Standardization
52. ASTM C837 (2009, revised 2019) Standard test method for methylene blue index of clay. ASTM International
53. NF P11-300 (1992) Classification des matériaux utilisables dans la construction des remblais et des couches de forme d'infrastructures routières. Association Française de Normalisation
54. Schroeder H (2018) The new DIN standards in earth building—the current situation in Germany. J Civ Eng Archit 12:113–120. https://doi.org/10.17265/1934-7359/2018.02.005
55. ISO 11464 (2006) Soil quality—pretreatment of samples for physico-chemical analysis. International Organization for Standardization
56. Ouellet-Plamondon C, Soro N, Nollet M-J (2017) Characterization, mix design, mechanical testing of earth materials, stabilized and unstabilized, for building construction. Poromechanics VI. ASCE, Paris, France. https://doi.org/10.1061/9780784480779.112

57. NZS 4298 (1998) New Zealand Standard—materials and workmanship for earth buildings. Standards New Zealand
58. HB 195 (2001) The Australian earth building handbook. Walker P. and Standards Australia
59. New Mexico Code (2006) New Mexico Earthen Building Materials Code 14.7.4. Construction Industries Division (CID) of the Regulation and Licensing Department
60. Lehmbau Regeln (2009) Begriffe, baustoffe, bauteile. In Vieweg & Teubner, 3, Überarbeitete Auflage, ed., E.V. Dachverband Lehm. Praxis, Wiesbaden, Germany
61. XP P94-047 (1998) Sols : reconnaissance et essais—Détermination de la teneur pondérale en matières organiques d'un matériau—Méthode par calcination. Association Française de Normalisation
62. XP P94-055 (1993) Sols : reconnaissance et essais—Détermination de la teneur pondérale en matières organiques d'un sol—Méthode chimique. Association Française de Normalisation
63. BS 1881-124 (2015) Testing concrete. Methods for analysis of hardened concrete. British Standards Institution
64. EN ISO 10390 (2020) 'Soil quality—Determination of pH'. European Committee for Standardization. EN ISO 10693 (1995) Soil quality—Determination of carbonate content—Volumetric method. European Committee for Standardization
65. NF P94-048 (1996) Sols : Reconnaissance et essais Détermination de la teneur en carbonate Méthode du calcimètre. Association Française de Normalisation
66. ASTM D4373 (2014) Standard test method for rapid determination of carbonate content of soils. ASTM International
67. EN 15933 (2012) Sludge, treated biowaste and soil; Determination of pH. European Committee for Standardization
68. ASTM D4972 (2001) Standard test method for pH of soils. ASTM International
69. Thomas GW (1982) Exchangeable cations. In: Page AL et al (eds) Methods of soil analysis: Part 2: Chemical and microbiological properties. Monograph Number, vol 9, 2nd edn. ASA, Madison, WI, pp 154–157
70. Bower CA, Reitemeier RF, Fireman M (1952) Exchangeable cation analysis of saline and alkali soils. Soil Sci 73:251–261
71. Gillman GP (1979) A proposed method for the measurement of exchange properties of highly weathered soils. Aust J Soil Res 17:129–139
72. Janek M, Lagaly G (2003) Interaction of a cationic surfactant with bentonite: a colloid chemistry study. Colloid Polym Sci 281:293–301. https://doi.org/10.1007/s00396-002-0759-z
73. Mantin I, Glaeser R (1960) Fixation des ions cobaltihexamine par les montmorillonites acides. Bulletin du Groupe français des Argiles 12:83–88
74. Morel R (1958) Observation sur la capacité d'échange et les phénomène d'échange dans les argiles. Bulletin du Groupe français des Argiles 10
75. Orsini L, Remy JC (1976) Utilisation du chlorure de cobaltihexamine pour la détermination simultanée de la capacité d'échange et des bases échangeables des sols. Science du Sol 4:269–275
76. Chhabra R, Pleysier J, Cremers A (1975) The measurement of the cation exchange capacity and exchangeable cations in soils: a new method. In: Proceedings of the international clay conference, pp 439–449
77. Bergaya F, Vayer M (1997) CEC of clays: Measurement by adsorption of a copper ethylendiamine complex. Appl Clay Sci 12:275–280. https://doi.org/10.1016/S0169-1317(97)00012-4
78. Meier LP, Kahr G (1999) Determination of the cation exchange capacity (CEC) of clay minerals using the complexes of copper(II) ion with triethylenetetramine and tetraethylenepentamine. Clays Clay Miner 47:386–388. https://doi.org/10.1346/CCMN.1999.0470315
79. ASTM 7503-18 (2018) Standard test method for measuring the exchange complex and cation exchange capacity of inorganic fine-grained soils. ASTM International
80. NF X31-130 (1999) Qualité des sols—Méthodes chimiques—Détermination de la capacité d'échange cationique (CEC) et des cations extractibles. Association Française de Normalisation

81. ISO 23470 (2018) Soil quality. Determination of effective cation exchange capacity (CEC) and exchangeable cations using a hexamminecobalt trichloride solution. International Organization for Standardization
82. ISO 11260 (2018) Qualité du sol—Détermination de la capacité d'échange cationique effective et du taux de saturation en bases échangeables à l'aide d'une solution de chlorure de baryum. International Organization for Standardization
83. Metson AJ (1956) Methods of chemical analysis for soil survey samples. New Zealand Soil Bureau Bulletin No. 12
84. Charlet L, Schlegel ML (1999) La capacité d'échange des sols. Structures et charges à l'interface eau / particule' Compte Rendu d'Académie d'Agriculture. Paris, France 85(2):7–24
85. Ciesielski H, Sterckeman T, Santerne M, Willery J (1997) A comparison between three methods for the determination of cation exchange capacity and exchangeable cations in soils. Agronomy 17:9–16. https://doi.org/10.1051/agro:19970102
86. ISO 11265 (1994) Qualité du sol—Détermination de la conductivité électrique spécifique. International Organization for Standardization
87. Rhoades JD (1982) Soluble salts. In: AL Page et al. (ed.) Methods of soil analysis: Part 2: Chemical and microbiological properties. Monograph Number 9 (Second Edition). ASA, Madison, WI, pp 167–179
88. ISO 11048 (1995) Qualité du sol—Dosage du sulfate soluble dans l'eau et dans l'acide. International Organization for Standardization
89. ISO 14256-2 (2007) Qualité du sol—Dosage des nitrates, des nitrites et de l'ammonium dans des sols bruts par extraction avec une solution de chlorure de potassium—Partie 2 : méthode automatisée avec analyse en flux segmenté. International Organization for Standardization
90. El Fgaier F, Lafhaj Z, Antczak E, Chapiseau C (2016) Dynamic thermal performance of three types of unfired earth bricks. Appl Therm Eng 93:377–383. https://doi.org/10.1016/j.applthermaleng.2015.09.009
91. Aubert JE, Gasc-Barbier M (2012) Hardening of clayey soil blocks during freezing and thawing cycles. Appl Clay Sci 65–66:1–5. https://doi.org/10.1016/j.clay.2012.04.014
92. Ammari A, Bouassria K, Cherraj M, Bouabid H, Charif D'ouazzane S (2017) Combined effect of mineralogy and granular texture on the technico-economic optimum of the adobe and compressed earth blocks. Case Stud Constr Mater 7:240–248. https://doi.org/10.1016/j.cscm.2017.08.004
93. Galán-Marín C, Rivera-Gómez C, Petric J (2010) Clay-based composite stabilized with natural polymer and fibre. Constr Build Mater 24:1462–1468. https://doi.org/10.1016/j.conbuildmat.2010.01.008
94. Laborel-Préneron A, Aubert JE, Magniont C, Maillard P, Poirier C (2017) Effect of plant aggregates on mechanical properties of earth bricks. J Mater Civ Eng 29:04017244. https://doi.org/10.1061/(ASCE)MT.1943-5533.0002096
95. Aubert JE, Marcom A, Oliva P, Segui P (2015) Chequered earth construction in south-western France. J Cult Herit 16:293–298. https://doi.org/10.1016/j.culher.2014.07.002
96. Costi de Castrillo M, Philokyprou M, Ioannou I (2017) Comparison of adobes from pre-history to date. J Archaeol Sci Rep 12:437–448. https://doi.org/10.1016/j.jasrep.2017.02.009
97. Millogo Y, Morel JC, Aubert JE, Ghavami K (2014) Experimental analysis of Pressed Adobe Blocks reinforced with Hibiscus cannabinus fibers. Constr Build Mater 52:71–78. https://doi.org/10.1016/j.conbuildmat.2013.10.094
98. Millogo Y, Aubert JE, Sere AD, Fabbri A, Morel JC (2016) Earth blocks stabilized by cow dung. Mater Struct 49:4583–4594. https://doi.org/10.1617/s11527-016-0808-6
99. Dao K, Ouedraogo M, Millogo Y, Aubert JE, Gomina M (2018) Thermal, hydric and mechanical behaviours of adobes stabilized with cement. Constr Build Mater 158:84–96. https://doi.org/10.1016/j.conbuildmat.2017.10.001
100. Hakimi A, Yamani N, Ouissi H (1996) Results of mechanical strength tests on samples of compressed earth. Mater Struct 29:600–608
101. Dove CA, Bradley FF, Patwardhan SV (2016) Seaweed biopolymers as additives for unfired clay bricks. Mater Struct 49:4463–4482. https://doi.org/10.1617/s11527-016-0801-0

102. Venkatarama Reddy BV, Hubli SR (2002) Properties of lime stabilised steam-cured blocks for masonry. Mater Struct 35:293–300
103. Duarte I, Pedro E, Varum H, Mirao J, Pinho A (2017) Soil mineralogical composition effects on the durability of adobe blocks from the Huambo region Angola. Bull Eng Geol Environ 76:125–132. https://doi.org/10.1007/s10064-015-0800-3
104. Turanli L, Saritas A (2011) Strengthening the structural behavior of adobe walls through the use of plaster reinforcement mesh. Constr Build Mater 25:1747–1752. https://doi.org/10.1016/j.conbuildmat.2010.11.092
105. Uguryol M, Kulakoglu F (2013) A preliminary study for the characterization of Kültepe's adobe soils with the purpose of providing data for conservation and archaeology. J Cult Herit 14:117–124. https://doi.org/10.1016/j.culher.2012.12.008
106. Araki H, Koseki J, Sato T (2016) Tensile strength of compacted rammed earth materials. Soils Found 56:189–204. https://doi.org/10.1016/j.sandf.2016.02.003
107. Venkatarama Reddy BV, Prasanna Kumar P (2011) Cement stabilised rammed earth Part A: compaction characteristics and physical properties of compacted cement stabilised soils. Mater Struct 44:681–693. https://doi.org/10.1617/s11527-010-9658-9
108. Moore DM, Reynolds RC (1997) X-ray diffraction and the identification and analysis of clay minerals, 2nd edn. Oxford University Press
109. Thorez J (1975) Phyllosilicates and clay minerals: a laboratory hand book for their X-ray diffraction examination. Editions G. Lelotte, Belgium
110. Madejová J (2003) FTIR techniques in clay mineral studies. Vib Spectrosc 31:1–10. https://doi.org/10.1016/S0924-2031(02)00065-6
111. Van Olphen H, Fripiat JJ (1979) Data handbook for clay materials and other non-metallic minerals. Pergamon Press, Oxford and Elmsford, New York. https://doi.org/10.1346/CCMN.1980.0280215
112. Ouedraogo KAJ, Aubert JE, Tribout C, Escadeillas G (2020) Is stabilization of earth bricks using low cement or lime contents relevant? Constr Build Mater 236. https://doi.org/10.1016/j.conbuildmat.2019.117578
113. Maskell D, Heath A, Walker P (2010) Laboratory scale testing of extruded earth masonry units. Mater Des 45:359–364. https://doi.org/10.1016/j.matdes.2012.09.008
114. Wouatong ASL, Djukem WDL, Ngapgue F, Katte V, Beyala VKK (2017) Influence of random inclusion of synthetic wicks fibers of hair on the behavior of clayey soils. Geotech Geol Eng. https://doi.org/10.1007/s10706-017-0267-z
115. Deer WA, Howie RA, Zussman J (2013) In: Wilson MJ (ed) Rock-forming minerals, sheet silicates: clay minerals, vol 3C, 2nd edn. The Geological Society, London, 724 pp. ISBN: 978-1-86239-359-2
116. ASTM D2488 (2017) Standard practice for description and identification of soils (visual-manual procedures). ASTM International
117. Neves C, Faria O, Rotondaro R, Cevallos PS, Hoffmann M (2010) Seleção de solos e métodos de controle na construção com terra—práticas de campo. Rede Ibero-americana PROTERRA. Available at: http://www.redproterra.org
118. Jayasinghe C, Kamaladasa N (2007) Compressive strength characteristics of cement stabilized rammed earth walls. Constr Build Mater 21:1971–1976. https://doi.org/10.1016/j.conbuildmat.2006.05.049
119. Sitton JD, Zeinali Y, Story BA (2017) Rapid soil classification using artificial neural networks for use in constructing compressed earth blocks. Constr Build Mater 138:214–221. https://doi.org/10.1016/j.conbuildmat.2017.02.006
120. ASTM E2392/E2392M-10 (2016) Standard guide for design of earthen wall building systems. ASTM International
121. Silva RA, Oliveira DV, Miranda T, Cristelo N, Escobar MC, Soares E (2013) Rammed earth construction with granitic residual soils: the case study of northern Portugal. Constr Build Mater 47:181–191. https://doi.org/10.1016/j.conbuildmat.2013.05.047
122. Bruno AW, Gallipoli D, Perlot C, Mendes J (2017) Mechanical behaviour of hypercompacted earth for building construction. Mater Struct 50:160. https://doi.org/10.1617/s11527-017-1027-5

123. Donkor P, Obonyo E (2015) Earthen construction materials: assessing the feasibility of improving strength and deformability of compressed earth blocks using polypropylene fibers. Mater Des 83:813–819. https://doi.org/10.1016/j.matdes.2015.06.017
124. Eko RM, Offa ED, Ngatcha TY, Minsili LS (2012) Potential of salvaged steel fibers for reinforcement of unfired earth blocks. Constr Build Mater 35:340–346. https://doi.org/10.1016/j.conbuildmat.2011.11.050
125. Hakimi A, Fassi-Fehri O, Bouabid H, Charif D'ouazzane S, El Kortbi M (1999) Non-linear behaviour of the compressed earthen block by elasticity-damage coupling. Mater Struct 32:539–545
126. Leitão D, Barbosa J, Soares E, Miranda T, Cristelo N, Briga-Sá A (2017) Thermal performance assessment of masonry made of ICEB' stabilised with alkali-activated fly ash. Energy Build 139:44–52. https://doi.org/10.1016/j.enbuild.2016.12.068
127. McGregor F, Heath A, Fodde E, Shea A (2014) Conditions affecting the moisture buffering measurement performed on compressed earth blocks. Build Environ 75:11–18. https://doi.org/10.1016/j.buildenv.2014.01.009
128. Mesbah A, Morel JC, Olivier M (1999) Clayey soil behaviour under static compaction test. Mater Struct 32:687–694
129. Muntohar AS (2011) Engineering characteristics of the compressed-stabilized earth brick. Constr Build Mater 25:4215–4220. https://doi.org/10.1016/j.conbuildmat.2011.04.061
130. Venkatarama Reddy BV, Gupta A (2005) Characteristics of soil-cement blocks using highly sandy soils. Mater Struct 38:651–658. https://doi.org/10.1617/14265
131. Zine-Dine K, Bouabid H, El Kortbi M, Charif-d'Ouazzane S, Hakimi A, El Hammoumi A, Fassi-Fehri O (2000) Rheology of walls in compressed earth blocks in uniaxial compression: study et modelling. Mater Struct 33:529–536
132. Achenza M, Fenu L (2005) On earth stabilization with natural polymers for earth masonry construction. Mater Struct 39:21–27. https://doi.org/10.1617/s11527-005-9000-0
133. Ashour T, Korjenic A, Korjenic S, Wu W (2015) Thermal conductivity of unfired earth bricks reinforcedby agricultural wastes with cement and gypsum. Energy Build 104:139–146. https://doi.org/10.1016/j.enbuild.2015.07.016
134. Balkis AP (2017) The effects of waste marble dust and polypropylene fiber contents on mechanical properties of gypsum stabilized earthen. Constr Build Mater 134:556–562. https://doi.org/10.1016/j.conbuildmat.2016.12.172
135. Binici H, Aksogan O, Shah T (2005) Investigation of fibre reinforced mud brick as a building material. Constr Build Mater 19:313–318. https://doi.org/10.1016/j.conbuildmat.2004.07.013
136. Dhandhukia P, Goswami D, Thakor P, Thakker JN (2013) Soil property apotheosis to corral the finest compressive strength of unbaked adobe bricks. Constr Build Mater 48:948–953. https://doi.org/10.1016/j.conbuildmat.2013.07.043
137. Fratini F, Pecchioni E, Rovero L, Tonietti U (2011) The earth in the architecture of the historical centre of Lamezia Terme (Italy): characterization for restoration. Appl Clay Sci 53:509–516. https://doi.org/10.1016/j.clay.2010.11.007
138. Illampas R, Ioannou I, Charmpis DC (2014) Adobe bricks under compression: Experimental investigation and derivation of stress–strain equation. Constr Build Mater 53:83–90. https://doi.org/10.1016/j.conbuildmat.2013.11.103
139. Miccoli L, Müller U, Fontana P (2014) Mechanical behaviour of earthen materials: a comparison between earth block masonry, rammed earth and cob. Constr Build Mater 61:327–339. https://doi.org/10.1016/j.conbuildmat.2014.03.009
140. Parisi F, Asprone D, Fenu L, Prota A (2015) Experimental characterization of Italian composite adobe bricks reinforced with straw fibers. Compos Struct 122:300–307. https://doi.org/10.1016/j.compstruct.2014.11.060
141. Piattoni Q, Quagliarini E, Lenci S (2011) Experimental analysis and modelling of the mechanical behaviour of earthen bricks. Constr Build Mater 25:2067–2075. https://doi.org/10.1016/j.conbuildmat.2010.11.039
142. Quagliarini E, Lenci S (2010) The influence of natural stabilizers and natural fibres on the mechanical properties of ancient Roman adobe bricks. J Cult Herit 11:309–314. https://doi.org/10.1016/j.culher.2009.11.012

143. Wu F, Li G, Li HN, Jia JQ (2013) Strength and stress–strain characteristics of traditional adobe block and masonry. Mater Struct 46:1449–1457. https://doi.org/10.1617/s11527-012-9987-y
144. Cagnon H, Aubert JE, Coutand M, Magniont C (2014) Hygrothermal properties of earth bricks. Energy Build 80:208–217. https://doi.org/10.1016/j.enbuild.2014.05.024
145. Fouchal F, Gouny F, Maillard P, Ulmet L, Rossignol S (2015) Experimental evaluation of hydric performances of masonry walls made of earth bricks, geopolymer and wooden frame. Build Environ 87:234–243. https://doi.org/10.1016/j.buildenv.2015.01.036
146. Maillard P, Aubert JE (2014) Effects of the anisotropy of extruded earth bricks on their hygrothermal properties. Constr Build Mater 63:56–61. https://doi.org/10.1016/j.conbuildmat.2014.04.001
147. Maskell D, Heath A, Walker P (2014) Inorganic stabilisation methods for extruded earth masonry units. Constr Build Mater 71:602–609. https://doi.org/10.1016/j.conbuildmat.2014.08.094
148. Arrigoni A, Grillet AC, Pelosato R, Dotelli G, Beckett CTS, Woloszyn M, Ciancio D (2017) Reduction of rammed earth's hygroscopic performance under stabilisation: an experimental investigation. Build Environ 115:358–367. https://doi.org/10.1016/j.buildenv.2017.01.034
149. Bui QB, Morel JC, Venkatarama Reddy BV, Ghayad W (2009) Durability of rammed earth walls exposed for 20 years to natural weathering. Build Environ 44:912–919. https://doi.org/10.1016/j.buildenv.2008.07.001
150. Bui QB, Morel JC, Hans S, Walker P (2014) Effect of moisture content on the mechanical characteristics of rammed earth. Constr Build Mater 54:163–169. https://doi.org/10.1016/j.conbuildmat.2013.12.067
151. Bui TT, Bui QB, Limam A, Maximilien S (2014) Failure of rammed earth walls: from observations to quantifications. Constr Build Mater 51:295–302. https://doi.org/10.1016/j.conbuildmat.2013.10.053
152. Cristelo N, Glendinning S, Miranda T, Oliveira D, Silva R (2012) Soil stabilisation using alkaline activation of fly ash for self compacting rammed earth construction. Constr Build Mater 36:727–735. https://doi.org/10.1016/j.conbuildmat.2012.06.037
153. da Rocha CG, Consoli NC, Dalla Rosa Johann A (2014) Greening stabilized rammed earth: devising more sustainable dosages based on strength controlling equations. J Clean Prod 66:19–26.https://doi.org/10.1016/j.jclepro.2013.11.041
154. Gerard P, Mahdad M, McCormack AR, François B (2015) A unified failure criterion for destabilized rammed earth materials upon varying relative humidity conditions. Constr Build Mater 95:437–447. https://doi.org/10.1016/j.conbuildmat.2015.07.100
155. Hall MR (2007) Assessing the environmental performance of stabilised rammed earth walls using a climatic simulation chamber. Build Environ 42:139–145. https://doi.org/10.1016/j.buildenv.2005.08.017
156. Michiels T, Napolitano R, Adriaenssens S, Glisic B (2017) Comparison of thrust line analysis, limit state analysis and distinct element modeling to predict the collapse load and collapse mechanism of a rammed earth arch. Eng Struct 148:145–156. https://doi.org/10.1016/j.engstruct.2017.06.053
157. Tripura DD, Singh KD (2015) Axial load-capacity of rectangular cement stabilized rammed earth column. Eng Struct 99:402–412. https://doi.org/10.1016/j.engstruct.2015.05.014

Chapter 3
Hygrothermal and Acoustic Assessment of Earthen Materials

Antonin Fabbri, Jean-Emmanuel Aubert, Ana Armanda Bras, Paulina Faria, Domenico Gallipoli, Jeanne Goffart, Fionn McGregor, Céline Perlot-Bascoules, and Lucile Soudani

Abstract Thanks to their microstructure which allows both exchange of gas with their surrounding environment and internal water vapour sorption phenomena, earthen materials are highly hygroscopic. If no material is used as a barrier or retardant to the diffusion into the envelope between the earth and the indoor environment of a building, they have a great potential to enhance the thermal comfort and to regulate indoor air quality. In addition, even if few studies have been realised on that point, a high acoustic absorption can be anticipated due to their open porous structure. However, notably due to the lack of standardized procedure to measure their performances, these multi-functional capabilities of earthen walls are almost not considered in the design and rehabilitation operations. In that context, in the framework of the RILEM Technical committee TCE 274, this chapter aims at presenting a critical bibliographic review related to the assessment of hygrothermal and acoustic performance of earthen structures. It is a first necessary step in order to define performance-oriented tests to properly assess their hygrothermal and acoustic performances. In particular, the analysis of collected information allowed to underline some consensus on the protocols that should be used to measure some of the key parameters, while the necessity to perform some additional investigations on others was clearly identified.

A. Fabbri (✉) · F. McGregor · L. Soudani
LTDS-ENTPE, CNRS, University of Lyon, UMR 5513, Vaulx-en-Velin, France
e-mail: antonin.fabbri@entpe.fr

J.-E. Aubert
Université de Toulouse, UPS, INSA, LMDC, Toulouse, France

A. A. Bras
Liverpool John Moores University, Liverpool, UK

P. Faria
CERIS and NOVA School of Science and Technology, NOVA University of Lisbon, Caparica, Portugal

D. Gallipoli · C. Perlot-Bascoules
Laboratoire SIAME, Fédération IPRA, Université de Pau Et Des Pays de L'Adour, Anglet, France

J. Goffart · L. Soudani
Université de Savoie Mont Blanc, LOCIE, Chambéry, France

A. Fabbri et al. (eds.), *Testing and Characterisation of Earth-based Building Materials and Elements*, RILEM State-of-the-Art Reports 35,
https://doi.org/10.1007/978-3-030-83297-1_3

Keywords Earthen materials · Hygroscopic behaviour · Internal air quality · Hygrothermal performances · Acoustic performances

3.1 Introduction

The broadest definition of "hygroscopicity" is the ability of a material to adsorb (and release) water vapour molecules from (to) the surrounding atmosphere. The adsorption kinetic is not mentioned in this definition, and without additional details it can encompasses most porous materials.

In civil engineering, the hygroscopicity of a material can be used to regulate the indoor variation of humidity within a dwelling. The characteristics of the inside air are very important for building's inhabitants because it can significantly influence their comfort, health, and productivity. Extremely low levels of relative humidity (RH) (below 30%) may cause eye or skin irritations and dry the nasal mucous membranes, resulting in a higher risk of respiratory infections. On the other hand, high levels of RH may lead to the development of fungi, which can cause allergies as manifested by asthma and rhinitis [12, 48], and the emission of volatile organic compounds is favoured [41]. Today there is consensus that indoor RH should remain between 40 and 60%.

With this in mind, it may be more convenient to adopt a more restrictive definition of hygroscopicity, i.e., the ability of a material to be used as a passive humidity regulation system. For that purpose, the velocity at which the water molecules are adsorbed is at least as important as the total amount of water molecules that can be adsorbed. As a consequence, a proper estimation of hygroscopicity requires good knowledge of both vapour/liquid water transfer and phase change in porous media.

It is also important to underline that the development of bio contaminants (moulds, mites, ...) is conditioned upon the level of RH: the capacity of earthen materials to limit bio contaminants could be evaluated through the characterisation of its hygroscopic behaviour.

The term "hygrothermal" is commonly used to denote the couplings between mass transfers of water phases, including their phase changes, and heat transfer. An increasing number of research publications focus on assessing these phenomena within hygroscopic walls and modelling them.

The hygrothermal properties of earthen materials can be explained by their composition and the resulting microstructure. Whatever the construction technique, they are composed of clays, silts, sands, and possibly gravels and fibres. Since the connections between these constituents are not perfect, pores are embedded within the solid material. The resulting porous network, which enables fluids (either liquid or gas) to flow through the material, makes it quite permeable. In addition, since some of the constituents of the solid matrix such as clays and fibres are themselves porous media, earthen material is a double-porosity medium. Its morphological description could be summarized by the illustration shown in Fig. 3.1.

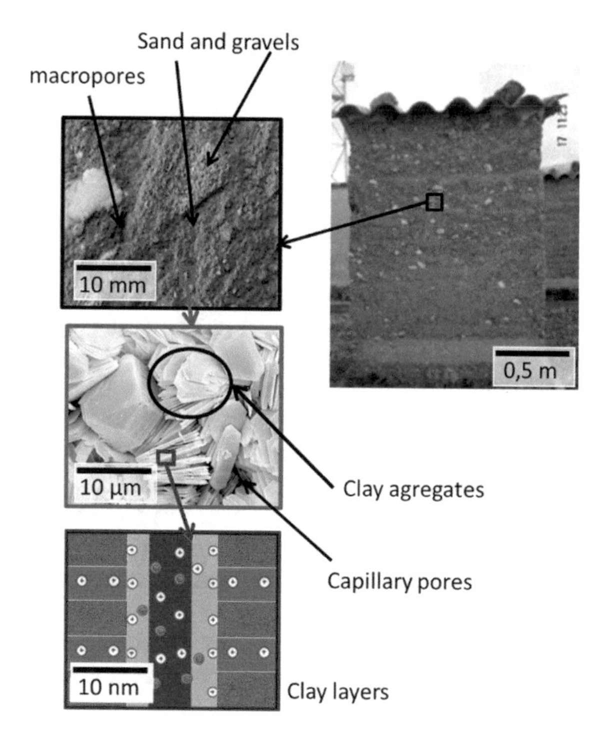

Fig. 3.1 Representation of the multi-scaled nature of earthen materials

The size distribution of the pores within an earthen material is particularly broad. Indeed, clay aggregates are commonly formed of interacting particles, which are themselves made of three to ten coupled layers, each several nanometres thick (7 nm for kaolinite, between 14 and 17 nm for Montmorillonite). This results in an interlayer porosity on the order of 1–10 nm [90], while, depending on the type of soil, the larger pores can extend to 1 mm. The presence of this nano-porosity, as well as the strong affinity between the clay particles and the water molecules, allows a relatively substantial retention of water within the porous network even when the RH of the surrounding air is far below 100%.

Under this context, this paper aims at drawing a critical review of the tools which have been developed in order to assess the hygrothermal performance of earthen materials. For that purpose, the main conservations equations and the resulting key parameters are succinctly presented in the first section, while the laboratory investigations which have been made to measure them are presented in the second section. The third section aims at describing the existing monitoring setups on buildings that can be found in the literature. It does not mean to be exhaustive but to provide a global overview relying on a certain amount of experimental setups over the last decade. First, the main physical parameters usually measured are described, and then a focus on case study of rammed earth houses, for which the positioning of sensors need specific placements. Finally, the last part of the paper is dedicated to the acoustic performances of earthen materials and buildings.

3.2 Key Material Parameters Involved in Hygrothermal Couplings

3.2.1 General Diagram of Heat and Mass Transfers

Earthen walls are subjected to spatial and temporal variations of temperature, vapour pressure and possibly air pressure. As mentioned above, these variations lead to migration of in-pore liquid water and water vapour, internal heat transfers as well as phase change phenomena. A diagram of the couplings between thermal and hydrodynamic processes is reported in Fig. 3.2, where RH stands for the relative humidity.

The main equations of hygrothermal couplings are nowadays quite well known by the scientific community. Detailed and comprehensive descriptions on the hygrothermal couplings are provided in numerous papers, for example [69] or [101]. Today, one of the main goals within this topic concerns the development of accurate and user-friendly software that can predict, with enough accuracy, the coupling between mass and heat transfers and their impact on the overall performance of a building. This research activity gives rise to commercial software such as WUFI (Fraunhofer, s.d. IBP/WUFI. [on line] Available at: http://www.wufi.de), which can provide reliable results on a wide range of materials and climatic loads but cannot always accurately reproduce the hygrothermal behaviour of unconventional materials such as earth when they are submitted to hygrometry and temperature variations. To overcome some of the restrictions of the WUFI code (for example, but not limited to, the lack of sorption–desorption hysteresis and dependence on temperature, the quite simple form imposed for the variation of the transport parameters with water content, etc.), a fairly high number of codes and procedures have been developed by

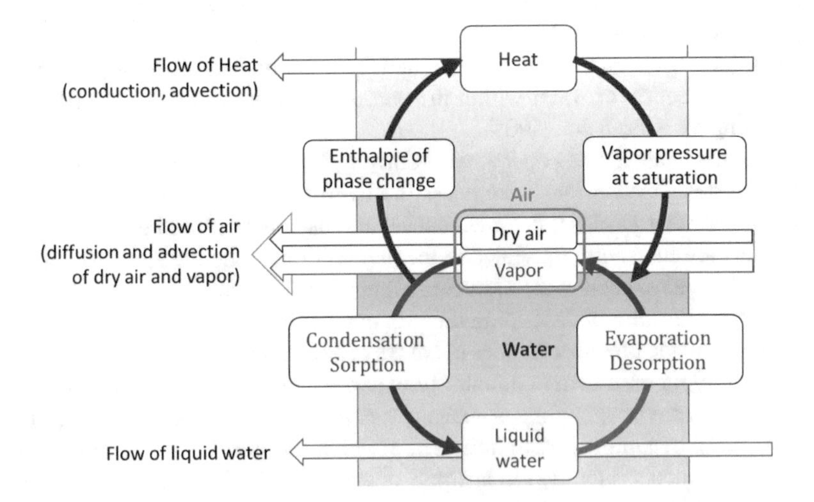

Fig. 3.2 Diagram of heat and moisture transfers within an earthen wall

the scientific community. One of the more visible studies in this field was conducted within the HAMSTAD project (see for example [50]).

At equilibrium, the mass conservation equations that drive the hygrothermal couplings can restrict to a system of three partial differential equations, two of them based on mass balance equations, and one on the heat balance. There are several ways to write these equations, for example:

$$\frac{\partial m_w}{\partial t} = \underline{\nabla} \cdot \left[D_L \left(\underline{\nabla} P_L - \rho_L \underline{g} \right) + \frac{p_v}{P_G} \frac{M_L}{M_G} D_G \underline{\nabla} P_G + P_G \delta_p \underline{\nabla} \left(\frac{p_v}{P_G} \right) \right] \tag{3.1}$$

$$\frac{\partial m_G}{\partial t} = \underline{\nabla} \cdot \left[D_G \underline{\nabla} P_G \right] + \dot{m}_{\rightarrow v} \tag{3.2}$$

$$\rho_d c \frac{\partial T}{\partial t} = \underline{\nabla} \cdot \left(\langle \lambda \rangle \underline{\nabla} T \right) - \left(\sum_{i=v,a,L} c_i \underline{\omega}_i \right) \cdot \underline{\nabla} T - \Delta h_v \dot{m}_{\rightarrow v} \tag{3.3}$$

In these equations, m_w and m_G are respectively the mass of water (both liquid and vapour) and of air per unit of material initial volume, $\dot{m}_{\rightarrow v}$ is the mass rate of water desorption per unit of material initial volume, ρ_d is the dry density, c is the average specific heat capacity, $\langle \lambda \rangle$ is the equivalent thermal conductivity, Δh_v is the enthalpy of desorption, $\underline{\omega}_i$ is the mass flow of the phase i, P_L and P_G are respectively the liquid and wet air total pressures (all the pressures are expressed in Pa), T is the temperature (all the temperatures are expressed in Kelvin) and p_v is the partial pressure of vapour. This latter is linked to the relative humidity, denoted by (RH), through the relation:

$$p_v = (RH) p_v^{sat} \tag{3.4}$$

where p_v^{sat} is the partial pressure at saturation, whose values are quite well tabulated (cf. [53], for example) and which is, at first order, only function of the temperature. The gas, liquid and vapour pressures are linked together by the Kelvin's law, which traduces the chemical equilibrium between the liquid water and its vapour:

$$P_G - P_L = -\frac{\rho_L R T}{M_L} \ln \left(\frac{p_v}{p_v^{sat}} \right) \tag{3.5}$$

where ρ_L is the density of liquid water, M_L its molar mass, R the perfect gas constant and T the temperature. The other parameters of these equations are presented in the following of this section.

3.2.2 Mass Conservation of Water

The first equation stands for the mass conservation of water, either in liquid or vapour forms (water solidification is not considered here). Assuming that the mass variation of liquid water is strongly higher than the mass variation of vapour (which seems quite reasonable since the density of water is at least a thousand time higher than the density of vapour), the left side of this equation can be directly linked to the water content, denoted by w, and defined as the ratio between the mass of water within the material and its dry mass, through the relation:

$$\frac{\partial m_w}{\partial t} = \rho_d \frac{\partial w}{\partial t} = \rho_d \left(\xi \frac{\partial (RH)}{\partial t} + \chi \frac{\partial T}{\partial t} \right) \tag{3.6}$$

where ξ is the variation of water content with RH at constant temperature (that is the slope of the sorption or desorption curves) while χ is the variation of water content with temperature at constant humidity.

The right side of the Eq. (3.1) is divided in three terms. The first one, $\underline{\nabla} \cdot \left[D_L \left(\underline{\nabla} P_L - \rho_L \underline{g} \right) \right]$, denotes the transport of liquid water through the porous network. $\rho_L \underline{g}$ is the gravity term, and D_L is the coefficient of water permeability, which is linked to the intrinsic permeability, denoted by κ_0 (in m^2), through the relation:

$$D_L = \rho_L \frac{\kappa_0 \kappa_r^L}{\eta_L} \tag{3.7}$$

where ρ_L is the density of water, η_L is its dynamic viscosity and is κ_r^L liquid water's relative permeability coefficient, which is a nondimensional quantity that varies between 0 (at a low saturation ratio) and 1 (when the material is fully saturated).

The second term, $\underline{\nabla} \cdot \left[\frac{p_v}{P_G} \frac{M_L}{M_G} D_G \underline{\nabla} P_G \right]$, denote the advective transport of vapour by the wet air phase. M_L and M_G are the molar mass of water and air while D_G is the coefficient of gas permeability, which is linked to the intrinsic permeability through the relation:

$$D_G = \rho_G \frac{\kappa_0 \kappa_r^G}{\eta_G} \tag{3.8}$$

Similar to the expression (3.7), κ_r^G is the gas's relative permeability, which varies between 0 (at saturation) and 1 (when the material is dried), and η_G is the gas's dynamic viscosity.

The last term, that is $\underline{\nabla} \cdot \left[P_G \delta_p \underline{\nabla} \left(\frac{p_v}{P_G} \right) \right]$ is the diffusion of vapour molecules within the air phase. δ_p is a form of the water vapour diffusion coefficient, which can be linked to the vapour resistance factor, μ, through:

$$\delta_p = \frac{\delta_a}{\mu} \tag{3.9}$$

δ_a is the diffusion coefficient of vapour in free air. It is a function of temperature and gas pressure, and it can be evaluation through the relation given by Künzel [65], which gives similar results that the one proposed by De Vries in the late 1960's [112]:

$$\delta_a = 2.10^{-7}\frac{T^{0.81}}{P_G}\left[\mathrm{kg/(m\,s\,Pa)}\right] \tag{3.10}$$

The vapour resistance factor, μ, stand for the impact of interactions between the vapour molecules and the porous network on their diffusion kinetic though the material. Theoretically, it may be a function of either water content, temperature and gas pressure.

3.2.3 *Mass Conservation of Air*

The Eq. (3.2) stands for the mass balance equation of air, composed by either dry air and water vapour. Under the perfect gas assumption, the left side of this equation, that is the mass variation of air, is directly linked to the water content and total gas pressure through the relation:

$$\frac{\partial m_G}{\partial t} = \frac{\partial}{\partial t}\left[\left(\phi - \frac{\rho_d}{\rho_L}w\right)P_G\frac{M_G}{RT}\right] \tag{3.11}$$

where ϕ is the porosity.

The right side of the Eq. (3.2) is composed by the Darcean transport of total air, $\underline{\nabla}\cdot\left[D_G\underline{\nabla}P_G\right]$, and by the mass variation due to sorption/desorption processes ($\dot{m}_{\rightarrow v}$ which is positive in desorption). Assuming that the mass exchange of vapour between the sample and the surrounding air is strongly higher than the mass variation of vapour within the sample, this last term can be written in the form:

$$\dot{m}_{\rightarrow v} \approx \underline{\nabla}\cdot\left[P_G\delta_p\underline{\nabla}\left(\frac{p_v}{P_G}\right)\right] \tag{3.12}$$

The range of validity of this last relation have been estimated by and it seems that its use is most of the time acceptable for earthen materials.

3.2.4 *Heat Balance*

Finally, the Eq. (3.3) is a form of the heat balance equation. The left member, that is $\rho_d c\frac{\partial T}{\partial t}$ denotes the heat associated to an increase (or decrease) of temperature. It is the sum of the contribution of each constituent. c is the total mass heat capacity, which is the amount of heat stored in an element of earth of unit dry mass when its temperature is increased by one degree. Another option would be to use the volumetric heat capacity, $C = \rho_d c$, which is the amount of heat stored in an element of earth of unit volume when its temperature is increased by one degree.

The right member of the Eq. (3.3) is divided in three main terms. The first one, $\underline{\nabla} \cdot (\lambda \underline{\nabla} T)$, is the thermal conduction through the material. λ is the equivalent thermal conductivity, which is typically measured in W/m/K. It is defined as the proportionality constant that relates the heat flux to the area traversed by the flux and the thermal gradient normal to the area. Similar to the heat capacity, the thermal conductivity of raw earth depends on the conductivities of the solids, water and air components that constitute the material.

The second term, $-(c_v \underline{\omega}_v + c_a \underline{\omega}_a + c_L \underline{\omega}_L) \cdot \underline{\nabla} T$, denotes the heat which is convectively transported by the fluid phases, where $\underline{\omega}_v$, $\underline{\omega}_a$ and $\underline{\omega}_L$ are respectively the mass flow of vapour, dry air and liquid water. Assuming that the velocity of in-pore phases is small, the Péclet number remains much lower than 1 and this term can be neglected. This might not be true when the earthen walls are submitted, for example, to capillary rises or to strong winds.

The last term, $-\Delta h_v \dot{m}_{\rightarrow v}$, is the gain (or loss) of heat associated to the desorption or sorption of moisture. Δh_v is the integral specific enthalpy of sorption, also called the latent heat of sorption, and it is defined as the amount of heat released to convert a unit mass of vapour into adsorbed water without change of temperature. The minus sign of the above expression is because desorption (positive $\dot{m}_{\rightarrow v}$) corresponds to an inflow of heat to the material while a gain of moisture (negative $\dot{m}_{\rightarrow v}$) correspond to an outflow of heat from the material. If the value of the latent heat of condensation is quite well tabulated for unconfined water (for example [53]), it is not yet the case for in-pore adsorbed water, and some additional studies are currently in progress on this topic.

3.2.5 Boundary Conditions and Interfaces

3.2.5.1 Boundary Conditions

Hagentoft et al. [50] presented a thorough overview of boundary conditions used in hygrothermal analyses. Here, we summarise the main points.

At the internal surfaces, it is generally admitted that the main process is the mass exchange of vapour (and dry air) between the ambiance and the earthen wall. This mass exchange is depicted by a Fourier's like boundary:

$$\underline{\omega}_{V,Si} = \beta_i \left(p_V^{Si} - p_V^i \right) \tag{3.13}$$

where $\underline{\omega}_{V,Si}$ is the mass flow at the internal surface, p_V^i is the vapor pressure in the indoor atmosphere and p_V^{Si} is the vapor pressure at the internal surface of the wall. β_i is the internal moisture surface exchange coefficient whose value may depend on surface roughness, indoor air agitation, temperature... but no clear methodology has been yet developed to properly assess it.

The heat exchange at the internal surface is composed by a convective term, a latent term and a mass exchange term. The convective term is quite classical and it is taken into account by the Fourier's like boundary:

$$\underline{q}_{c,Si} = h_i(T_{Si} - T_i) \tag{3.14}$$

where $\underline{q}_{c,Si}$ is the heat flow caused by heat convection at the internal surface, T_i is the temperature in the indoor atmosphere and T_{Si} is the temperature at the internal surface of the wall. h_i is the internal heat surface exchange coefficient.

The latent term makes it possible to take into account the evaporation/condensation processes occurring at the boundary surface in order to keep the equilibrium condition between the liquid water and vapour. In particular, a flow of liquid water toward the surface must be counterbalanced by evaporation, and thus a heat consumption. On the other hand, water condensation is imposed at the surface to keep the liquid–vapour equilibrium if the flow of liquid water is directed towards the material core, which leads to a heat supply. The simplest way to express this term, denoted by $\underline{q}_{l,Si}$, is to link it to the mass flow of liquid at the surface:

$$\underline{q}_{l,Si} = \Delta h_v \underline{\omega}_{L,Si} \tag{3.15}$$

Finally, the mass exchange term takes into account the heat convectively transported at the surface due to the exchange of matter (vapour and dry air) between the indoor atmosphere and the earthen wall. This last term is however most of the time negligible.

At the external surfaces, the problem is even more complex. For the mass balance equations, it becomes necessary to consider the mass income of water due to rain and of air due to the wind. But, no sound methodology exists to properly assess these two contributions, even more that all the processes which occur at the external surface are coupled.

Concerning the heat balance equation, in addition to the effect of these matter exchanges on the sensitive heat, and especially rain, we need to consider the impact of solar radiation.

3.2.5.2 Interfaces and Discontinuities

The hygrothermal transfers in enclosures composed of multiple layers has to bridge interfaces at the contact planes between the layers. Three types of contact are distinguished: ideal (perfect), real (imperfect) and free (no) contact. According to the contact type, interface resistances can exist for each of the heat and mass transfers. The interface modelling is done by taking into account these contact conditions (boundary conditions at interfaces between layers). This implies considering the interface as a layer with an equivalent thickness and properties.

Usually, ideal contact is considered, and interfaces are neglected in simulations. The heat and mass transport potentials are supposed to be unambiguously known and the fluxes in both layers are assumed equal at the contact plane. This approach can be valid for some particular configuration. For example [107] have shown that mortar joints between bricks can be neglected. However, De Freitas, Abrantes and Crausse [45] have shown that the interface may significantly affect mass transfer. Anyway, as it is underlined by [96], the risk on interstitial condensation can be reduced in cases of a good contact between the wall and the insulation.

3.2.6 Summary on the Key Parameters used for Hygrothermal Simulations

The set of equation presented below allows to identify the key parameters that should be determined to correctly assess the hygrothermal behaviour of earthen materials. They can be divided in three main groups.

The first one is the thermal parameters, which are identified as the thermal conductivity and the heat capacity (either mass of volumetric). However, for convenient purpose, these parameters can be replaced by the thermal diffusivity (a) and effusivity (e).

Thermal diffusivity α, which is typically measured in m^2/s, is defined as the ratio of the apparent thermal conductivity λ and volumetric heat capacity C:

$$a = \frac{\lambda}{C} \tag{3.16}$$

and is an inverse measure of the rapidity with which a material exposed to a heat source changes of temperature during the transient phase before attaining steady state. The smaller is the diffusivity, the greater is the thermal inertia of the material (i.e. the slower the material will tend towards thermal equilibrium when subjected to a change of boundary conditions). This is because, the smaller is the conductivity λ, the slower will be the heat flux to/from the material while, the greater is the capacity C, the larger will be the cumulative flux that is necessary to produce an increase of temperature.

The thermal effusivity e, which is typically measured in $J/(s^{1/2}m^2 {}^\circ C)$, is defined as the square root of the product of thermal conductivity λ and volumetric heat capacity C:

$$e = \sqrt{\lambda C} \tag{3.17}$$

and is a direct measure of the rapidity with which a material exposed to a different temperature exchanges heat during the transient phase before attaining steady state.

The larger is the effusivity, the grater is the rate of heat transfer at the boundary of the material. This is because, the greater is the conductivity λ, the faster is the heat flux to/from the material while, the greater is the capacity C, the smaller is the resulting temperature change which preserves the thermal gradient and hence the rate of transfer.

The second one would be the transport parameters, that is the vapour conductivity as well as the liquid and gas permeabilities and their evolutions with water content and, eventually, temperature.

Finally, the last one would be the sorption–desorption curves, which will give the water content of the material in function of the temperature and the RH, that allows to determine the coefficients ξ and λ (defined at Eq. (3.5)). In addition, most of all parameters strongly depends on water content, thus the knowledge of this latter is of main importance.

3.3 Laboratory Measurement of Hygrothermal Parameters

3.3.1 Measurement of Thermal Properties

Heat capacity and conductivity are the two most important parameters to predict the thermal behaviour of a material. In the following the focus will be on the main techniques to measure these two parameters.

3.3.1.1 Methods

Test methods are differentiated between steady-state and transient ones. In steady-state methods, measurements are made after a stationary temperature distribution has been attained in the earth sample. The attainment of such condition requires time and therefore allows migration of vapour under a spatial gradient of temperature. This means that the moisture distribution of an unsaturated specimen at the time of measurement might be significantly different from the initial one. Instead, in transient methods, measurements are relatively fast (of the order of minutes), which reduces vapour movements and therefore minimizes disturbance of the initial moisture distribution.

The "Guarded Hot Plate" is a steady-state method for the measurement of thermal conductivity which is recognized by a number of international standards such as, for example, [2, 60] and [31]. During tests, one or two specimens of given thickness are sandwiched between a hot and a cold plate while the heat flux across the material is continuously measured. The hot plate includes a "measuring" area surrounded by a "guard" area, both at the same temperature. The purpose of the guard area is to reduce horizontal heat losses inside the sample and therefore to ensure a flux that is as one-dimensional as possible, in the direction perpendicular to the two plates, over the measuring area. Once steady-state conditions are attained, the thermal conductivity λ is calculated from the heat transfer equation by using the measured heat flux $\dot{q}_\lambda$, the imposed thermal difference ΔT, the sample thickness ΔL and the measuring hot plate area A:

$$\lambda = \frac{\dot{q}_\lambda / A}{\Delta T / \Delta L}$$

The Guarded Hot Plate test may provide different values of conductivity depending on whether heat flows upwards or downwards. This is due to the effect of gravity on moisture transfers between the hot and cold plates.

The "Line Source" or "Hot Wire" test (Fig. 3.3) is a transient method for measuring thermal conductivity described in the international standards [3] and [55]. The test consists in inserting inside a sample a thin cylindrical heating element with a temperature sensor positioned at mid-height. The heating element provides a constant heat flux $\dot{q}$ while the temperature sensor records the corresponding increase of temperature at the interface between the probe and the tested material. The recorded temperature increase is then interpreted according to theory of line heat sources in semi-infinite homogeneous isotropic media. This theory indicates that, at relatively large times, the plot of the temperature change against logarithm of time follows the following linear relationship:

$$T_2 - T_1 = \frac{\dot{q}_\lambda}{4\lambda\pi} \ln \frac{t_2}{t_1}$$

where T_2 and T_1 are the temperatures measured by the sensor at times t_2 andt_1, respectively. Therefore, as time increases, measurements tend towards a straight line in the semi-logarithmic plane of temperature versus time. The slope of this line is then determined to infer the thermal conductivity λ.

Measurements of thermal conductivity from the "Hot Wire" test compare well with those from the Guarded Hot Plate test, though the latter method yields slightly

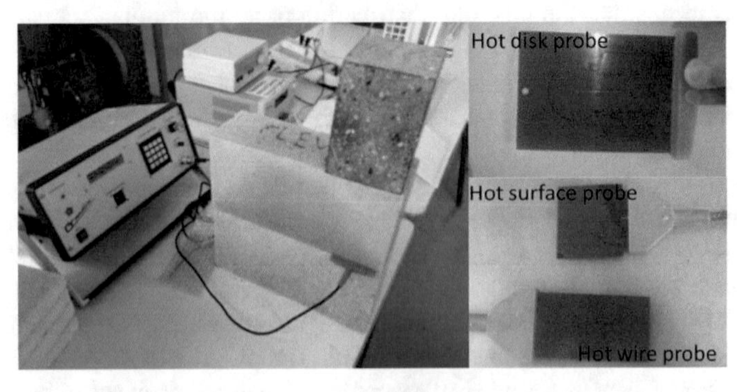

Fig. 3.3 Photos of acquisition system and probes used for "hot wire", "hot disk" and "hot surface" tests

smaller values of thermal conductivity probably because of the effects associated to moisture convection. One of the advantages of the “Hot Wire” test with respect to the Guarded Hot Plate test is the relatively short duration of measurements, which limits disturbance to the original moisture distribution in partly saturated porous materials. Moreover, the small size of the “Hot Wire” probe.

The “Hot Disk” test (Fig. 3.3) is a transient method for measuring both thermal conductivity and specific heat capacity and is described in the international standard [59]. Similarly to the Needle Probe test, this test allows rapid measurements of material properties, thus minimizing disturbance of initial moisture inside unsaturated samples. The test consists in sandwiching a flat probe with the shape of a disk between the planar surfaces of two identical samples. The disk probe incorporates a continuous double spiral of electrically conducting nickel enclosed between two Kapton films, which provide electrical insulation. During measurements, a fixed electrical current is imposed across the spiral to increase the temperature of the probe and therefore to produce dissipation of heat into the tested material. By measuring the increase of temperature at the interface between the probe and the samples, the rate of heat dissipation is measured and the thermal properties are then inferred.

The “Hot Disk” test can also be performed in a single-sided configuration where the flat disk probe is sandwiched between the tested sample on one side and a known insulation material on the other side. Single-sided testing requires no additional equipment, only the use of an insulating material to support the probe.

On other probe that it is quite often used is the “hot surface” (Fig. 3.3). This latter can be used to estimate the thermal effusivity.

The “Differential Scanning Calorimetry” is one of the most common techniques to measure the heat capacity of materials and is described in international standards [4, 30, 56]. The technique is also used to measure the exchanges of latent heat during thermal transitions between material phases or crystalline structures. Two types of Differential Scanning Calorimeter exist, namely the “heat flux calorimeter” and the “heat flow calorimeter” (or “power compensation calorimeter”). The latter is the most common device and consists of two identical pans, i.e. an empty pan used as reference and a filled pan containing the sample. These two pans are heated at the same rate so that their temperatures remain identical throughout the test. The principle of the test resides in the fact that the heat supplied to produce a given increase of temperature is different between the two pans because one pan is empty while the other one contains the sample. This difference is measured and related to the heat capacity or the latent heat of the sample.

3.3.1.2 Typical Findings

Typical values of massic heat capacity, conductivity and density of different earth constituents as reported in the literature, are given in Table 3.1. The volumetric heat capacity can be calculated as the product of the massic heat capacity and density.

The values of the dry thermal conductivity of earthen material reported in literature is quite scattered, and no clear tendency can be found with dry density (cf. Fig. 3.4).

Table 3.1 Typical values of heat capacity, thermal conductivity and density for different earth constituents at constant temperature of 20 °C and pressure of 1 atm [10, 27, 28, 62, 78, 80, 82, 83]

Material	Massic heat capacity, c [J/(g K)]	Thermal conductivity, λ [W/(m K)]	Density, ρ [g/cm^3]
Silicon dioxide (quartz)	0.732–0.875	6.5–11.3	2.65
Calcium carbonate (calcite)	0.800	3.2–3.7	2.71
Calcium sulfate (gypsum)	0.816	2.1–3.7	2.45
Aluminium oxide (alumina)	0.908	33	3.70
Diverse clay minerals	0.757–1.13	1.8–2.9	2.60–2.65
Kaolin	0.975–1.02	2.8	2.60
Humus (organic matter)	1.854–1.996	0.25	1.30
Water (liquid)	4.186	0.6	1.00
Water (vapour)	1.910	Depends on humidity	Depends on humidity
Dry air	1.000	0.025	0.00125

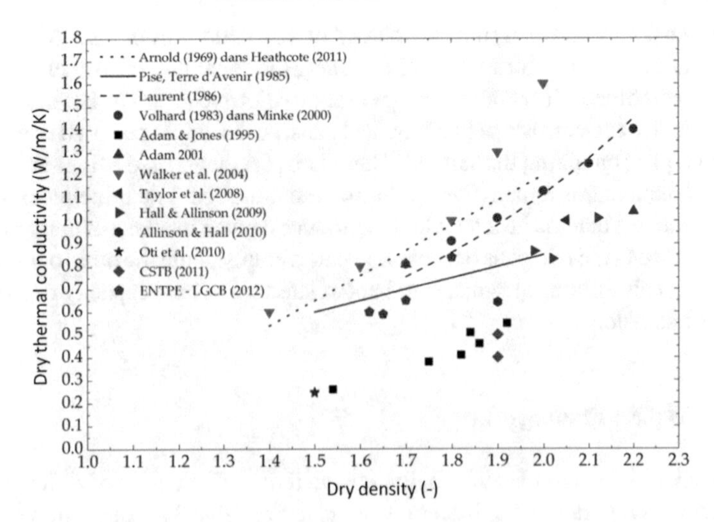

Fig. 3.4 Dry thermal conductivity in function of dry density for several earthen materials. Graph extracted from [25], with the data of [6, 5, 8, 11, 25, 20, 51, 71, 85, 105, 108, 113]

Similarly, the measured incensement of thermal conductivity with water content seems strongly dependant on the tested material and on the test protocol. For these results, the question of a proper measurement of the thermal conductivity of the material appears to be of main importance.

The thermal conductivity of earthen materials also strongly depends on water content. If the constituents were ideally arranged in parallel, the thermal conductivity of the earth, denoted by $\lambda_{||}$ would be calculated as the average conductivity of the constituents weighted by volume:

$$\lambda_{||} = (1-\phi)_d + \frac{\rho_d}{\rho_L} w_L + \left(\phi - \frac{\rho_d}{\rho_L} w\right)_G \tag{3.18}$$

where λ_d, λ_L and λ_G are the thermal conductivities of solids, water and air, respectively.

If the constituents were instead ideally arranged in series, the thermal resistivity of the earth $\frac{1}{<\lambda>_\perp}$ (i.e. the reciprocal of conductivity) would be calculated as the average resistivity of the constituents weighted by volume:

$$\frac{1}{\lambda_\perp} = (1-\phi)\frac{1}{d} + \frac{\rho_d}{\rho_L} w \frac{1}{\lambda_L} + \left(\phi - \frac{\rho_d}{\rho_L} w\right)\frac{1}{\lambda_G} \tag{3.19}$$

The theoretical conductivities $\lambda_{||}$ and $\lambda_\perp$ provide an upper and lower limit to the real values because, in reality, earth constituents are arranged neither in parallel nor in series but rather in a configuration between the two.

An option is therefore to be closer to the real apparent conductivity λ is to use the arithmetic mean of the two limits:

$$\lambda = \frac{\lambda + \lambda_\perp}{2} \tag{3.20}$$

or, alternatively, to estimate λ as the geometric mean of the conductivities of each constituent weighted by volume [1]:

$$\lambda = \lambda_d^{(1-\phi)} \lambda_L^{\left(\frac{\rho_d}{\rho_L} w\right)} \lambda_G^{\left(\phi - \frac{\rho_d}{\rho_L} w\right)} \tag{3.21}$$

However, there is no clear consensus on the applicability of these formula for earthen materials. Indeed, if Eqs. (3.20) and (3.21) were found to provide realistic predictions in [47, 111], it is not the case in [71]. Deeper investigations are thus necessary to draw a definitive conclusion on that point.

Fewer studies can be found on the measurement of heat capacity. Similarly to the thermal conductivity, it varies with water content. At first order, this variation is equal to:

$$c = c_d + w c_L; C = (1-\phi)C_d + \frac{\rho_d}{\rho_L} w C_L \tag{3.22}$$

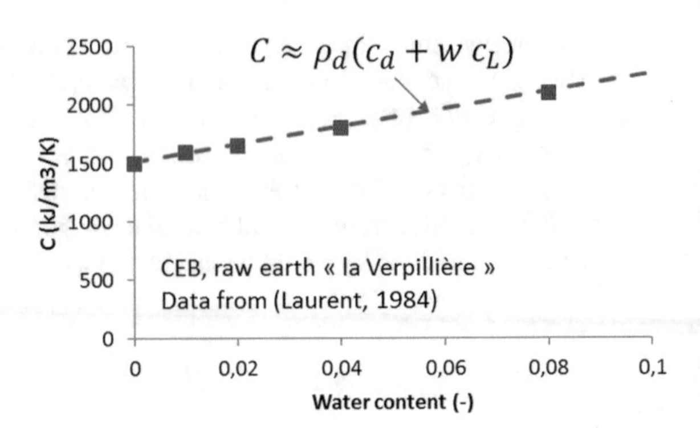

Fig. 3.5 Volumetric heat capacity measured by [71] on unstabilized compacted earth blocks at several water content

where c_d and c_L are respectively the massic heat capacity of the dry material and of liquid water (in J/kg/K) while C_d and C_L are respectively the volumetric heat capacity of the dry material and of liquid water (in J/m^3/K). A good consistency is generally observed between the relation (3.22) and experimental data (cf. Fig. 3.5).

3.3.2 *Transport Properties*

3.3.2.1 Vapour Permeability

Water vapour permeability is commonly measured according to the "wet cup" or "dry cup" methods using the standard EN ISO 12572 [58]. The experimental protocol used for these two tests consists in placing the sample on top of a cup whose RH is controlled by a saline solution. For the wet cup, a potassium chloride solution is used (for example RH level, 85% at 23 °C). For the dry cup, silica gel may be used, but potassium acetate solution (RH level, 25% at 23 °C) was found to provides better stability of RH within the dry cup [74]. To seal the samples to the cup, a vapour-tight aluminium tape is often used, because it does not adsorb a significant quantity of moisture itself [104]. The diagram of the wet cup method is presented in Fig. 3.6.

From the experimental mass variation of the cup assembly (wet or dry) a regression line $G = \Delta mass/\Delta time$ is determined when the permanent state is reached. Assuming that the mass flow of liquid water is negligible towards the vapour water one, and that the total gas pressure remains homogeneous and equal to the atmospheric pressure, the Eq. (3.1) provides the following relation between G and δ_p:

$$\delta_p = \frac{Gd}{A\Delta p_v} \tag{3.23}$$

where d is the thickness of the sample, A its cross section and Δp_v is the difference of partial pressure of vapour between the two sides of the samples, which is not

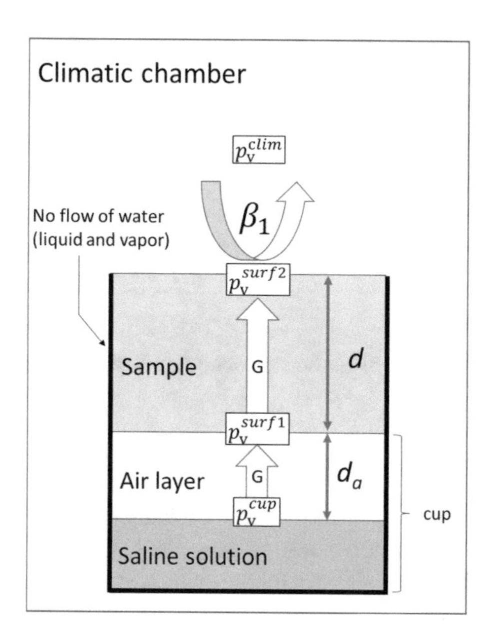

Fig. 3.6 Schematic diagram of the wet cup test

directly equal to the difference in partial pressure of vapour within the cup and the climatic chamber. To take into account this effect, a first correction that can be made is to consider, as it is depicted in the EN ISO-12572, the resistance of the air layer between the sample and the salt solution by assuming that the transport of vapour within the cup is only made by diffusion (no convection). It will be referred to as the "ISO correction" and it leads to the:

$$\delta_p^{iso} = \frac{Gd}{A\Delta p_v - G\frac{d_a}{\delta_a}} \tag{3.24}$$

where d_a is the thickness of the air layer between the salt solution and the sample.

However, many studies had investigated that the water vapour permeability obtained from such experiment, even after the ISO correction, show a significant dependency on the sample thickness [44, 74], while this latter should be intrinsic. To avoid this problem, and thus to determine the "real" water vapour permeability it is necessary to do a second correction which consider the effect of film moisture resistances at sample surfaces. The general expression of the vapour diffusion coefficient then becomes:

$$\delta_p = \frac{Gd}{A\Delta p_v - \frac{G}{\beta}} \tag{3.25}$$

where β stands for the cumulate effect of the surface films and air layer resistances.

An experimental procedure to determine $1/\beta$ is provided by [110]. It consists in assuming that only the film resistance of the top surface (that is the one in contact

Table 3.2 Values of some classical construction materials (data from [65]

Material	$\mu[-]$
Cellular concrete	7.7–7.1
Lime silica brick	27–18
Solid brick	9.5–8
Gypsum board	8.3–7.3
Concrete	260–210
Lime plaster	7.3–6.4

with the climatic chamber's atmosphere) has to be taken into account. It leads to:

$$\frac{1}{\beta} = \frac{d_a}{\delta_a} + \frac{1}{\beta_1} \tag{3.26}$$

where β_1 is the moisture exchange coefficient at the top surface.

Another option, proposed by [74], rather consider that both surfaces are submitted to film resistance effects, which leads to the following expression for β :

$$\frac{1}{\beta} = \frac{1}{\alpha}\left(\frac{d_a}{\delta_a} + \frac{1}{\beta_c}\right) \tag{3.27}$$

where β_c is the "air layer effect" at the top surface, which can be estimated from the evaporation rate of a cup of water which have the same diameter as the sample, while α takes into account the interactions between water molecules and the sample surface that are not considered in the evaporation of the cup filled with pure water and in vapour diffusion through the air thickness.

These two methods need the empirical estimation of a parameter (β_1 for Eq. (3.26) and α for Eq. (3.27)), which requires to test at least samples of three different thicknesses. What is more, these previous developments where made under the assumption that no vapour advection and no liquid water transport occur within the material. It is quite clear that these two assumptions are not true, and the clear assessment of their impact on the estimation of δ_p is an important topic that should be treated in the near future. Instead of δ_p the ability of the vapour to go through the material is commonly expressed using the vapour resistance factor, denoted by μ. The link between δ_p and μ is given by the Eq. (3.9). μ -values commonly obtained for earthen materials, and for some other classical construction materials, are reported in the Tables 3.2 and 3.3.

3.3.2.2 Liquid Water and Air Permeability

At first, even if intrinsic permeability should not depend on the fluid which filled the porous material, it is not practically the case. Indeed, a difference up to one order of magnitude can be observed between gas and liquid water intrinsic permeabilities. It

Table 3.3 Some typical values of μ of earthen materials

Material	μ [–]	References
Light earth	2–7	[68, 109]
Earth plasters	7–10	[74]
Rammed earth Compressed earth block	6–14	[8, 25, 38, 73]
Adobe/Cob/Wattle and Daub	3–9	[21]

is commonly attributed to the differences between water/solid and gas/solid interactions, and slip effects during gas permeability measurement [14]. Measurement of intrinsic permeabilities can be done with several methods, mainly transient methods (pulse-test) 17 or steady-state methods [84], and these methods can be applied with no main problems to earthen materials.

Several authors have pointed out the practical troubles to measure the liquid water permeability of unsaturated soils, even if some experimental set up have been designed to overcome these difficulties [29, 33, 46, 84, 115]. Still, they remain quite sophisticated and not widespread. Therefore, many formulations to evaluate the liquid water transport properties have been developed, based on different measurements. In [24], the Hazen formula, provides an evaluation of the intrinsic permeability (κ) from the particle size distribution; in 22, the multiscale network approach derives D_L from the water vapour permeability; in [66], the capillary transport coefficient (similar toD_L) is approximated from the water absorption coefficient and the free water saturation.

Concerning earth-based material, the use of the absorption coefficient, commonly called the A-Value, and defined by the total amount of water absorbed (in kg) per surface unit in contact with water (in meter square) and per square root of the immersion time (in seconds), is the most widespread method to assess the behaviour of the material in highly saturated states. However, even though the protocol of measurement is provided by European standards (BS-3921 for example), the reliability of this method have been pointed out by many inter-laboratory investigations [44, 95]. In addition, this test does not allow a direct measurement of the liquid transport coefficient. For that purpose, the use of indirect analysis is required. Classically, the relation proposed by [66] is used, even if no clear validation of it has been made for earthen construction materials. At last, these kinds of laboratory tests are commonly made on homogeneous samples, which have been dried and cured under controlled external conditions. The similarity between the tested material and the one manufactured on site is thus questionable.

Finally, few studies have been yet realized on the estimation of the gas relative permeability for earthen materials. According to Fabbri et al. [38], this latter can correctly estimate through the well-known Corey's law if the material remains within the hygroscopic range of saturation.

3.3.3 Water Vapour Sorption Curves

The sorption isotherms denote the variation of water content resulting from a variation of air relative humidity when the steady state is reached. Some typical adsorption–desorption curves obtained on earthen materials and hemp concrete (for matter of comparison) are reported in Fig. 3.7. In these curves, the water content can be either expressed in percent (that is $w_{\%} = 100w$, $w = m_w/m_d$) or in kg/m^3 (that is $w_\rho = \rho_d w$). Because it is rather the product $\rho_d w$ that is involved in the mass conservation equation, for a matter of comparison between material, $w_\rho(RH)$ should be rather used than $w_{\%}(RH)$ or $w(RH)$. In Fig. 3.7, for example, the use of w instead of w_ρ might lead to the wrong conclusion that the sorption properties of hemp concrete elements is more important than CEB ones.

For a given relative humidity, a difference in water content can be observed if the material is subjected to an increase or a decrease of the air relative humidity. This phenomenon, called hysteresis, is quite common and has been widely studied by many authors for a large variety of materials, including hemp concrete [72]. The general conclusion that may be drawn from these study is that it could be explained by the complex shape of the porous network [22], the difference of the wetting angle between wetting and drying [114], the occurrence of metastable processes [87] and/or local heterogeneity in behaviour between the numerous sorption sites that form the porous network [34]. To take into account this effect, some hygrothermal models have been developed [64, 72]. But no clear evidence of the necessity to use such level of complexity to properly model the hygrothermal behaviour of earthen materials has been provided yet.

Along the same line, no clear consensus exists on the importance of taking into account the impact of temperature on sorption/desorption curves. The study from [86] tends to indicate stronger variations than the ones anticipated by the Clapeyron's relations while [40] concludes on a material dependent but limited effect if we remain in the range of temperature commonly experienced in building applications.

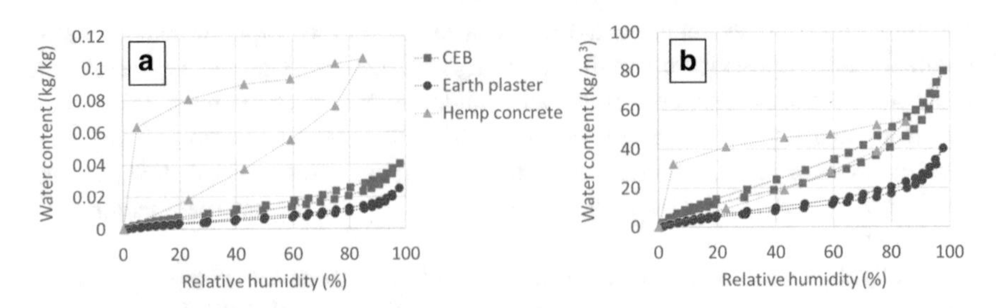

Fig. 3.7 Example of typical adsorption–desorption curves of an earth plaster, a compacted earth block (data from [40]) and a hemp concrete (data from 39), with the water content expressed either in kg/kg (**a**) or in kg/m^3 (**b**)

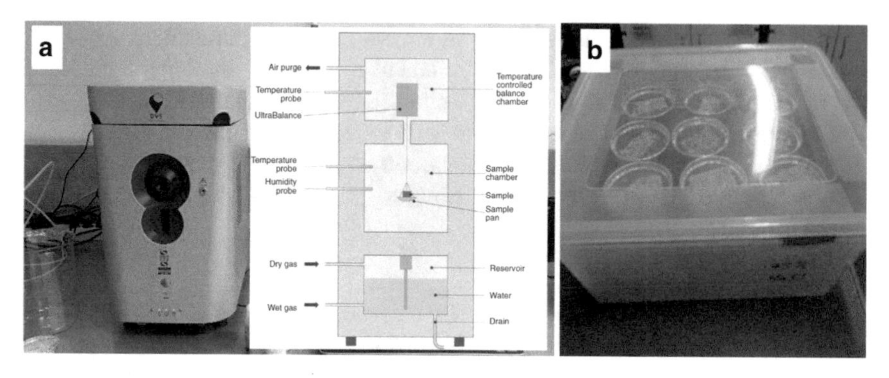

Fig. 3.8 Example of devices used to measure the sorption curves through dynamic gravimetric vapour sorption (**a**) or dessicator (**b**) methods

Several methods exist to estimate the isothermal sorption–desorption curves, but the two most widely used are the desiccator and dynamic gravimetric vapour sorption methods (cf. Fig. 3.8).

The experimental protocol of the desiccator method is precisely defined in the international standard ISO 12751 [57]. The sorption stage consists in successively putting a previously dried sample in several environments of increasing RH and constant temperature. The sample is periodically weighed, and it stays within a given environment until mass constant. The desorption stage consists in successively putting a sample previously equilibrated at 95%RH (at least) in several environments of decreasing RH until mass constant and at constant temperature. The RH of the environments is fixed by equilibrium with saturated saline solutions.

The dynamic gravimetric sorption method, commonly called the DVS (dynamic vapour sorption) method, consists in measuring uptake and loss of moisture by flowing a carrier gas at a specified RH (or partial pressure) over a small sample (from several milligrams to several grams depending on the device used) suspended from the weighing mechanism of an ultrasensitive recording microbalance. Variations in the gas's RH are automatically calculated by the device when the target condition in mass stability is reached. A sorption–desorption loop can thus be made in approximatively 1–2 weeks for earthen materials, while a period of 2–4 months is necessary if the desiccator method is used. On the other hand, the desiccator method can test several specimens at the same time, and it is the only way to test specimens with high levels of heterogeneity like earth-fibres mixtures. Whatever method which is used, the isothermal sorption–desorption curves should be intrinsic to the material tested. However, when direct comparisons are made between the curves obtained by the DVS or the desiccator methods, certain significant differences may be observed. According to [40, 68] these differences are mainly due to the difference of the protocol used to measure the dry mass of the sample between these two methods.

For practical reasons, the use of single parameter can be preferred than the use of the sorption curves in order to give an idea of the sorption capability of a material.

For that purpose, even if there is not yet a consensus, two parameters appears to be commonly used: the hydric capacity, equal to $\rho_d\xi$ (that is the slope of the linear part of the desorption curve in kg/m^3), and $w_{\rho,80}$ which is the water content for a relative humidity of 80%. For compacted earth blocks, $\rho_d\xi$ is generally in the range of 30–100 kg/m^3 and $w_{\rho,80}$ is in the range of 60–120 kg/m^3. For earth plasters, $\rho_d\xi$ is rather in the range of 3 –30 kg/m^3, while $w_{\rho,80}$ is in the range of 5–30 kg/m^3 [26, 40, 81].

3.3.4 Assessment of the Hygroscopic Buffering and Hygrothermal Potential

A common way to quantify the hygroscopic potential of a porous medium is to perform dynamic sorption–desorption tests, also called the MBV (Moisture Buffering Value) test. The protocol of this test has been originally defined within the framework of the NordTest project [94]. It indicates the amount of moisture transported in or out of a sample during isothermal daily cycles of RH between 33 and 75% at 23 °C, with 8-h time steps at high RH and 16-h time steps at low RH (other RH cycles can be chosen, but to facilitate comparisons, this tends to be the reference cycle). The sample tested is isolated on all its sides except one, and the MBV, in kg/m^2/%RH, is calculated as the mass variation per unit of surface area of the open surface (denoted by A):

$$MBV = \frac{\Delta mass}{A(RH_M - RH_m)} \tag{3.28}$$

where $\Delta mass$ is the total mass variation of the sample during the cycle, while RH_M and RH_m are the maximum and the minimum relative humidities of the cycle (in %). The value obtained for a daily cycle with $RH_M = 75\%$ and $RH_m = 33\%$ allows to classify the tested material into one of the five categories defined by [94] and summarized in the Table 3.4.

A recent review on the buffering capacity of earth based building materials has presented a compilation of results based on the Nordtest type measurement [77]. Those results are presented in Fig. 3.9, notice that variable humidity conditions

Table 3.4 MBV categories of materials [10, 27, 28, 62, 78, 80, 82, 83]

MBV [g/(m^2 %RH)]	Hygroscopic potential
0–0.2	Negligible
0.2–0.5	Limited
0.5–1	Moderate
1–2	Good
> 2	Excellent

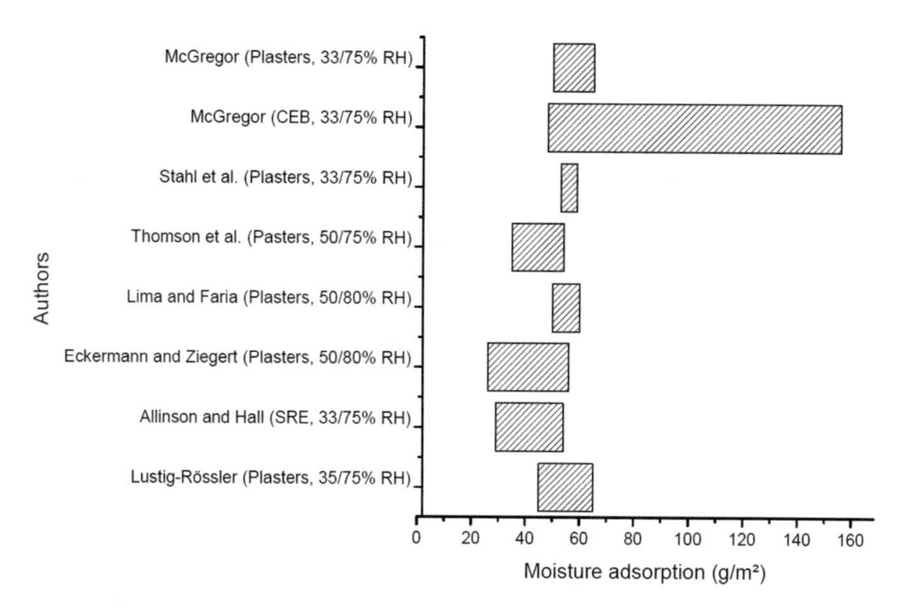

Fig. 3.9 Moisture Buffering Value of earth based building materials [77]

have been used. However, a general tendency appears with values of a moisture buffer capacity around 60 g/m^2, comparable to the values of earth based materials in Figure. It has been demonstrated [75], that the composition of compressed earth blocks (CEB) can be tailored through the clay content and clay nature to reach very high moisture adsorption around 150 g/m^2. According to the classification proposed by the Nordtest protocol earth materials present in most cases a good to excellent moisture buffering value.

The moisture buffering capacity presented through the moisture buffering test depends on the humidity conditions used during the cycles [76]. If the same conditions are respected the test can give an indication on the buffering potential of the material. This test is realised in isothermal laboratory conditions, no direct comparison with in-situ performance is possible. The transient nature of this test does not allow direct assessment of intrinsic material properties. To overcome this, an inverse modelling approach was described to determine the intrinsic properties such as the vapour diffusion coefficient or the moisture capacity [36]. If such a method can be validated for several materials it would bring considerable benefits in terms of duration to determine intrinsic material properties.

The MBV is determined under isothermal conditions therefore neglecting the influence of temperature. It is commonly accepted and recent studies confirm this for earth based materials, that the temperature has a little influence on the sorption and diffusion properties [40], but it will strongly influence the results of the MBV tests, due to the strong increase of the vapour pressure at saturation with temperature.

Currently the main limitation of the laboratory investigation of the buffering potential is that there is no method to relate measured performance in the laboratory to in-situ performance of the materials. In that sense, there is a total lack of data sets which compare laboratory and in-situ buffering performances of earth based building materials.

A comparison between the moisture buffer capacity of some bio-based and earthen-based materials have been realized by [42]. Instead of the MBV, this study has considered the maximal adsorption value defined in the DIN 18947 (2018) standard. It consists in the measurement of the mass uptake after 12 h at 80%rh of samples initially stabilized at 50%rh. The results, presented in Fig. 3.10, stand out the higher capacity of earthen-based materials.

Let us underline that water vapour is not the only gaseous component which can be adsorbed at the pores surfaces of earthen material. Indoor pollutants are mainly gaseous or solubilised compounds: CO_2, NO_x, volatile organic compounds (VOCs), formaldehyde, phthalates, polycyclic aromatic hydrocarbons (PAHs), tetrachloroethylene. The experimental methods to evaluate the contribution of earthen materials to IAQ are typical developed in the analytic chemistry field. The retention capacity could be evaluated on samples exposed to the different sources of pollutants (alone or in mixture) through the characterisation of kinetics and adsorption isotherms (chromatography), retention factor, diffusion and emission. Furthermore, as concentration of pollutants depend on the temperature, these characteristics have to be considered relating to the thermal properties of earthen materials.

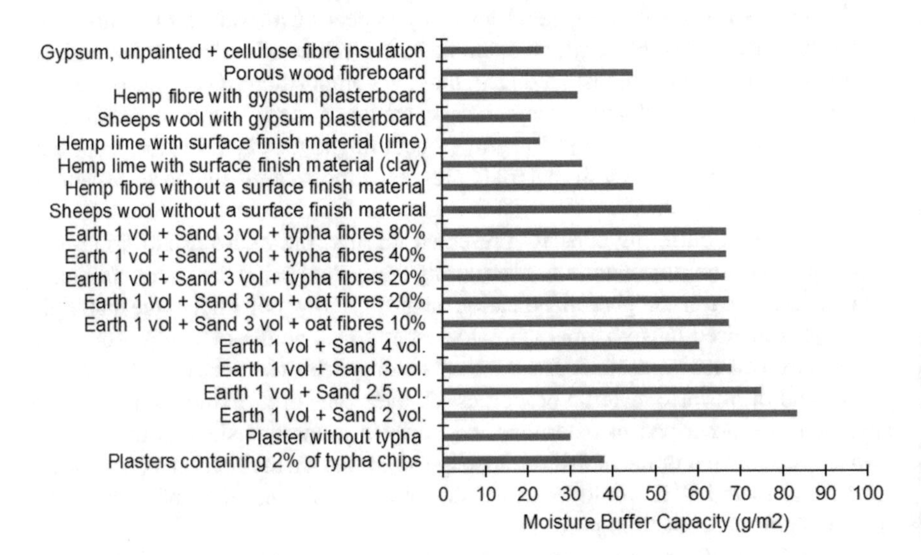

Fig. 3.10 Maximal adsorption value of some bio-based materials [42]

3.4 Assessment of In-Situ Thermal and Hygrothermal Performance

From this bibliographic review, it is quite clear that if numerous laboratory studies have been realized on the hygrothermal assessment of earthen materials, only few papers present works conducted at building scale. In addition, whenever data is available, it remains difficult to draw any conclusions, for different reasons.

A first one relies on the in-situ evaluation of material properties, potentially very different from those estimated in the laboratory. Tested samples can have different properties and the surrounding environment is less controlled on site. A second one can be the major difficulty to analyse properly data when coming from occupied dwellings, which brings uncertainties of the same order of magnitude as the sensor precision. Finally, experimental setup can create bias likely to introduce distorted phenomena.

In order to avoid these kind of issues, it seems important to clarify some setup techniques and to identify precautions to consider when monitoring hygroscopic materials at building scale. The first part of this section gathers few publications dealing with rammed earth houses and exposing problems and solutions found for different cases. The second part explores more deeply some questions (sensors types and location, etc.) thanks to laboratory measurement on hemp concrete samples.

3.4.1 Description of the Instrumented Houses

A focus is realized on instrumented earthen houses to investigate the specific measures taken regarding this material. For this purpose, this review gathers 22 publications/thesis, among them 6 using earth, and exposes the main characteristics of monitored buildings and the instrumentation set in place. It does not mean to be exhaustive but to provide a global overview relying on a certain amount of experimental set-ups over the last decade. The summary of all studied buildings, with the source and their main characteristics is provided in the Table 3.5 for the earthen houses and experimental cells and on Appendix 3.1 for the other buildings. Different building types have been studied, from the real occupied house, to experimental cells from 8 to 10 m^2. In between, unoccupied small houses are also listed. Real houses have a living area from 70 to 215 m^2; small houses usually have the size of a main room in a house, from 12 to 30 m^2.

The majority of studied buildings are located in Europe (western, eastern and northern), fewer are located in Australia, USA, and China. This covers areas with different climates: cold, rainy, Mediterranean, arid, etc. The majority of the buildings have a wooden frame, with different types of insulating materials (mineral wool, OSB panels, wood fibres, cellulose fibres, hemp, etc.); these building types are usually highly insulated. Some of the buildings are made with earthen materials without any insulating material.

Table 3.5 Characteristics of earthen houses and experimental cells

References	Location	Size	Material and insulation	Thickness (m)	Goal of the study
[8]	UK	8 m^2	Stabilized rammed earth with PS	0.23	Hygrothermal performance Numerical validation with WUFI Plus
[16, 15]	Australia		Rammed earth	0.30	Thermal monitoring to assess the good thermal behaviour of rammed earth walls
[100]	Australia	104	Rammed earth	0.33	Hygrothermal performance and comparison between insulated and non-insulated rammed earth walls
		96 m^2			
		175		0.11	
[102]	France	150 m^2	Wooden frame/rammed earth	0.50	Energy performance of the walls for different seasons and orientation
[106]	Australia		Rammed earth		Measure real thermal behaviour of rammed earth walls to highlight their high thermal mass
[93]	Spain	5 6	Rammed earth	0.50 0.29	Thermal performance and materials´ influence
[23]	Portugal	8	Air lime stabilized cob with reed fibres and reeds	0.40	Assessment of technology and hygrothermal monitoring

The thickness of the walls depends on the material: it ranges from 20 cm for light weights structures, to 50 cm for rammed earth walls. The insulating layer varies from 10 to 30 cm.

3.4.2 *Instrumentation Protocol*

The experimental houses and cells are highly instrumented, but fewer sensors are places into occupied houses, even if their number remains acceptable. Generally, very few information about the exact sensor used (brand, model) is provided in the papers. The summary of the instrumentation scheme is provided in Table 3.6 for the earthen

Table 3.6 Description of the instrumentation scheme of the 5 buildings with earth

References	Duration	Variable	Weather station	Specificity
[8]	10 months	T, RH	T, RH, wind, sun, pressure, precipitation (7 km)	
[16, 15]	1 year	T, RH	T, RH, precipitation. Solar and atmospheric data from the local airport	Embedded thanks to a box and a pipe placed during compaction
[100]		T	T, RH – sun, wind (nearby)	
[102]	5 years	T, RH, WC	T, RH – sun, wind (100 km)	Placed in the walls during compaction, at the middle and 10 cm from each surface
[106]	4 days	T, RH, heat flux	T, RH, sun	Embedded at the surface with mortar on them
[93]	8 + 8 days	T	T	Indoor sensors
[23]	28 days	T, RH (non-ventilated cell)	T, RH (nearby)	Indoor sensors

houses and experimental cells and on Appendix 3.1 for the other buildings (Tables 3.7 and 3.8).

Temperature sensors are always placed in the indoor and the close outdoor environments. They can be located near the walls or at the centre of the room(s). They are often located at mid-height, but also near the ceiling and the potential thermal bridges. Different heights can be also monitored in order to study temperature stratification.

Thermocouples are usually used to measure the surface temperature and can also be positioned inside the walls. When insulating materials are used, some of them can be placed between the layers. The sensors are set in place during the construction or can be added later with a drilling, in materials with a low density, mostly insulating materials.

Flux meters are sometime placed at the surface of the wall. Even if they are almost always used for experimental dwellings, they are not used for occupied houses. Among the sources describing precisely the sensors, the band "Hukseflux" is often used.

The thermal monitoring of these buildings is always coupled with other measurements: the RH is always recorded, and the use of a single sensor for temperature and RH is common. Otherwise, depending on the main objective of the study, the CO_2 concentration, air speed or the electric consumption can also be recorded.

The use of a weather station in order to fully take advantage of the measurement appears to be essential as most of the experimental set-up uses one. However, they can be more or less sophisticated. The external temperature and RH are always recorded; follow solar irradiance and wind speed. The most precise ones measure

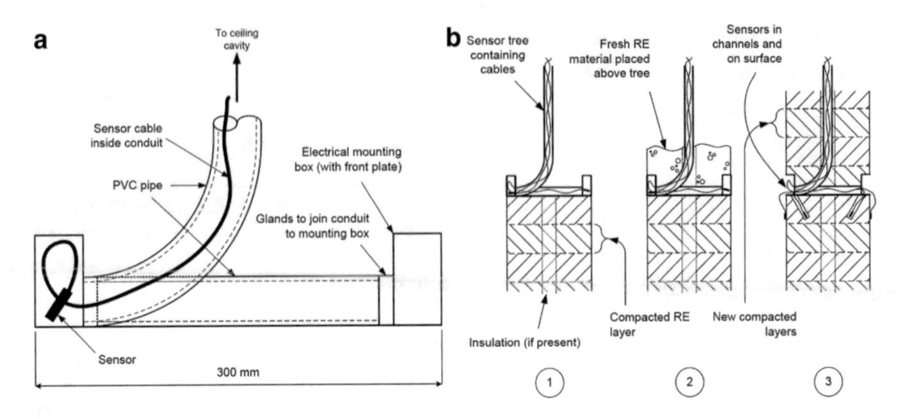

Fig. 3.11 **a** Sensor tree components and layout; **b** installation of sensors trees [16]

wind speed and direction, global and diffuse solar radiation, atmospheric pressure and precipitation.

The equipment used during the monitoring strongly depends on the type of house and the aim of the experimental set-up. Heaters, ventilation systems and/or cooling systems are used when dealing with passive houses, but natural ventilation and/or cooling is also investigated with some materials such as raw earth. Whatever the chosen equipment, it is usually well monitored.

In [16], the sensors were either placed at the surface of the wall or embedded (a box and pipe are placed during compaction, see Fig. 3.11). This system enables to easily replace the sensor in case of damage. However, it can also create a thermal bridge and thus flaw the results. To avoid any problem, the sensor is placed 15 cm below the thermal bridge created by the pipe spanning the wall, and this latter was filled with insulated foam once the sensors were installed.

In [102], sensors were placed in the walls during manufacture at three different locations (middle, and 10 cm from each surface), and for two orientations (south and west). The heat flux through the wall surfaces were estimated using the temperature measurements within the wall trough the Fourier's law rather than with flux-meters, since these latter can hardly be used on south surfaces because the sun radiations would have created too important local perturbation. The applicability of this instrumentation scheme to measure the heat flow is based on a quasi-stationary assumption within the 10 firsts centimetres of the wall. The applicability of this assumption was found to be acceptable for daily and seasonal studies [102].

3.4.3 Results Obtained on Earthen Buildings

The monitoring at building scale differs from the experimental set-up to the occupied house: the number and location of the sensors are more intrusive and loads are easier to analyse when the house is empty. In particular, the paper of David Allinson and Hall [8] which deals with the investigation on hygrothermal behaviour of an experimental cell of $8m^2$, built in cement-stabilized rammed earth walls, and with a numerical analysis using WUFI Plus (v1.2), underlines a quite good consistence between experimental data and numerical simulations (a difference of less than 5%) when the earthen material is precisely characterized. However, other studies, on real buildings, rather tend to the conclusion that thermal simulations (the Australian software Accurate (v2.26) in [16], BERS Pro v4.3 [15] and ENERWIN-EC [100]) underestimate the potentialities of rammed earth constructions, especially for not considering the appropriate thermal comfort criteria.

In comparison with a much lighter earthen-based construction, as the case of reed-cob related by [23], although the cell walls have a much lower mass, was not occupied nor ventilated, the thermal inertia effect is still very significant, with a much higher temperature stabilization indoors.

Actually, even if data is lighter and more difficult to analyse, occupied houses are usually monitored for longer periods and provide a better insight on long term behaviour. For example, the study of Taylor and Luther [106] demonstrates that the large thermal capacity of rammed earth walls improves their thermal properties above the expectation by consideration of R-value alone. Moreover, the weather station is often stated as a key element for a better analyse of the indoor measurements. This point is underlined [103], which has observed strong time lags and decrement factors along with a release of heat from the wall to the indoor environment, during the night, after being heated up by the sun. But other complexities like wind-induced advection-diffusions problems or discontinuities between the buildings elements may be of main importance to correctly assess the hygrothermal behaviour of earthen buildings.

The following chart presents the comparison of indoor air temperatures between an earthen building, constructed with 50 cm thick adobe block (left) and concrete building, constructed of 10 cm thick precast concrete (right) in Cairo, [67], which subsequently was used by Minke in [81], to highlight ways in which to improve the indoor climate using a natural material.

From analysing Fig. 3.12; the Indoor temperature over the 24 h test period within the earthen structure showed no real fluctuation in temperature change and remained within the human comfort zone, represented by the black horizontal band. However, Labarta [67] and Minke [81] stated that the temperature variation within the concrete structure over the same period fluctuated like the outdoor temperature, concluding that the earthen structure could maintain and regulate that comfortable indoor temperature despite similar climate conditions.

'Decrement factor' and 'time lag' as stated by, Labarta [67] and Minke [81] refer to the way in which an exterior wall responds to humidity at a specific time of day before the external temperature reaches the internal environment. The climate conditions related to the location were on average between 18 and 27 °C so, '*thermal capacity is important when trying to create a comfortable indoor temperature*' [81]. Finally, one promising way to highlight the benefits of earth in constructions could be to directly analyse the performance of earthen refurbishment solutions like it was made in [18].

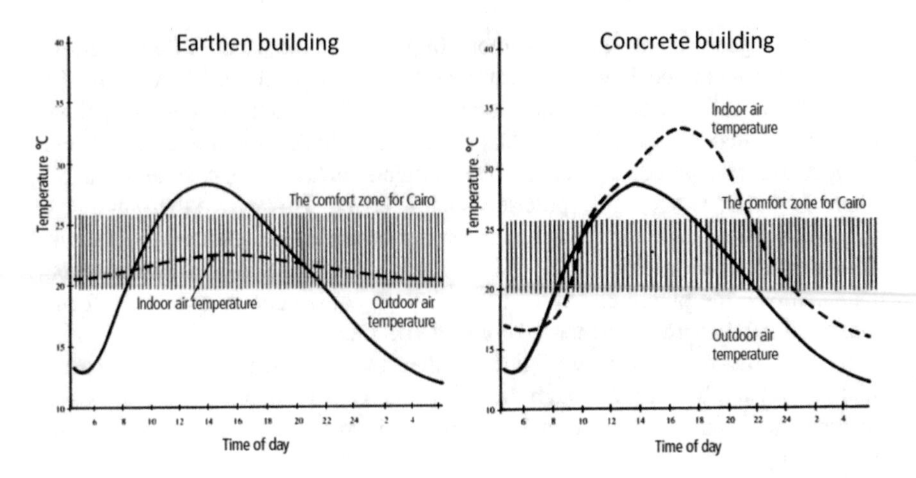

Fig. 3.12 Comparison of indoor air temperatures between earthen and concrete buildings after [67]

3.5 Acoustic Properties

The acoustical properties of earth building materials have been poorly studied and only few references could be found, similar conclusion were made in [52]. The following references found in a literature survey are taken mainly from proceedings or more general books on earth buildings. Most often those studied the direct measurement of acoustic attenuation in earth buildings. Interestingly the theory of sound propagation in porous media such as soils is highly documented due the exploration techniques used in geophysics. The examination of this rich literature can only be done by taking a theoretical approach. Even though this would bring valuable information on the acoustical properties of earth building materials, the considerable task overcomes the scope of this review. The following review therefore only focus on references directly concerned with earth as building material.

The modelling of sound wave propagation in raw earth building materials can be linked to the general modelling of porous media. Where earth has however some specific particularities such as the hygroscopicity or variable pore size (swelling/shrinkage) depending on water content. The pore size seems to be a crucial parameter influencing the propagation of pressure variation.

The non-acoustical parameters in the model used by [54] includes the flow resistivity [kg Pa s/m^2], the porosity [–], the tortuosity [–] and the standard deviation in pore size. This model provides rather good agreement, however in this study the parameters were adjusted to obtain the best fit.

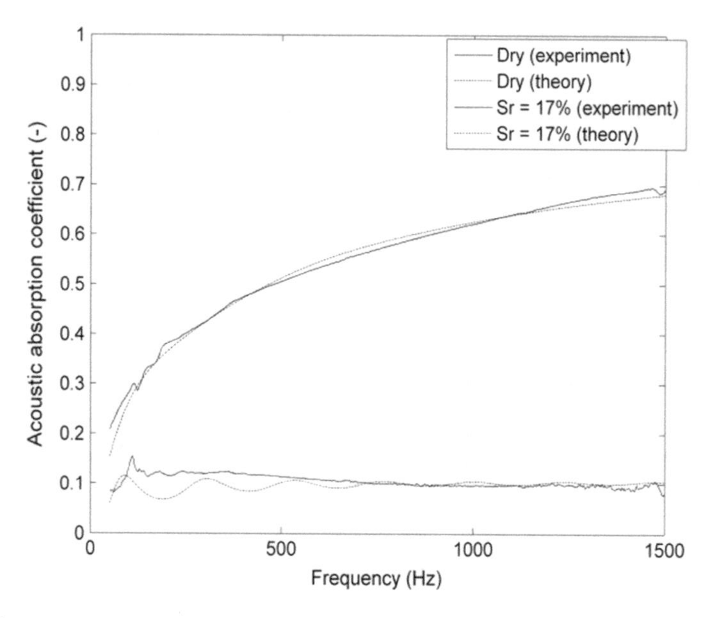

Fig. 3.13 Absorption coefficient of dry and 17% saturated clay soil

3.5.1 Measurement of Material Acoustical Properties

A conference paper that deals with noise reduction through green panels directly discusses the acoustical properties of soils taking in to account the particularity of the porous hygroscopic media. The study investigated the influence of the saturation ratio, Sr, on the absorption coefficient. The absorption coefficient is most often measured in laboratory with an impendence tube (Kundt tube). The absorption coefficient represents the amount of absorbed wave at a given frequency to the total incident wave and therefore is presented as a ratio. The Fig. 3.13 was taken from [54] it shows the absorption on a range of frequencies between 100 and 1500 Hz (typically measured range). The absorption coefficient is close to 0.1 which is low when compared to materials with larger pores (for example containing fibres).

Similar values were given in Hall et al. [52], where the absorption coefficient of earth bricks are compared with fibre board, plaster board, brickwork or concrete. Only fibre board shows a higher absorption, see Fig. 3.14.

3.5.2 Measurement of Building Elements Acoustic Properties

The acoustic behaviour of building components can be differentiated into either the performance regarding airborne sound or impact sound between rooms, or coming from the outside.

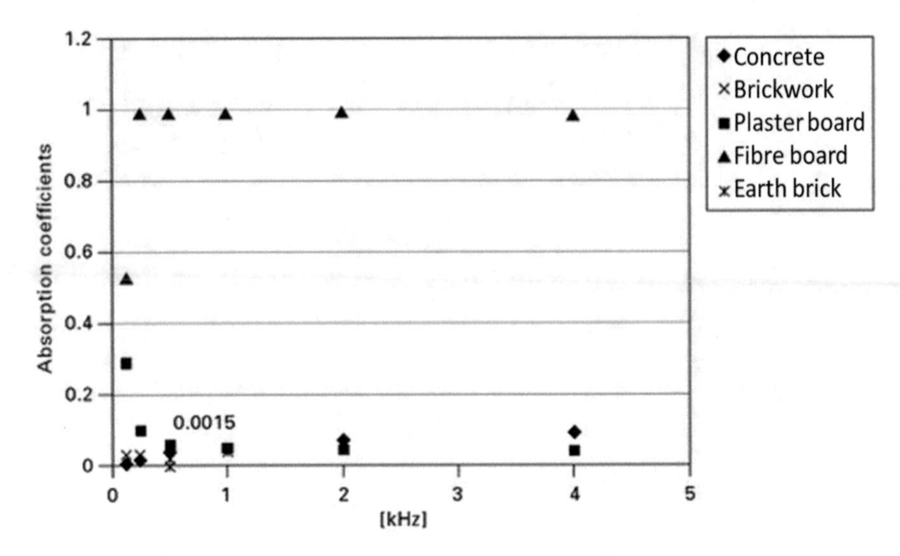

Fig. 3.14 Absorption coefficient of different building materials [52]

3.5.2.1 Attenuation of Airborne Sounds

A chapter in the book of [97] focuses on the attenuation of airbone sounds. This chapter is probably a good resume of the research that has been conducted on the subject in Germany. After the Second World War, in Germany, the attention was more driven to the mechanical performance of the earth building materials and hardly any attention was put on acoustics. The heavy weight construction was then the most common techniques. It is only around 1997 when light weight and dry wall techniques were used that the acoustical comfort became an issue. Thereafter most of the data on sound attenuation came from the producers of earth building materials.

To estimate the acoustical performance of a building element Weighted Sound Reduction Index, R_w, can be used. R_w, for building elements is a measure of the sound insulation performance. This index is measured in laboratory conditions. The Weighted Apparent Sound Reduction Index, R'_w, is on the contrary measured on site and includes flanking effects or other 'on-site' influence on the measurement. The general observation is that the earth building systems having high material densities also present the highest sound reduction index.

Reference values are given in the standards for R_w, in Schroeder [97] a reference value of 53 [dB] is given which is not reached in most cases for light weight earth building walls (p. 350, Lehmbau).

It is related in [32], that the Australian acoustic standard AS/NZS 1276-1979 is one of the few norms that indicate values for rammed earth walls. Research has demonstrated for a 300 mm rammed earth wall a sound reduction index of 58.3 dB (values taken from a rammed earth constructor). The standard classifies a 300 mm thick wall in the sound transmission class STC 57 dB whereas a 250 mm thick wall reaches a STC 50 dB. In this same paper a relation between the superficial density (m) and reduction values is given, sourced from the British Standard BS 8233:2014:

$$R_w = 21.65\log_{10}(m) - 2.3 \tag{3.29}$$

In-situ reverberation times and transmission loss were measured in [32]. The reverberation time is related to the geometry of the room but also on the acoustic absorption coefficient of the materials in the room. This absorption varies with the frequency of the sound wave. In this study no values of absorption coefficients are given, yet the reverberation times are compared to recommended values for conference rooms of similar size. The first room constructed with adobe and rammed earth presents a value below recommended values. In the second room only two walls are made of adobe and rammed earth and two others are masonry brick and double dry wall, I this case the reverberation time exceeds the recommended value.

On site measurement were performed in [19]. The measurements on a compressed earth block home and a wood framed home could be compared. Both houses had a similar configuration. The test consist of positioning a source (loudspeaker) at the outside of the residence and measuring at specific positions the transmission loss (cf. Fig. 3.15).

The study concludes on better performance of the CEB home were they have found a difference of about 8 dBA between the CEB home and wooden-framed home.

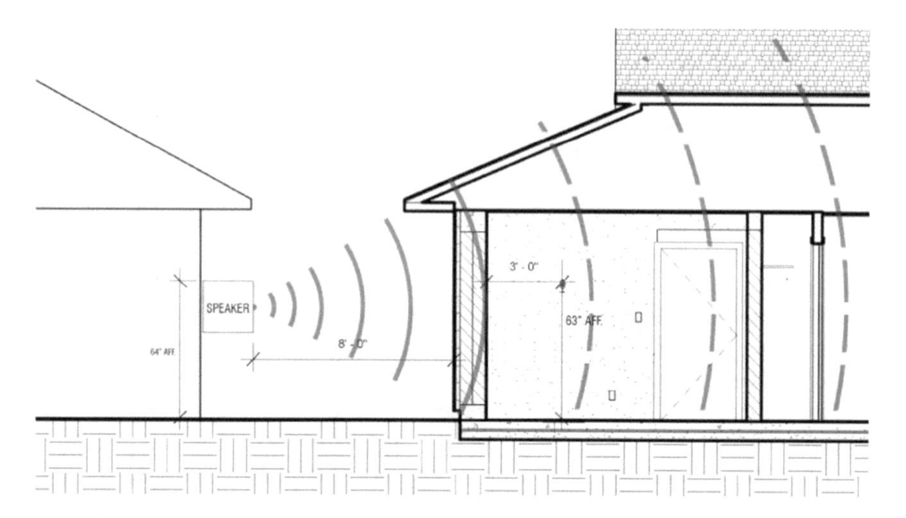

Fig. 3.15 Schematic diagram of the on-site measurement performed by [19]

3.5.2.2 Attenuation of Impact Sounds

The attenuation of impact sounds seems to be more related to the performance of ceilings in building acoustics. No specific studies could be found, however the science is relatively well understood, the understanding of sound wave propagation in soils is intensively studied in seismology and geophysical prospecting. It may however be interesting to share the experience of earth builders. In the case of rammed earth construction in France. To test the cohesion of a rammed earth wall, the builders refer to the impact sound of a hammer on the dry rammed earth wall. A low resonance may indicate pathology or too high moisture content in the wall.

3.6 Conclusion

This is a field open to further research. Acoustical methods already used in geophysics could potentially be applied on heritage buildings for detecting pathologies. Also new developments can be expected through the theory of sound propagation in porous hygroscopic media to extend these techniques to detect water content and transfers.

3.6.1 Concluding Remarks

This chapter has focused on the means to evaluate the hygrothermal and acoustic performances of earthen material. The main physical concepts that drive these performances are already well known by the scientific community. They were recalled in the first part of this chapter. However, it lacks of standard experimental protocols for assessing the engineering performance of earthen materials. In particular, this chapter underlines the lack of a procedure to measure dry mass in spite of the importance of this parameter for the determination of material characteristics such as sorption–desorption capacity, dry density, thermal conductivity, heat capacity, strength and stiffness. Similarly, different protocols exist for the determination of moisture conductivity, moisture buffering potential and water durability, which complicates comparison between measurements.

In addition, the existing laboratory procedures to assess the hygroscopic performance of the material, like the Moisture Buffering Value (MBV), may not be representative of field conditions. Indeed, the homogeneous samples which are tested may be quite different from the in-situ materials, and the tests conditions can be potentially very different from those estimated in the laboratory, in particular since the inhabitants' behaviour is difficult to predict.

On the other side, consensus is starting to emerge for some categories of laboratory tests. It is now acknowledged that the measurement of vapour diffusion should take into account the film moisture resistance at the sample surface and that the determination of moisture buffering capacity must follow standard wetting–drying protocols that do not damage the material.

In conclusion, if some final results have been identified, this chapter rather indicates the direction towards which the research activities should be oriented in order to allow the definition of accurate performance oriented test for earthen materials.

3.7 Appendix 3.1: Characteristics of Houses and Experimental Cells Built with Other Materials Than Earth

See Tables 3.7 and 3.8.

Table 3.7 Characteristics of houses with other hygrothermal material than earth

References	Location	Size	Material and insulation	Thickness (m)	Goal of the study
[7]	Estonia	18 m^2	Log house with mineral wool, cellulose fibre, reed, clay plaster	0.27	Study air leakage of a log house and validation of a simulation model for three insulation materials simulations with WUFI 5.1
[13]	Romania	140 m^2	brick	0.55	LCC analysis of a passive house
[35]	Belgium	12 m^2	Wooden and gypsum Mineral wool	0.27	Experimental data set for validation of HAM modelling
[37]	USA	single storey	Wooden frame		Effect of occupant's behaviour on energy consumption
[43]	France	16m^2	OSB VIP	0.20	Development and evaluation of hybrid façade
[49]	UK	99 m^2 76 m^2	Wooden frame with wood fibre, OSB	0.47	Actual performance of low energy dwellings performance tests: coheating test, tracer gas decay, in situ U value measurement, pressurisation, infrared thermography
[61]	Spain	85 m^2	Cross-laminated		Energy performance of modular structures to create adaptive passive houses

(continued)

Table 3.7 (continued)

References	Location	Size	Material and insulation		Thickness (m)	Goal of the study
[63]	France	36.5 m^3	Timber frame with wood fibre, cellulose, OSB		0.21	Numerical validation
[70]	UK	19 m^2	Wooden frame	Hemp cell	0.37	Energy performance comparison using in-situ coheating test on five buildings
				PIR	0.27	
				Wood fibre	0.36	
				Mineral wool	0.32	
[79]	France	27 m^2	OSB		0.38	Hygrothermal measurement on a test room made with wooden concrete Numerical modelling with Cast3M
[88]	China	17 m^2	Reinforced concrete Insulation mortar		0.30	In-situ method for the measurement of thermal resistance of buildings and comparison of their efficiency
[89]	France	20 m^2	Wooden frame with mineral wool, gypsum board, OSB, PS		0.23	Description of an experimental wooden frame house for the validation of HAM models and for a better understanding of hygrothermal behaviour of light weight structure
[91]	UK	101 m^2	Wooden frame heavily insulated			Passive House: comparison with design predictions
[92]	UK	87 m^2	Wooden frame heavily insulated		0.43	Monitoring of Passive House
		67 m^2				

(continued)

Table 3.7 (continued)

References	Location	Size	Material and insulation	Thickness (m)	Goal of the study
[99]	UK	27 m^2	Wooden frame hemp lime	0.20	Study the hygrothermal performance of a hemp-lime building test
[98]	UK	9.6 m^2	Wooden frame straw bale	0.49	Full scale testing
[103]	France	215 m^2	Mineral wool	0.40	Comparison between forecasts and actual performance simulations with EnergyPlus

Table 3.8 Description of the instrumentation scheme of the buildings with other HT material than earth

References	Duration	Variable	Weather station	Specificity
[7]	1 year	T, RH, heat flux	T, RH	
[13]			/	
[35]	18 months	T, RH, heat flux, pressure, air velocity	Yes	A bitumen impregnated wood fibreboard as coating in order to measure the accumulated moisture by weighting some specimen of the board
[37]	15 years	T, RH, heat flux, power	T, RH, wind, sun	Location of flux meters determined with thermographic image check for stratification with two sensors in the vertical position
[43]	Several months	T, RH, heat flux	Yes	Drill to insert thermocouples during construction
[49]	6 months	T, heat flux, CO_2	/	
[61]		T, RH		
[63]	Several months	T, RH, air speed, heat flux	T, RH, wind, sun, pressure, precipitation	Surface sensors seems to be settle in the first cm of the panel and are protected with a tube and a Gore-Tex membrane
[70]	Several months	/		
[79]	Several years	T, RH	T, RH, wind, sun	
[88]	2 years	T, heat flux	Yes	
[89]	Several months	T, RH, heat flux	T, RH, wind, sun	
[91]	1 year	T, RH, CO_2	T, RH, wind, sun, pressure, precipitation	
[92]	2 years	T, RH, CO_2		
[99]	Several months	T, RH		
[98]		T, RH	Yes	
[103]	Several months	T, RH, CO_2, COV	T, RH, sun, wind (38 km)	

References

1. **Farouki OT (1981) Thermal properties of soils**
2. ASTM-C177 (2013) Standard test method for steady-state heat flux measurements and thermal transmission properties by means of the guarded-hot-plate apparatus. ASTM International
3. ASTM-D5334 (2014) Standard test method for determination of thermal conductivity of soil and soft rock by thermal needle probe procedure. ASTM International
4. ASTM-E793-06 (2012) Standard test method for enthalpies of fusion and crystallization by differential scanning calorimetry. ASTM International
5. Adam EA, Jones PJ (1995) Thermophysical properties of stabilised soil building blocks. Build Environ 30(2):245–253. https://doi.org/10.1016/0360-1323(94)00041-P
6. Adam EA (2001) Compressed stabilized earth block manufacturing in Sudan—UNESCO—Technical Note No. 12. Organization, p 101
7. Alev U et al (2014) Air leakage and hygrothermal performance of an internally insulated log house. NSB 2014:55–61
8. Allinson D, Hall M (2010) Hygrothermal analysis of a stabilised rammed earth test building in the UK. Energy and Buildings. https://doi.org/10.1016/j.enbuild.2009.12.005
9. Allinson D, Hall MR (2010) Hygrothermal analysis of a stabilised rammed earth test building in the UK. Energy and Buildings 42:845–852
10. Aplin AC, Fleet AJ, Macquaker JHS (1999) Muds and mudstones: physical and fluid-flow properties. Geological Society of London
11. Arnold P (1969) Thermal conductivity of masonry materials. J Inst Heat Ventilating Eng 37:101–108
12. Arundel A, Sterling E, Biggin J (1986) Indirect health effect of relative humidity in indoor environments. Environ Health Perspect 65:351
13. Badea A et al (1994) A life-cycle cost analysis of the passive house "Politenica" from Bucharest. Energy Build 80(2014):542–555. https://doi.org/10.1016/j.enbuild.2014.04.044
14. Baroghel-Bouny V et al (2007) Assessment of transport properties of cementitious materials: a major challenge as regards durability? Eur J Environ Civ Eng 11(6):671–696
15. Beckett CTS, Cardell-Oliver R, Ciancio D, Huebner C (2018) Measured and simulated thermal behaviour in rammed earth houses in a hot-arid climate. Part A: Structural behaviour. J Build Eng 15:243–251
16. Beckett C et al (2014) Sustainable and affordable rammed earth houses in Kalgoorlie, western Autralia : development of thermal monitoring techniques. In: ASEC 2014
17. Brace WF, Walsh JB, Frangos WT (1968) Permeability of granite under high pressure. J Geophys Res 73(6):2225–2236
18. Bras A (2015) Repair of Quinta da Mina buildings belonging to a social neighbourhood (Barreiro, Portugal)—selection of retrofit solutions to improve the energy efficiency of a group of buildings. Technical report for the city council
19. Butko D et al (2014) Comparing the acoustical nature of a Compressed Earth Block (CEB) residence to a traditional wood-framed residence. In: Proceedings of meetings on acoustics. Acoustical Society of America, p 015002. https://doi.org/10.1121/2.0000083
20. CSTB (2011) Analyse des caractéristiques des systèmes constructifs non industrialisés
21. Cagnon H et al (2014) Hygrothermal properties of earth bricks. Energy Build 80:208–217
22. Carmeliet J, Descamps F, Houvenaghel G (1999) A mutliscale network model for simulating moisture transfer properties of porous media. Transp Porous Media 35:67–88
23. Carneiro P et al (2016) Improving building technologies with a sustainable strategy. Procedia Soc Behav Sci 216:829–840
24. Cassan M (2005) Les essais de perméabilité sur site dans la reconnaissance des sols. Presses de l'Ecole Nationale des Ponts et Chausées
25. Chabriac PA (2014) Mesure du comportement hygrothermique du pisé. ENTPE. Available at: https://tel.archives-ouvertes.fr/tel-01413611

26. Champiré F et al (2016) Impact of relative humidity on the mechanical behavior of compacted earth as a building material. Constr Build Mater 110:70–78. https://doi.org/10.1016/j.conbuildmat.2016.01.027
27. Chesworth W (2008) Encyclopedia of soil science. Springer
28. Cliffton AW, Wilson GW, Barbour SL (1999) Emergence of unsaturated soil mechanics. NRC Research Press
29. Corey A (1954) The interrelation between gas and oil relative permeabilities. Producers Monthly 19(19):38–41
30. DIN-51007 (1994) Thermal analysis; differential thermal analysis; principles. Deutsches Institut für Normung
31. DIN-52612-2 (1984) Tesing of thermal insulating materials; determination of thermal conductivity by means of the guarded hot plate apparatus; conversion of the measured values for building applications. Deutsches Institut für Normung
32. Daza AN, Zambrano E, Ruiz JA (2016) Acoustic performance in raw earth construction techniques used in Colombia. In: EuroRegio2016, European Association of Acoustics. Porto, pp 1–10
33. Delage P, Cui YJ (2000) L'eau dans les sols non saturés. Techniques de l'ingénieur, C301
34. Derluyn H et al (2012) Hysteretic behavior of concrete: Modeling and analysis. Cem Concr Res 42:1379–1388
35. Desta T, Langmans J, Roels S (2011) Experimental data set for validation of heat, air and moisture transport models of building envelopes. Build Environ 46:1038–1046
36. Dubois S et al (2014) An inverse modelling approach to estimate the hygric parameters of clay-based masonry during a Moisture Buffer Value test. Build Environ 81:192–203. https://doi.org/10.1016/j.buildenv.2014.06.018
37. Emery AF, Kippenhan CJ (2005) A long term study of residential home heating consumption and the effect of occupant behavior on homes in the Pacific Northwest constructed according to improved thermal standards. Energy 31(2006):677–693. https://doi.org/10.1016/j.energy.2005.04.006
38. Fabbri A, Al Haffar N, McGregor F (2019) Measurement of the relative air permeability of compacted earth in the hygroscopic regime of saturation. CR Mec 347:912–919
39. Fabbri A, McGregor F (2017) Impact of the determination of the sorption-desorption curves on the prediction of the hemp concrete hygrothermal behaviour. Constr Build Mater. https://doi.org/10.1016/j.conbuildmat.2017.09.077
40. Fabbri A et al (2017) Effect of temperature on the sorption curves of earthen materials. Materials and Structures/Materiaux et Constructions 50(6). https://doi.org/10.1617/s11527-017-1122-7
41. Fang L, Clausen G, Fanger PO (1999) Impact of temperatre and humidity on chemical and sensory emissions from building materials. Indoor Air 9:193–201
42. Faria P, Bras A (2017) Building physics. In: Jones D, Brischke C (eds) Performance of bio-based building materials. Woodhead Publishing Series in Civil and Structural Engineering, pp 335–344
43. Faure X (2008) Enveloppe hybride pour bâtiment à haute performance énergétique. Université Joseph Fourier, Grenoble I
44. Feng C et al (2015) Hygric properties of porous building materials : Analysis of measurement repeatability and reproducibility. Build Environ 85:160–172
45. De Freitas VP, Abrantes V, Crausse P (1996) Moisture migration in building walls—analysis of the interface phenomena. Build Environ 31:99–108
46. Gardner W (1958) Some steady state solutions of the unsaturated moisture flow equation with application to evaporation from a water table. Soil Sci 85:228–232
47. Gemant A (1952) How to compute thermal soil conductivities. Heat Piping Air Cond 24(1):122–123
48. Goncalves H et al (2014) The influence of porogene additives on the properties of mortars used to control the ambient moisture. Energy Build 74:61–68

49. Guerra-santin O et al (2013) Monitoring the performance of low energy dwellings : two UK case studies. Energy Build 64:32–40. https://doi.org/10.1016/j.enbuild.2013.04.002
50. Hagentoft C-EE et al (2004) Assessment method for numerical prediction models for combined heat, air and moisture transfer in building components: benchmarcks for one-dimensional cases. J Therm Envelope Build Sci 27(4):327–351. https://doi.org/10.1177/1097196304042436
51. Hall MR, Allinson D (2009) Assessing the effects of soil grading on the moisture content-dependent thermal conductivity of stabilised rammed earth materials. Appl Therm Eng 29:740–747
52. Hall M, Lindsay R, Krayenhoff M (2012) Modern earth buildings 1st Edition Materials, Engineering, Constructions and Applications
53. Haynes WM, Lide DR, Bruno TJ (2015) Handbook of chemistry and physics. CRC Press
54. Horoshenkov KV et al (2011) The effect of moisture and soil type on the acoustical properties of green noise control elements. In: Proceedings of forum acusticum 2011, (ii), pp 845–849
55. IEEE-442 (1981) IEEE Guide for soil thermal resistivity measurements. Institute of Electrical and Electronics Engineers
56. ISO 11357-1 (2016) Plastics—differential scanning calorimetry (DSC)—Part 1: General principles. International Organization for Standardization
57. ISO 12571 (2013) Hygrothermal performance of building materials and products—determination of hygroscopic sorption properties. International Organization for Standardization, Geneva, Switzerland, pp 1–22
58. ISO 12572 (2001) Determination of water vapour transmission properties. International Organization for Standardization, Geneva, Switzerland
59. ISO 22007-2 (2015) Determination of thermal conductivity and thermal diffusivity—Part 2: Transient plane heat source (hot disc) method. International Organization for Standardization
60. ISO 8302 (1991) Thermal insulation—determination of steady-state thermal resistance and related properties—Guarded hot plate apparatus. International Organization for Standardization
61. Irulegi O et al (2014) The Ekihouse : an energy self-sufficient house based on passive design strategies. Energy Build 83:57–69. https://doi.org/10.1016/j.enbuild.2014.03.077
62. Jacobson MZ (1999) Fundamentals of atmospheric modeling. Cambridge University Press
63. Kedowide Y (2015) Analyses expérimentales et numériques du comportement hygrothermique d'une paroi composée de matériaux fortement hygroscopiques. Université Savoie Mont-Blanc
64. Kwiatkowski J, Woloszyn M, Roux J-J (2009) Modelling of hysteresis influence on mass transfer in building materials. Build Environ 44:633–642
65. Künzel HM (1995) Simultaneous heat and moisture transport in building components. Fraunhofer IRB Verlag Suttgart
66. Künzel HM, Gertis EHMK (1995) Simultaneous heat and moisture transport in building components One-and two-dimensional calculation using simple parameters. Fraunhofer IRB Verlag Suttgart
67. Labarta G (2015) Rammed earth as a construction buiding material. Warsaw University of Technology
68. Labat M et al (2016) From the experimental characterisation of the hygrothermal properties of straw-cvlay mixtures to the numerical assessment of their bu ering potential. Build Environ 97:69–81
69. Labat M, Woloszyn M (2016) Moisture balance assessment at room scale for four cases based on numerical simulations of heat–air–moisture transfers for a realistic occupancy scenario. J Build Performance Simulation 9:487–509
70. Latif E et al (2016) In situ assessment of the fabric and energy performance of five conventional and non-conventional wall systems using comparative coheating tests. Build Environ 109:68–81. https://doi.org/10.1016/j.buildenv.2016.09.017
71. Laurent J (1986) Contribution à la caractérisation thermique des milieux poreux granulaires. Institut National Polytechnique, Grenoble

72. Lelievre D, Colinart T, Glouannec P (2014) Hygrothermal behavior of bio-based building materials including hysteresis effects: experimental and numerical analyses. Energy Build 84:617–627
73. Liuzzi S et al (2013) Hygrothermal behaviour and relative humidity buffering of unfired and hydrated lime-stabilised clay composites in a Mediterranean climate. Build Environ 61:82–92
74. McGregor F et al (2017) Impact of the surface film resistance on the hygric properties of clay plasters. Mater Struct 50:193
75. McGregor F, Heath A, Shea A (2014) The moisture buffering capacity of unfiered clay masonry. Build Environ 82:207–599
76. McGregor F et al (2014) The moisture buffering capacity of unfired clay masonry. Build Environ 82. https://doi.org/10.1016/j.buildenv.2014.09.027
77. McGregor F et al (2016) A review on the buffering capacity of earth building materials 169(5). https://doi.org/10.1680/jcoma.15.00035
78. McQuarrie M (1954) Thermal conductivity: VII, Analysis of variation of conductivity with temperature for Al_2O_3, BeO, and MgO. J Am Ceramic Soc 37:91–95
79. Medjelekh D et al (2014) Mesure et modélisation des transferts hygrothermiques d'une enveloppe en béton de bois. IBPSA 2014:1–8
80. Midttomme K, Roaldset E, Aagaard P (1998) Thermal conductivity of selected claystones and mudstones from England. Clay Miner 33:131–145
81. Minke G (2012) Building with earth: design and technology of a sustainable architecture. Birkhäuser—Publishers for Architecture
82. Mwaba MG et al (2006) Experimental investigation of $CaSO_4$ crystallization on a flat plate. Heat Transfer Eng 27(3):42–54
83. Osborn P (2013) Handbook of energy data and calculations. Butterworth
84. Osselin F et al (2015) 'Experimental investigation of the influence of supercritical state on the relative permeability of Vosges sandstone', Comptes Rendus - Mecanique. Elsevier Masson SAS 343(9):495–502. https://doi.org/10.1016/j.crme.2015.06.009
85. Oti JE, Kinuthia JM, Bai J (2010) Design thermal values for unfired clay bricks. Mater Des 31:104–112
86. Oumeziane YA et al (2016) Influence of temperature on sorption process in hemp concrete. Constr Build Mater 106:600–607
87. Pellenq RJM et al (2009) Simple model for phase transition in confined geometry. 2: Capillary condensation/evaporation in cylindrical pores. Langmuir 25:1393–1402
88. Peng C, Wu Z (2008) In situ measuring and evaluating the thermal resistance of building construction. Energy Build 40:2076–2082. https://doi.org/10.1016/j.enbuild.2008.05.012
89. Piot A et al (2011) Experimental wooden frame house for the validation of whole building heat and moisture transfer numerical models. Energy Build 43(6):1322–1328
90. Pusch R, Young R (2006) Microstructure of Smectite and engineering performance. Taylor & Francis Group, London
91. Ridley I et al (2013) The monitored performance of the first new London dwelling certified to the Passive House standard. Energy Build 63:67–78. https://doi.org/10.1016/j.enbuild.2013.03.052
92. Ridley I et al (no date) The side by side in use monitored performance of two passive and low carbon Welsh houses. Energy Build 82(2014):13–26. https://doi.org/10.1016/j.enbuild.2014.06.038
93. Rincón L et al (2015) Experimental rammed earth prototypes in Mediterranean climate. In: Mileto V, Soriano G (eds) Earthen architecture: past, present and future. Taylor & Francis, London, pp 311–316
94. Rode C et al (2005) Nordic Innovation Centre: moisture buffer value of building materials. Technical Report of the Technical University of Denmark
95. Roels S et al (2004) Interlaboratory comparison of hygric properties of porous building materials. J Thermal Envelope Build Sci 27(4):307–325. https://doi.org/10.1177/1097196304042119

96. Scheffler G, Grunewald J (2003) Material development and optimisation supported by numerical simulation for a capillary-active inside insulation material. In: Vermeir GLG, Hens H, Carmeliet J (eds) Research in building physics—proceedings of the 2nd CIB co-sponsored international conference on building physics. In-house Publishing
97. Schroeder H (2010) Lehmbau. Vieweg+Teubner, Wiesbaden
98. Shea A, Beadle K, Walker P (2010) Dynamic simulation and full-scale testing of a prefabricated straw-bale house. ICSBE 2010:101–107
99. Shea A, Lawrence M, Walker P (2012) Hygrothermal performance of an experimental hemp—lime building Construction and Building Materials. Constr Build Mater 36:270–275. https://doi.org/10.1016/j.conbuildmat.2012.04.123
100. Soebarto V (2009) Analysis of indoor performance of houses using rammed earth walls. In: Eleventh international IBPSA conference. Glasgow
101. Soudani L, Fabbri A et al (2016) Assessment of the validity of some common assumptions in hygrothermal modeling of earth based materials. Energy Build 116:498–511
102. Soudani L, Woloszyn M, Fabbri A, Morel JC, Grillet AC (2017) Energy evaluation of rammed earth walls using long term in-situ measurements. Sol Energy 141:70–80. https://doi.org/10.1016/j.solener.2016.11.002
103. Stefanoiu AM et al (2016) Comparison between the design phase and the real behavioral measurements of an Energy Efficient Building. In: IBPSA France, pp 1–8
104. Svennberg K (2006) Moisture buffering in the indoor environment. Lund University, LTH
105. Taylor P, Fuller RJ, Luther MB (2008) Energy use and thermal comfort in a rammed earth office building. Energy Build 40:793–800
106. Taylor P, Luther MB (2004) Evaluating rammed earth walls: a case study. Sol Energy 76(1–3):79–84. https://doi.org/10.1016/j.solener.2003.08.026
107. Vereecken E, Roels S (2013) Hygric performance of a massive masonry wall: how do the mortar joints influence the moisture flux? Constr Build Mater 41:697–707
108. Volhard F (1983) Leichtlehmbau: alter Baustoff - neue Technik. Karlsruhe
109. Volhard F (2016) Construire en terre allégée. Actes Sud
110. Vololonirina O, Perrin B (2016) Inquiries into the measurement of vapour permeability of permeable materials', Construction and Building Materials. Constr Build Mater 102:338–348. https://doi.org/10.1016/j.conbuildmat.2015.10.126
111. De Vries DA (1952) The thermal conductivity of soil. Mededelingen van de Landbouwhogeschool te Wageningen 52(1):1–73
112. Vries DA, Kruger AJ (1966) On the value of the diffusion coefficient of water vapour in air. Phénoménes de transport avec changement de phase dans les milieux poreux ou colloïdaux, pp 561–572
113. Walker P et al (2004) Rammed Earth: design and construction guidelines. In: Innovation Project: "Developing Rammed Earth for UK Housing"
114. Zhou A-N (2013) A contact angle-dependent hysteresis model for soil-water retention behaviour. Comput Geotech 49:36–42
115. Zillig W et al (2006) Liquid water transport in wood : towards a mesoscopic approach. In: Fazio P, Ge H, Rao J, Desmarais G (eds) Research in building physics engineering. CRC Press

Chapter 4
Mechanical Behaviour of Earth Building Materials

H. N. Abhilash, Erwan Hamard, C. T. S. Beckett, Jean-Claude Morel, Humberto Varum, Dora Silveira, I. Ioannou, and R. Illampas

Abstract Earth based building materials having low or negligible carbon footprint are looked upon as a sustainable alternative building material in the construction sector. The confidence of using any building material is augmented with through understanding of its mechanical properties. A brief review on the mechanical properties of the building materials such as Rammed Earth, Earth Blocks (Adobe, Compressed Earth Block, and Extruded Blocks) and Cob, which are manufactured using raw earth or by adding very little additives are presented in this chapter. The mechanical behaviour of earth based building material is highly dependent on raw material, manufacturing technique and testing conditions. Therefore it is highly recommended to conduct through experimental campaign for every soil mix. This chapter also presents various experiments recommended to study the mechanical properties of the materials.

Keywords Rammed Earth · Earth Blocks · Cob · Mechanical properties · Mechanical tests

---

H. N. Abhilash · J.-C. Morel (✉)
Coventry University, Coventry, UK
e-mail: Jean-claude.morel@entpe.fr

E. Hamard
University Gustave Eiffel MAST/GPEM, Bouguenais, France

C. T. S. Beckett
The University of Edinburgh, Edinburgh, UK

H. Varum
CONSTRUCT-LESE, Faculty of Engineering, University of Porto, Porto, Portugal

D. Silveira
ADAI-LAETA, Itecons, Coimbra, Portugal

I. Ioannou · R. Illampas
University of Cyprus, Nicosia, Cyprus

A. Fabbri et al. (eds.), *Testing and Characterisation of Earth-based Building Materials and Elements*, RILEM State-of-the-Art Reports 35,
https://doi.org/10.1007/978-3-030-83297-1_4

4.1 Rammed Earth

4.1.1 Introduction

The moist earth compacted in layers within a formwork is known as rammed earth. This is an ancient construction technique, which is commonly found in Europe, Asia, Australia, Africa, from Millennia, earliest examples can be dated to 2000 BCE [10, 49]. The main advantage of this technique is that, it makes use of the locally available raw earth in its natural state, thereby reducing the carbon footprint produced in comparison to other conventional building materials. Rammed earth in ancient days were constructed with the help of only raw earth. But various stabilisers are also in use to produce stabilised rammed earth to increase its mechanical and durability characteristics, and thus assimilating to conventional building materials. The rammed earth produced with the help of stabilisers are generally known as stabilised rammed earth (SRE) and are not in the scope of this chapter.

4.1.2 Mechanical Properties

4.1.2.1 Compressive and Tensile Properties

The compressive strength is one of the most important mechanical parameter that dictates the choice of material for building construction. This key performance indicator in rammed earth material can be affected by many other inter related parameters such as manufacturing/moulding water content, compaction energy, particle size distribution, clay content and dry density. Specifying a standard or generalised compressive strength for rammed earth is next to impossible considering the variability in specification of the raw earth from region to region. Therefore, ideal scenario is to study the compressive strength of rammed earth experimentally for each instance the soil is used for construction.

The compressive strength of unstabilised rammed earth is directly proportional to its dry density. The maximum dry density for any soil is related to the optimum water content (OWC) required for the compaction energy adopted. Note that, the variation of soil properties from region to region, makes it next to impossible for arriving at a generalised soil mix that is recommendable. Also the relationship between the dry density and the compressive strength is unique for the soil mix used. Therefore the soil properties such as particle size distribution (PSD), Atterberg's limits shall not be considered as sole criteria to choose the soil suitability [24, 80].

Table 4.2, at the end of this section, provides the compressive strength and related properties reported in the literature of the rammed earth. The compressive strength of rammed earth varies between 0.3 and 7 MPa, depending on the dry density, moisture at test and clay content & type. The linear elastic region in rammed earth is generally seen up to 30% of the failure load, in most literature, this is termed as initial tangent modulus (ITM). Each time ITM or secant modulus is reported, it is ideal to indicate the load or strain at which the calculation is made.

4.1.2.2 Effect of Clay

In rammed earth, clay acts as a binder, hence the amount of clay present will be crucial in most cases to enhance mechanical strength. It is understood that the clay content of 5–30% is considered acceptable [24, 39, 50, 56], excess clay content will increase the possibility of shrinkage. In addition to clay content, the study of Champiré et al. [22] suggests that the activity of clay, qualified by Methylene Blue Value, have more impact on the mechanical behaviour of compacted earth, whereby clays with higher activities induce greater amounts of shrinkage or swelling with changes in water content.

In the study of Hall and Djerbib [41], an attempt to study the influence of particle size distribution effect on compressive strength was carried out. It was interesting to observe that the compressive strength remained low (0.7–1 MPa) when the binder (silt and clay) was 20%, whereas the strength was higher (1.4–1.5 MPa), when the binder ratio was 30%, contrastingly, when the binder was 40%, the strength decreased (1–1.35 MPa). It may not be appropriate to conclude the compressive strength characteristics only on the basis of binder percentage, as the other parameters such as binder/aggregate ratio and dry density of each series was varying. As authors specifies hypothetically the binder/aggregate ratio is perhaps more important factor, which has not been proven yet. In addition, the study of Beckett et al. [18] suggests that the higher internal friction would provide relatively higher compressive strength even if the cohesion and clay content of the soil is relatively less (Table 4.1). To make any conclusive statement, the dry density and testing moisture content of the specimens in question should be the same. At this point, it can be said that, currently there is no conclusive statement to suggest what clay content and type would be ideal to achieve the higher compressive strength.

4.1.2.3 Effect of Testing Moisture Content

There are two important timeframe when the water or moisture content of the material is taken into consideration. The first one is called initial water content or manufacturing/moulding water content, this is important for achieving the near maximum dry density of the material, for a given energy of compaction. The second one is called as testing moisture content, this is important as suction values will come into play. The testing moisture content may be higher or lower than the manufacturing moisture content, depending on whether the material has absorbed water through its surfaces or lost water through evaporation.

One of the problems related to rammed earth material is its sensitivity towards moisture ingress when the surface is exposed to different environmental conditions. In reality these conditions can vary from −40 to +50 °C and 0% relative humidity to

Table 4.1 Compressive strength (MPa) of 100 mm cube rammed earth specimens conditioned at different temperature and relative humidity (RH), (source: [18])

Soil		Soil: 4-5-1 (Clay 19.9%) Dry density 1.94 g/cc				Soil: 2–7-1 (Clay 9.9%) Dry density 1.96 g/cc			
Temperature, °C		15°	20°	30°	40°	15°	20°	30°	40°
RH	30%	1.18	1.41	1.16	1.60	1.44	1.46	1.56	1.76
	50%	1.00	1.18	1.40	1.13	1.12	1.16	1.55	1.39
	70%	0.75	1.10	1.09	1.12	0.95	1.24	1.11	1.23
	90%	0.74	0.67	0.71	0.70	0.85	0.87	0.95	0.87

Table 4.2 Compressive strength properties of rammed earth reported in literature

Author(s)	Sample(s)	Clay (%)	Testing moisture content (%)	Compressive strength (MPa)	Dry density (g/cc)	ITM (MPa)	Secant modulus (MPa)	Tensile strength, MPa	Specimen dimension, mm		
									L/D	W	H
Araki et al. [6]	Cylinder	16		2.8–4.2	1.95–1.99	2500	1000	10% of fc	50	–	100
Bui et al. [21]	Cylinder	5 9 10	2	1.95 1.75 1.95	1.92	950 750 –			160	–	300
Bui et al. [21]	Cylinder	8		1.9	1.92	500		10% of fc	160		300
	Wallette		1.80	1.22	1.92				1000	300	1000
Champire et al. [22]	Cylinder	16 15 8	1.2 0.9 0.7	4.80 4.10 3.20	1.97 1.95 1.98	2200 5300 3600			64.4		140
Ciancio et.al. [24]	Cylinder	5 30 15 30 40	Dry	0.30 0.56 0.34 0.42 0.54	1.97 1.97 2.01 1.79 1.76				100		200
Gerard et al. [38]	Cylinder	13	2 3.5 4 5.5 6.0 8.0 14.5	7.00 5.20 4.20 3.50 3.75 2.00 0.75	2.00			13% of fc	36		72

(continued)

Table 4.2 (continued)

Author(s)	Sample(s)	Clay (%)	Testing moisture content (%)	Compressive strength (MPa)	Dry density (g/cc)	ITM (MPa)	Secant modulus (MPa)	Tensile strength, MPa	Specimen dimension, mm		
									L/D	W	H
Hall and Djerbib [41]	Cube	20		0.90	2.14				100	100	100
		20		1.00	2.14						
		20		0.77	2.06						
		20		1.10	2.01						
		30		1.40	2.15						
		30		1.37	2.13						
		30		1.45	2.13						
		30		1.15	2.06						
		40		1.35	2.07						
		40		1.10	2.09						
Maniatidis and Walker [56]	Cylinder	12		2.46	1.85	160			100		200
	Cylinder	12		1.9					300		600
	Prisms	12	5.1	0.62	1.76	60			300	300	600
			7.5	0.84	1.97	65					
			7.0	0.97	2.03	70					
	columns	12	6.0	0.52	1.73				300	300	1800
			6.3	0.73	1.95				300	300	2400
			6.3	0.65	2.05				300	300	3000
Miccoli et al. [61]	Wallette	11		3.7	2.19*	4143			500	110	500
Bui et al. [20]	Prisms	4	3	0.84	1.9	74			400	400	650
Abhilash and Morel [6]	Cylinder	17	1.5	2.3	1.89	825			160		300
				2.6	1.95	909					
				3.2	2.02	630					

(continued)

Table 4.2 (continued)

Author(s)	Sample(s)	Clay (%)	Testing moisture content (%)	Compressive strength (MPa)	Dry density (g/cc)	ITM (MPa)	Secant modulus (MPa)	Tensile strength, MPa	Specimen dimension, mm		
									L/D	W	H
		20	1.80	2.17 2.3 4.6	1.79 1.83 2.02	1668 1311 3261			110		220

100% relative humidity. There are multiple combinations that can be tried to simulate these conditions in a laboratory.

The effect of testing moisture content on the mechanical strength of rammed earth was presented by Bui et al. [20]. In this study, different samples with great variation of moisture content from the moment of manufacturing (11–13%) to dry states (1–2%) were tested in unconfined compression at a different state of moisture contents. From the interpretation of the experimental results, the compressive strength of the specimens increased with the decrease in moisture content, but the rate of increase in strength varied with change in soil. However, in the study of Bui et al. [20], the samples were not conditioned in the controlled environment. After manufacturing the samples, during its drying process, the samples were tested to extract their mechanical properties at different moisture content. The distribution of moisture throughout the specimens may therefore not have been uniform. Whereas [22, 34, 38], conditioned the samples in a relative humidity controlled environment using salt solutions. Champiré et al. [22] studied three different soils having different clay content and type, statically compacted. We can assume that there is not much difference in sample internal structure with a dynamic compaction process (as is usually used for rammed earth) leading to the same dry density [40]. The compressive strength of rammed earth samples at three different testing moisture content conditioned at 25%, 75%, 95% relative humidity in 24 °C ± 2 °C were studied. The compressive strength of rammed earth samples monitored at 95% RH was found to be 60–80% of the samples monitored at 25% RH. From the study reported in [34, 38], the reduction of compressive strength in the order of 50% for the samples monitored at 97% RH (6% moisture) with respect to samples tested at 40% RH (2% moisture) can be found. [38] also succeeds in extracting the compressive strength of rammed earth in its saturated condition (14.5% moisture), the compressive strength at saturation is 0.75 MPa, which is 10% of the samples monitored at 40% RH (2% moisture). The results of [22, 34, 38] show that there is a great variation of compressive strength in between the moisture content of 0% to 6%, which are in contrast with the results of [20]. The variation of compressive strength with resepect to the moisture content is pictorically shown in the Fig. 4.1.Though these studies present the extreme conditions of the RH, in reality, the moisture present in the rammed earth wall would be varying in between 0.5 and 2% [93].

Earlier studies were carried out by changing only the relative humidity at constant temperature. Whereas Beckett et al. [18] carried out experimental investigation on specimens conditioned at different temperatures (15°, 20°, 30° and 40°) and different relative humidities (30%, 50%, 70%, and 90%). The compressive strength (values in MPa) of rammed earth specimens at different climatic condition is given in Table 4.1. As temperature increases and with reduction in relative humidity, the moisture content in specimen would reduce, thereby increasing the suction values. As quoted in earlier studies, Beckett et al. [18] also reports the increase of compressive strength with reduction in relative humidity. The compressive strength at 30% relative humidity was almost twice that of at 90% relative humidity. Also, the compressive strength of rammed earth was found to be increasing with the increase in temperature for the respective relative humidity.

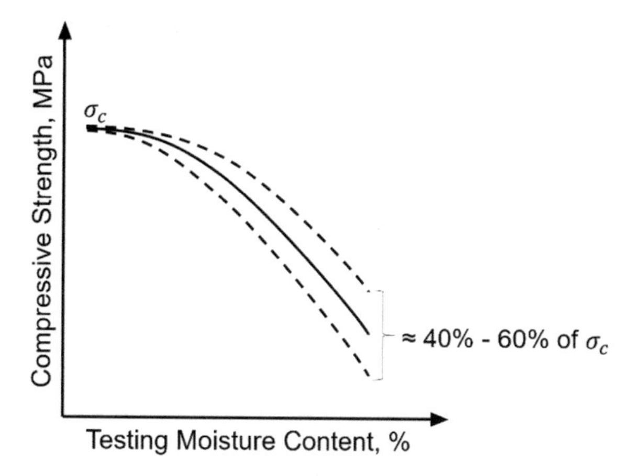

Fig. 4.1 Variatrion of compressive strength with respect to the testing moisture content

4.1.2.4 Effect of Dry Density

The compressive strength is a direct function of dry density of the material, applicable to the soil in use only. It has to be clearly understood that the dry density of two different soils should not be compared to correlate the compressive strength. Abhilash and Morel [6] studied the compressive strength of the rammed earth samples manufactured for the same soil at different dry densities. The different dry densities were achieved by altering the compaction energy and manufacturing water content. The compressive strength was found to increase with the increase in the dry density of the material, this was found to be consistent with the two soils tested in the study.

4.1.2.5 Effect of Layer Thickness

In general practice the layer thickness of the rammed earth walls after compaction is found to vary from 80 to 150 mm thick. During replicating the rammed earth sample in the laboratory, attention should be given to the layer thickness along with the sample geometry and dimension. The study of Raju and Venkatarama Reddy [76] present an interesting work related the optimum layer thickness of stabilised rammed earth for achieving the maximum compressive strength. We believe that their conclusions can be extended to unstabilised rammed earth. The study suggests that the layer thickness of 100 mm is found to be optimum, as the layer thickness increases, the compressive strength is found to be decreasing.

4.1.3 Full Scale Behaviour and Shear Properties

To design a load bearing wall, the material shear properties such as shear strength, cohesion and angle of internal friction should be known. Table 4.3 presents some of the shear strength properties of rammed earth reported in the earlier works. At the same time, literature highlights the importance of understanding or incorporating the shear strength parameters of both within the layer and interface of the rammed earth [31, 62, 87]. There are very few works that report shear parameters of both within the layer and interface of the rammed earth [23, 30]. Along with the compressive characteristic of the rammed earth, the shear strength characteristics shall be obtained through experimental investigation. The experiments such as tri-axial test, large shear box test, diagonal tension test, pushover test, shall be used for obtaining shear strength properties of layered rammed earth. While the experiments such as triplet test, large shear box test, and interface shear test shall be used for obtaining shear strength properties of the rammed earth interface. Testing procedures are discussed in Sect. 4.1.4.

4.1.3.1 Shear Strength Properties of Layered Rammed Earth

The cohesion and angle of internal friction of rammed earth reported from tri-axial compression test by Araki et al. [10] was 626 kPa and 48.9° respectively, and by Gerard et al. [38] was 6.2 kPa and 36.5° respectively. [10] used 10 layered cylindrical specimens with dimensions 50 mm (diameter), and 100 mm (height), while Gerard et al. [38] used 3 layered cylindrical specimens with dimensions 36 mm (diameter) and 72 mm (height). The dry densities reported by both the studies were 1.99 g/cc and 2 g/cc respectively. Araki et al. [10] carried out tri-axial test in drained and unsaturated condition, and the moisture of specimens at test was reported to be 1.46–1.65%, whereas Gerard et al. [38] carried out tri-axial test in undrained and saturated condition.

In the study of El-Nabouch et al. [30], large shear box was used to study the shear strength properties of the rammed earth. The cohesion and angle of internal friction was reported to be 24 kPa and 37.3° respectively. The author(s) also make an argument that the high moisture content (4%) at test could have impacted the cohesion and angle of internal friction. It highlights that the cohesion and angle of internal friction could me susceptible to the moisture content, which is backed by Jaquin et al. [48].

As said earlier the shear strength and shear modulus of rammed earth can be calculated with the help of diagonal tension test (DTT) [10]. From the experimental investigation carried out by Miccoli et al. [62], the average shear strength and shear modulus of five rammed earth wallettes were reported to be 0.7 MPa and 1582 MPa respectively. The soil used had 11% of clay and the testing moisture content of the wallettes were found to be 2–3%. In another study by Silva et al. [89], 11 rammed earth wallettes having an average dry density of 2.02 g/cc were tested in DTT and the

Table 4.3 Shear properties of rammed earth reported in the literature

Author(s)	Soil type	Test type	Dry density, g/cc	Moisture at test	Shear strength, MPa	Shear Modulus, MPa	Cohesion, kPa	Friction angle, °	Remarks
Liu et al. [55]	USRE	Push over	–	–	73 kN	–	–	–	
Nabouch et al. [67]	USRE	Push over	–	3%	37 – 53 kN	–	–	–	
Silva et al. [89]	USRE	DTT[a]	1.97–2.06	–	0.11–0.19	340–1036	–	–	
Miccoli et al. [62]	USRE	DTT[a]	2.19(bulk)	2–3%	0.7	1582	–	–	
Corbin and Augarde [26]	USRE	Shear box	–	–	–	–	55–80	23°–65°	
Araki et al. [10]	USRE	Tri-axial	1.99	1.45–1.65%	–	–	626	49°	
Gerard et al. [38]	USRE	Tri-axial	2	–	–	–	6.2	36.5°	
El-Nabouch et al. [30]	USRE	Shear box	–	4%	–	–	30.9	37.3°	Intra-layer
					–	–	24	34.8°	Interface
Cheah et al. [23]	CSRE	DTT[a]	–	–	0.73	–	–	–	
		Tri-axial	2.04	2.4–3.5%	–	–	724	48°	
		Triplet	2.15	2.4–3.5%	–	–	328	45°	
Abhilash et al. [5]	USRE	Interface shear test	1.7–1.9	1–2%	–	–	56	43°	Interface
			1.7–1.85	1–2%	–	–	118	37°	Interface

Note [a]DTT—Diagonal Tension Test; USRE—Unstabilised Rammed Earth; CSRE—Cement Stabilised Rammed Earth average shear strength and shear modulus were reported to be 0.15 MPa and 640 MPa respectively. The clay content in the soil used was reported to be 14% and the testing moisture content was 1.04%. It is interesting to note that the shear strength reported in the above two studies were only 18.8% and 7.14% of their respective compressive strengths reported. In the study of Miccoli et al. [62] wallettes were used to study the compressive strength, whereas Silva et al. [89] used cylindrical specimens. At this stage, it may be premature to state the generic shear strength value of rammed earth in terms of its compressive strength. But the interesting aspect is, in both the studies, the authors observe delamination of rammed earth interface along with the diagonal cracks within the layer and suggests the importance of knowing interface shearing properties.

4.1.3.2 Shear Strength Properties of Rammed Earth Interface

Along with the intra-layer shearing properties, the interface properties of the rammed earth was extracted in the study of El-Nabouch et al. [30], using large shear box. The cohesion and angle of internal friction of rammed earth interface was reported to be 24 kPa and 34.8° respectively. The interface cohesion and angle of internal friction is naturally lower in comparison to intra-layer. The cohesion and angle of internal friction of interface is 77.7% and 93.3% of their intra-layer properties respectively. The authors also suggests that the correlation might not be accurate owing to the higher moisture content at the centre of the specimen. This suggests that the cohesion and angle of internal friction of the intra-layer might be even higher at lower testing moisture contents.

In the study of Abhilash et al. [5], the interface shearing properties of two soils were reported by interface shear test. The rammed earth interface cohesion of two soils were reported to be 55 kPa and 118 kPa, and the angle of internal friction to be 43° and 37°. The clay content of the two soils were 17% and 20% respectively. At this stage, there is very little study and evidence on influence of clay role on the rammed earth interface properties.

Establishing the relation of cohesion and angle of internal friction of rammed earth interface and within the layer would be useful in future. Along with this the relation of the cohesion within the rammed earth layer, shear strength and compressive strength of rammed will also be helpful.

In the absence of triplet test on rammed earth, the literature from cement stabilised rammed earth has been taken for interpretation. By carrying out triplet and tri-axial test on cement stabilised rammed earth, the study of Cheah et al. [23] reports that the interface cohesion is 45% of the intra-layer cohesion, and the angle of internal frictions are 45° and 48° respectively for interface and intra-layer. Assuming that the cement stabilised and unstabilised rammed earth would account similar relation for interface and intra-layer properties, it can be seen that there is a clear distinction in the shearing properties of interface and intra-layer of rammed earth, and this has to be addressed in the design as and when required (Table 4.3).

4.1.4 Experimental Procedure

A range of experimental procedures exists for rammed earth, owing to the variety of factors affecting its mechanical properties at the single layer, multiple layer and building scale. These procedures are presented below.

4.1.4.1 Compression Test

Unconfined compression test (UCT) is the standardly accepted test to study the compressive strength and the material behaviour under compressive loads. Figure 4.2 illustrates the UCT of a rammed earth cylinder. The specimens are usually of 100 mm or 150 mm diameter with a 2:1 height:diameter ratio. The other important parameters that can be measured in the UCT are stiffness parameters of the material. The stiffness parameters such as initial tangent modulus (ITM) and coefficient of Poisson's can be calculated with the help of axial and lateral strain's measured using extensometer and linear variable differential transducer's (LVDT) respectively.

Advantage

- Compression testing equipment can be found in all conventional civil engineering laboratories

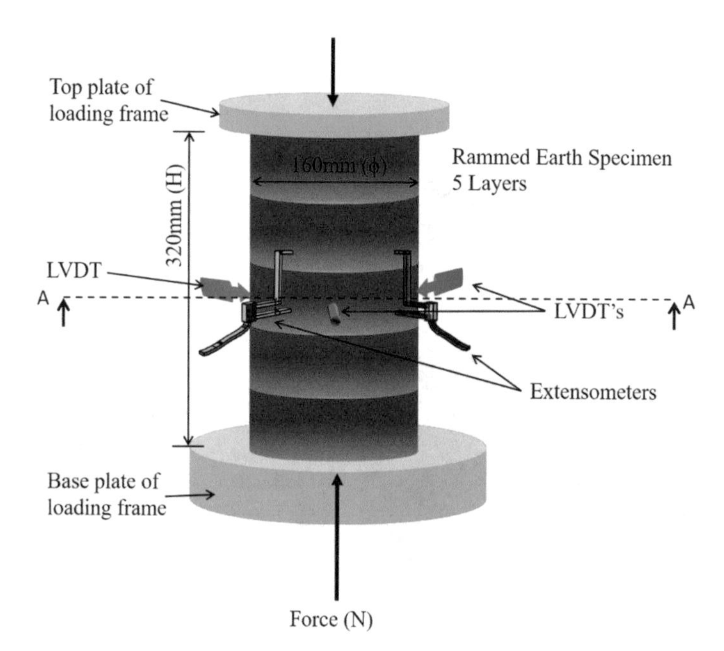

Fig. 4.2 Unconfined compression test set up, the sample has an aspect ratio of 2

- Specimens can be instrumented with extensometers to measure strain and elastic moduli.

Disadvantage

- Specimens may be difficult to manufacture for non-rammed materials
- Different layer thicknesses may affect results.

4.1.4.2 Tension Tests

Tensile strength can be measured directly or indirectly. The direct test employs a 'butterfly' specimen, broken across its midpoint. However, this test has met with little success due to the difficulty in preparing specimens with such confined geometries. Instead, the indirect tensile test is the most commonly used test for studying the tensile strength of the material. It is also widely known as the "split tensile" test or "Brazilian" test. Figure 4.3 illustrates the indirect tensile test performed on the rammed earth.

Advantage.

- Testing equipment can be found in all conventional civil engineering laboratories.

Disadvantage

- Specimens may be difficult to manufacture for non-rammed materials
- Converting between indirect tensile strengths and true tensile strength is subjective and poorly understood.

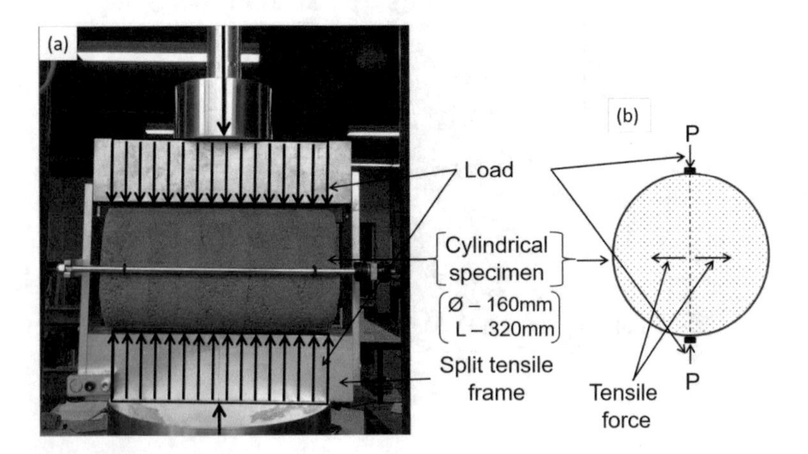

Fig. 4.3 Split tensile test set up

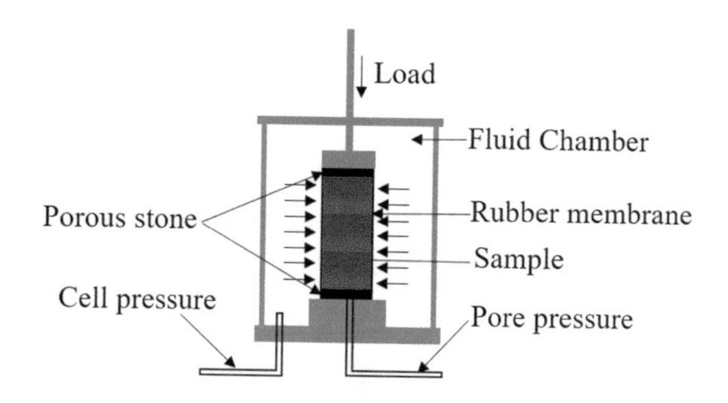

Fig. 4.4 Schematic diagram of tri-axial compression test

4.1.4.3 Shear Test

In general practice, the experiments adopted to study shear parameters of earthen building materials are the 'Tri-axial compression test', 'Shear box test', 'Diagonal tension test', 'Triplet test' and 'Push over test'. Though rammed earth is considered as monolithic wall, its methodology of construction leaves behind the layers. Under compression these layers would not have much impact on the mechanical behaviour. But under shear or lateral loading, the layer phenomena should be incorporated in the shear properties of the rammed earth [62, 88]. The shear parameters such as cohesion and angle of internal friction was found to be higher within the layers than at the interface [23]. Owing to this, shear parameters within the layers (stack of layers) and at the interface shall be extracted for incorporating in design considerations.

Tri-axial compression test is generally adopted to extract the shear parameter of rammed earth (stack of layers). Shear box test, triplet test, and interface shear test, are used for extracting the shear parameters of interface. Pushover test and diagonal tension test can be used extract the shear parameters of Wallette's (small wall specimens with dimension of 1.5 m (Length) × 1.0 m (Height) × 0.3 m (Thickness).

Tri-Axial Compression Test

One of the classic geotechnical experiment to study the shearing properties of cohesive soil is Tri-axial compression test. A schematic diagram of Tri-axial compression test is shown in Fig. 4.4. This test is generally carried out according to [4, 16]. The diameter of the cylindrical specimen shall not be less than 33 mm [4], or 34 mm [16], and the height of the specimen shall be maintained to satisfy an aspect ratio of 2. An examples of tri-axial compression test on rammed earth can be found in [98]. In case of rammed earth, where specimen with higher dimensions (>100 mm in diameter) shall be tested depending of the size of the biggest grains, customised tri-axial chamber shall be manufactured.

Advantages:

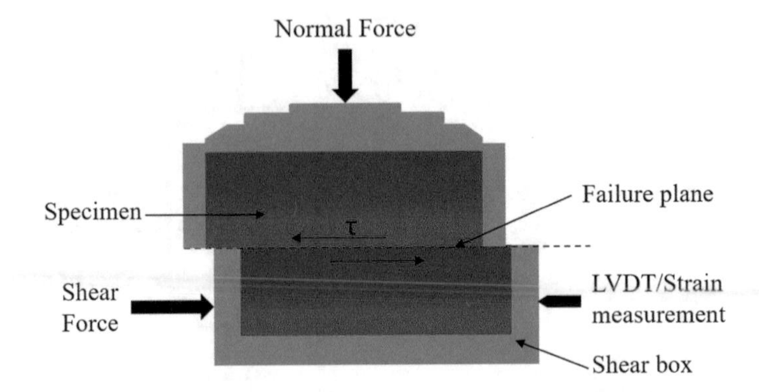

Fig. 4.5 Schematic diagram of direct shear test

- Tri-axial test can be found in all conventional civil engineering laboratories, hence it is easy to perform the experiments and obtain the basic shear properties of rammed earth on small cylindrical specimens.
- In tri-axial test, the shear strength, cohesion and angle of internal friction of the rammed earth can be extracted.

Disadvantage:

- Rammed earth is manufactured by dynamically compacting the top surface of the layer, this will induce a variation in density profile within the layer. Also since rammed earth is layered structure, the influence of the layered phenomena should be taken into account. To incorporated the layered phenomena and bring close replication to in-situ layer thickness, medium or large rammed earth cylindrical specimens should be tested. Therefore to facilitate cylindrical specimens having 100 mm diameter or 150 mm diameters, customised tri-axial setup should be prepared.

Shear Box Test (Direct Shear Test)

Shear box test is also commonly known as direct shear test. In shear box test for a pre induced normal stress, the specimen is sheared by introducing incremental lateral force on the bottom or top half the shear box. The direct shear test shall be carried out in accordance with [3, 13]. A schematic diagram of direct shear test using shear box is shown in Fig. 4.5. Shear box can be classified into small and large based on the size of the specimen tested. The regular or small shear box is of dimension 60 mm square or 100 mm square, and having a height of about 40 or 50 mm. The large shear box can accommodate a specimen with dimension 305 mm square and having a height of 150 mm [13]. In practice, the layer thickness of rammed earth vary from 60 to 120 mm. Therefore, specimens with thickness of 60 mm and above shall be tested in the large shear box.

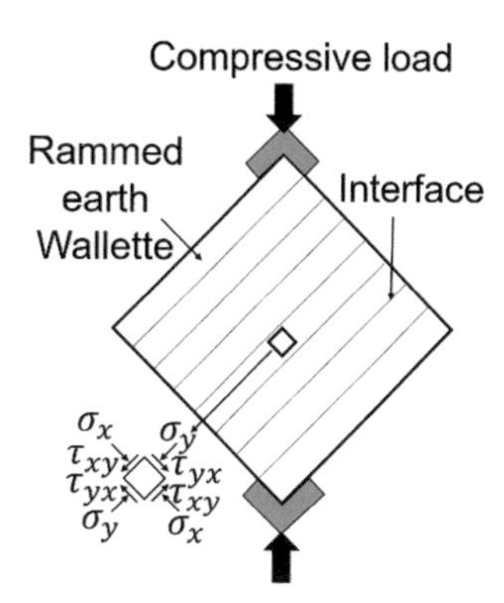

Fig. 4.6 Schematic representation of Diagonal Tension Test (DTT)

Since shear box will shear the specimen horizontally along the mid height, using rammed earth specimens comprising two even layers will enable to extract the shearing parameters at the layer interface, while using three-layered rammed earth specimens will enable to extract the shearing properties within the layer. If required, to incorporate the specimens with higher dimensions, a customised shear box with the required dimension shall be manufactured and the procedure mentioned in the ASTM or BS shall be adopted. Some examples of 500 mm square × 310 mm in height shear box [29], and 500 mm square × 450 mm in height shear box [30], can be found in the literature.

Advantage:

- Similar to tri-axial compression test, shear box is also one of the conventional testing facility that exist in all the basic civil engineering laboratories. Carrying out test in classical shear box should be very convenient.
- With the help of large shear box cohesion and angle of internal friction of interface and also within the layer can be extracted.

Disadvantage:

- Like said earlier, to facilitated specimens with layer thickness of 60 mm – 100 mm, a large shear box should be manufactured.

Diagonal Tension Test

Diagonal tension test (DTT) is carried out on Wallette's, which are larger specimens with dimension of 750 to 1000 mm (length) × 750 to 1000 mm (height) × 100 to 150 mm (thick). DTT shall be carried out in accordance to [10], a schematic diagram of DTT is shown in Fig. 4.6. From DTT, shear strength and strain of the rammed earth Wallette's (panels) can be extracted. The results obtained from this test is very useful in anticipating the shear behaviour of the wall [62, 71].

Advantage:

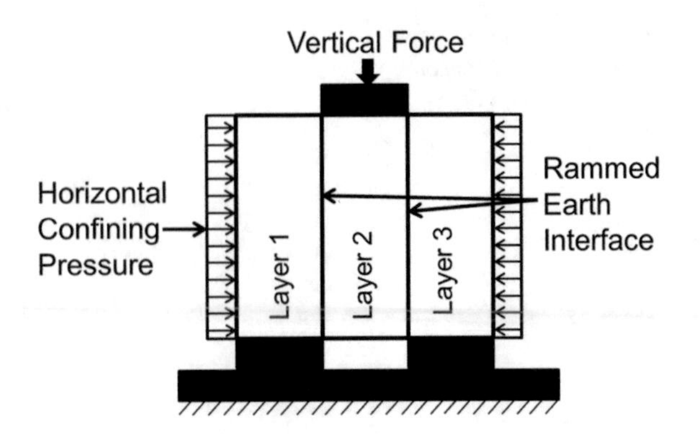

Fig. 4.7 Schematic representation of triplet test

- Some standards suggest the shear strength obtained from the DTT experiment shall be used for design consideration.
- The shear strength of the Wallette's is directly obtained.

Disadvantage:

- The specimen size is quite large and DTT facility is not readily available in most of the conventional civil engineering laboratory.
- Specimens are vulnerable to damage due to its positioning technic and size.
- Conditioning of specimens will take time and require some additional special equipment's.
- Owing to the dimension of specimen and the amount of raw material required, the sample size would be restrictive.
- Only shear strength of the specimen can be extracted, the cohesion and angle of internal friction cannot be obtained.

Triplet Test

Triplet test is one of the commonly used experiment to study the properties of the masonry joints. As the name suggests triplet test will have specimens with three layers (course) that is two joints (interface). The shearing properties such as cohesion and angle of internal friction of the interface shall be extracted from this test. To plot Mohr–Coulomb failure criteria, a minimum of three normal stress and shear stress combination shall be generated. Therefore in this test two actuators (horizontal and vertical) are required. The specimen shall be mounted with their interface perpendicular to the base plate as shown in Fig. 4.7, and specimen shall be positioned such that the extreme layers is resting on the base plate, leaving the interfaces and middle layer free to displace downwards. The specimen is then restrained using two horizontal actuators on opposite sides to induce pre-determined horizontal confining pressure (Fig. 4.7). The incremental vertical force (Shear force) shall be applied using vertical actuator on the middle layer until the interface fails. This method will help to extract

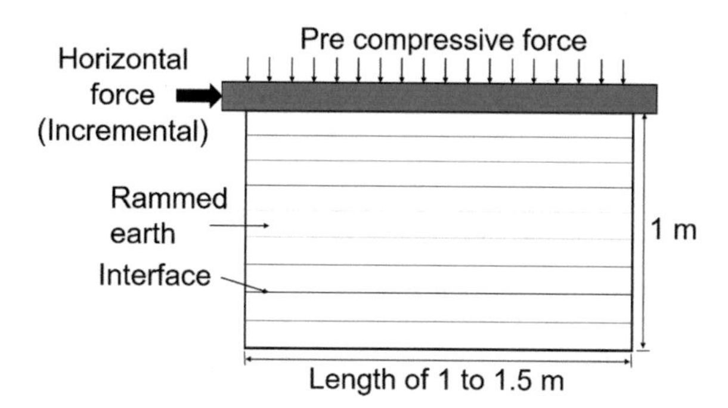

Fig. 4.8 Schematic representation of pushover test

the interface shearing properties of the rammed earth. Some examples of triplet test carried out on rammed earth can be found in [23, 71, 77, 84] literature.

Advantage:

- Triplet test enables to investigate the interface properties of the rammed earth.
- It is easy to represent specimens with layer thickness of 60–100 mm.

Disadvantage:

- Not easily found in conventional laboratory.
- It might be difficult to position multiple actuators, which will also consume lot of space.

Pushover Test

As per the definition of BS EN 1998-1:2004 + A1:2013 [17], 'pushover analysis is a non-linear static analysis carried out under conditions of constant gravity loads and monotonically increasing horizontal loads'. Pushover test is also called as 'In-plane shear compression test' [60]. Pushover test is recommended to study the seismic behaviour of the wall. To replicate seismic behaviour, monotonic or cyclic lateral loads will be simulated on the wall or Wallette along with the pre-compressive load. In other words, in pushover test a full scale wall or Wallette can be subjected to bi-directional loading and study their behaviour due to lateral forces acting on them. Figure 4.8, shows a simple schematic representation of the pushover or in-place compression test. The shear strength, stiffness parameters and failure pattern can be extracted from this experiment.

Advantage:

- Pushover test will enable to replicate the in-situ wall and its boundary conditions.
- The shear strength of the wall and the failure pattern can be identified.
- Along with shear strength of the wall, with different normal force (pre-compressive load), cohesion and angle of internal friction of the wall can be extracted.

- With the help of cyclic lateral loads, the seismic conditions can be induced on the wall.

Disadvantage:

- The experimental set up is huge and dedicated space is required.
- The sample dimension is huge and require additional time to condition the specimen before testing.
- For conditioning specimens, large facility and additional equipment's may be required.

Interface Shear Test

Considering some drawback of large shear box and triplet test, interface shear test was developed which can be used in any conventional civil engineering laboratory. The shearing properties such as cohesion and angle of internal friction of the interface can be extracted from the interface shear test.

To plot Mohr–Coulomb's failure criteria and obtain shear properties a minimum of three combination of normal and shear forces are required. To achieve this, the interface shear test uses the wedge technique to position the specimens in different inclination with respect to the axial load in a conventional axial compression press. Three pair of different wedges are required, which helps to orient the specimens at 20°, 30° and 45° with respect to vertical axis of the press [5]. Each pair of wedge will have identical bottom and top wedge. These three pair of wedges will help to provide the three combination of normal and shear force on the interface. The wedges can be custom designed for the specimen layer thickness that shall be tested.

Since the wedges create an angle with the vertical axis of the compression press, the length of the specimens tested at different inclination should be altered such that the vertical symmetry of the whole test setup is maintained. The length of the specimens tested at 45° inclination will be same as the height of the specimen. The length of the specimens keep increasing with decrease in angle of inclination. Figure 4.9 shows a schematic representation of interface shear test on two layered rammed earth specimen. This interface shear test can also accommodate three layered rammed earth specimens.

Advantage:

- Interface shear properties can be extracted with existing uniaxial compression press.
- Wedges can be customised to the requirement, and doesn't consume space.

Disadvantage:

- Requirement of new pair of wedge for new combination of normal and shear force.
- The length of the specimen should vary for each inclination.

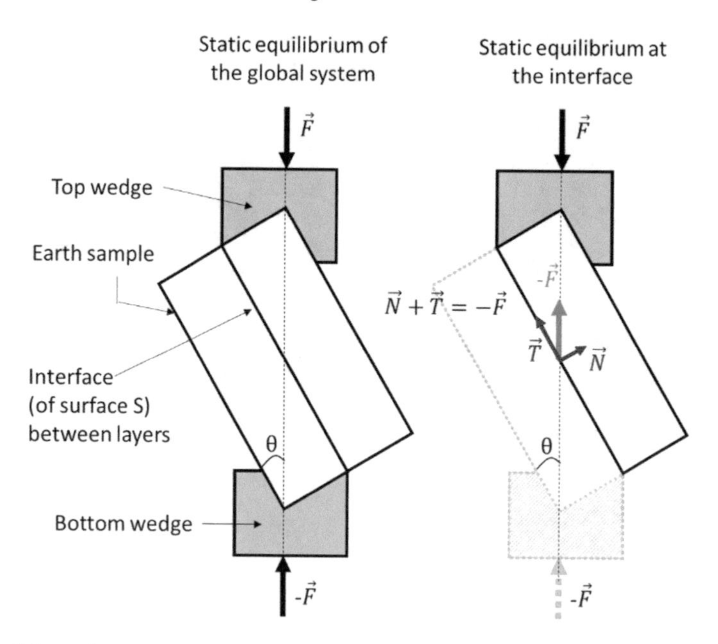

Fig. 4.9 Schematic representation of interface shear test (where N = Normal force, T = Tangent force, F = Force, θ = angle of inclination)

- Only interface shear properties can be investigated.
- The mounting technique requires some modifications to avoid fixation of top wedge to the press plate.

Flexure Test

To obtain the flexural strength and modulus of rupture of rammed earth the test procedure mentioned in [7] shall be followed. The rammed earth short beam specimens both with only one layer and multi-layers shall be tested for observing their flexural behaviour. Figure 4.10 shows a figure of single layered rammed earth beam being tested under four point bending test.

Advantage

- Flexural testing is one of the conventional testing facility that exist in all the basic civil engineering laboratories.

Disadvantage

- Specimens are large and must be manufactured with horizontal layer interfaces (if present).

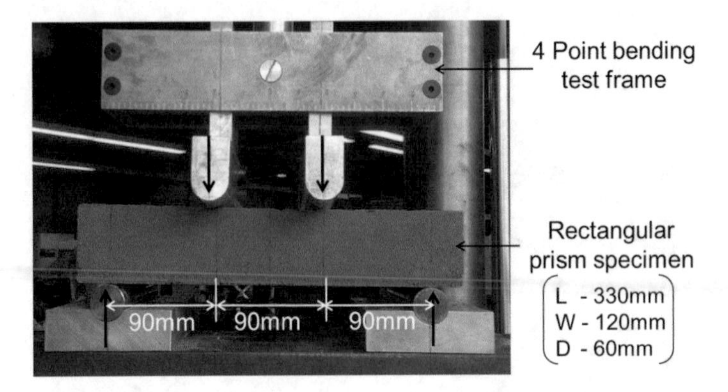

Fig. 4.10 Four point bending test set up

4.2 Earth Blocks

4.2.1 Introduction

"Earth blocks" may refer to a range of materials and construction techniques, depending on the literature. Common interpretations are discussed below.

4.2.1.1 Adobe

Adobe is one of the oldest building material man has made using raw earth. In ancient period (dated back to BC), Adobe could have potential been one of the primary choice as a building material. Adobe is a semi-liquefied raw earth mix poured into a mould and dried under sun. This simplicity in manufacturing of Adobe is one of the key reasons for its wide reach across the globe. As there is high clay and water present in the mix, there is high chance of shrinkage. In order to contain shrinkage, the agricultural (bio) products such as straw, coir, etc., are added, along with the raw earth. There are examples of dung usage in the Adobe, to enhance mechanical properties of the Adobe.

4.2.1.2 Compressed Earth Blocks (CEB)

The moist earth mix which is compacted within a mould by using manual or mechanical press is called compressed earth blocks. The compaction of CEB can be done with a single ram press of double compaction plate press. This is a static compaction process, the CEB's compacted with single ram would develop a varying density within the block due to the compaction mechanism. The density within the block will be higher towards the surface of moving ram and lower towards the static ram.

On the other hand, with the help of double plate compaction ram, there would be more uniform density distribution within the block. Thanks to mechanical moulds and press, this will help in having a proper, plane and smooth geometric faces of CEB. As the compaction effect increases the dry density of the block can also be increased, thereby increasing the compressive strength of the block. The other means to increase the mechanical properties of the CEB is by adding cement, lime or alkaline solution as a stabiliser.

The CEB's are allowed to dry under the sun or shaded area or at any ambient temperature, they can also be dried with the help of oven. When the two consecutive weight measurement of the CEB's are similar, then the drying process of CEB is completed.

4.2.1.3 Stabilised Compressed Earth Blocks (SCEB)

This is a similar product that of a compressed earth block with a stabiliser such as cement/lime/alkaline solution added to the moist earth mix. The stabilisers are added to enhance the mechanical and durability performance of the materials. If cement or lime is used as a stabiliser, then the SCEB shall be allowed to cure in burlap curing, similar to the cement based products. If alkaline solution is used as a stabiliser, then the SCEB shall be subjected heat (high temperature) curing process for the specified period. The dry density of the CEB/SCEB are generally in the range of 1.8–2.0 g/cm^3. Some blocks were dried in the sun [7], others had water sprinkled on them during the curing process (exposed to sunshine for 2–3 weeks, and to air for 1 week) [69] and others were simply stored for 28 days before testing. The references dealing with stabilized blocks containing plant aggregates or fibres are the most numerous. Several types of binders were used during these studies, such as: cement (with or without mineral additions), lime and organic stabilizer (alginate, and beetroot and tomato polymer). The size of the plant aggregates or fibres used in stabilized blocks was comparable to that observed in other types of blocks (only flax fibres were a little bit longer (8.5 cm)) but the plant particle content in some blocks could be very significant, especially for wood aggregates: 40% in [9], 37.5% in [19] and 29% in [53]. Such high plant aggregate or fibre contents would certainly lead to significant problems of strength but these seem to have been solved by using high binder content (cement for the references on wood aggregates). However, the amounts of binder used in other references are often very high, which could raise questions on the environmental impact of such materials when cement and/or lime are used. It is important to note that the cement content in concrete blocks is below 7% (150 kg/m^3) and these blocks are hollow. This means that the comparison with a solid earth block stabilized with cement is even more disadvantageous for the earth block regarding cement content. To date there are still few studies dealing with the use of natural organic stabilizers but this is certainly the most sustainable solution and should be developed in the future.

4.2.1.4 Extruded Earth Blocks/Bricks

Extruded blocks are manufactured with the earth in a plastic state. Generally produced in an industrial process, these blocks can present perforations and are dried in an oven (105 °C [28]).

4.2.2 Mechanical Properties

4.2.2.1 Strength and Stiffness

The results obtained by different authors following mechanical tests on adobe specimens and adobe masonry walls (considering non-stabilized or air lime stabilized adobes) are summarized in, Tables 4.4, 4.5 and 4.6.

It can be noted that, even though there is significant variability between the results of different authors, in general, these are of the same order of magnitude. Considering the differences in composition of the adobes and testing procedures adopted, this variability of results was expected. It is worth noting that the compressive and tensile strengths of adobe specimens depend greatly on fibre content and soil composition [46]. Furthermore, insufficient compaction of the soil mixture during the hand-moulding production process can lead to increased deformability of the end-product, whilst the preparation of samples under controlled laboratory conditions generally leads to superior mechanical properties in relation to materials available on the market that have been produced using empirical techniques [44]. The stiffness values obtained for the adobes and adobe masonry of Aveiro region (Portugal) [91, 92], in particular, are much greater than those obtained by other authors, and this difference is especially high for the adobe specimens. These higher values must be due to the differences in composition of the adobes and mortars, construction methods, and testing procedures [92]. In fact, the adobes of Aveiro region were made with sandy soils that could include some gravel in their composition and were stabilised with lime binder, while the adobes used in the other studies were made with finer soils and were not stabilised. Moreover, the way in which the deformation of specimens is measured—i.e. directly on the specimens or on the loading system—may also lead to different results [91]. When measurement is carried out on the loading system, higher values of deformation and, consequently, lower values of modulus of elasticity, are generally obtained. This was the case in at least four of the studies on adobe specimens (Quagliarini and Lenci [35, 46, 73]. In the studies on adobe walls, measurement of the deformation directly on the specimens seems to have been the customary practice.

Table 4.4 Strength of Adobe obtained by different studies (adapted from [90])

References	Location	Adobe bricks		Compression		Tension	
		Composition	Condition	Specimens	Comp. strength (MPa)	Test	Tensile strength (MPa)
Gavrilovic et al. [37]	Mexico	Clayey soil	Not indicated	Not indicated	1.18	Flexural	0.27
Meli [59]	Mexico	Clayey soil	New (produced in different regions of the country)	Not indicated	0.51–1.57	Flexural	0.20–0.43
Rivera Torres and Muñoz Díaz [78]	Colombia	Clayey soil	Collected from existing construction	Bricks	3.04	Flexural	0.41
Liberatore et al. [54]	Italy	Silty sand	Collected from existing constructions	Bricks and half bricks	0.29–1.56	Flexural	0.17–0.40
Baglioni et al. [17]	Morocco	Silty or clayey soil	Some collected from existing constructions and some new	… [a]	2.83	Flexural	0.18–0.35
Silveira et al. [90]	Portugal	Sandy soil and lime binder	Collected from existing constructions	Cylinders	1.17	Splitting	0.19
Illampas [45]	Cyprus	Lean clay, Lean clay with sand	New (produced in different regions of the country)	Cylinders and cubes with aspect ratio ~ 1	0.60–1.75	Flexural	0.10–0.95
Illampas et al. [47]	Cyprus	Sandy silty clay	New (produced in the laboratory)	Prisms with aspect ratio ~ 2	0.93–4.50 [b]	Direct tension	0.29–0.80 [b]

[a] Results obtained from in situ sclerometer tests
[b] Depending on the fibre content of samples which ranged from 0 to 5% w/w

4.2.3 Experimental Procedures

A review of the indications of earthen construction codes and standards and The Australian Earth Building Handbook [96] for the mechanical testing of adobe and adobe masonry (non-stabilized or stabilized) is presented in Table 4.8 and Table 4.9. From our literature review it was understood that, at present, there is no specific national or international code or standard for CEB's and extruded blocks. Since the dimension and function of CEB's or extruded blocks are similar to the fired clay

Table 4.5 Modulus of elasticity of Adobe obtained from different studies (adapted from [91])

References	Location	Adobe bricks		Test specimens	Measurement of deformations	Modulus of elasticity (MPa)
		Composition	Condition			
Gavrilovic et al. [37]	Mexico	Clayey soil	Not indicated	Not indicated	Not indicated	1471
Quagliarini and Lenci [73]	Italy	Clayey soil, straw and coarse sand, in variable proportions	New (produced for the study)	Blocks (bricks cut into 4 parts): $0.15 \times 0.23 \times 0.13$ m^3	Measurement of the relative displacement of testing platens	98 - 211
Fratini et al. [35]	Italy	Gravel clay, with a proportion of clay ranging from 16% up to 40%	Collected from existing constructions	Cubes: $0.05 \times 0.05 \times 0.05$ m^3	Measurement of the relative displacement of testing platens	15 - 87
Eslami et al. [32]	Iran	Clayey soil	New (produced for the study)	Bricks: $0.19 \times 0.19 \times 0.05$ m^3	Not indicated	≈ 85 [a]
Silveira et al. [91]	Portugal	Sandy soil and lime binder	Collected from existing constructions	Cylinders: $H \approx 0.15$–0.18 m $D \approx 0.08$–0.09 m	Performed directly on test specimens	13,214
Illampas et al. [46]	Cyprus	Lean clay	New (sampled from a local manufacturer)	Cubes: $0.05 \times 0.05 \times 0.05$ m^3 Cylinders: $H \approx D \approx 0.05$ m	Measurement of the relative displacement of testing platens	11–92
Illampas et al. [47]	Cyprus	Sandy silty clay	New (produced for the study)	Prisms (cut from bricks): $0.05 \times 0.05 \times 0.12$ m^3	Measurement of the relative displacement of testing platens	271–1239 [b]

Notation: H - Height; D - Diameter

[a] Estimated from the stress–strain curve presented by [32]

[b] Depending on the fibre content of samples which ranged from 0 to 5% w/w

Table 4.6 Results obtained by different studies in simple and diagonal compression tests conducted on Adobe walls (adapted from [92])

References			Meli [59]	San Bartolomé and Pehovaz [83]	Torrealva and Acero [94]	Liberatore et al. [54]	Yamín et al. (2007)	Wu et al. [97]	Silveira et al. [92]	Illampas et al. [45]
Location			Mexico	Peru	Peru	Italy	Colombia	China	Portugal	Cyprus
Adobe bricks used in the walls	Composition		Clayey soil	Soil, coarse sand and straw, in proportion 5:1:1	Soil, coarse sand and straw, in proportion 5:1:1	Silty sand	Clayey soil, with or without natural fibres	Soil with 44% clay-silt and 56% sand, and straw (0.5%)	Sandy soil and lime binder	Lean clay with straw fibres (35% vol/vol)
	Condition		New (produced for the study)	New (produced for the study)	New (produced for the study)	Collected from existing constructions	Collected from existing constructions	New (produced for the study)	Collected from existing constructions	New (sampled from a local manufacturer)
	Comp. strength (MPa)		0.51–1.57	2.94	Not indicated	0.29–1.56	2.84	1.66	0.47	1.28
	Tensile strength (MPa)		0.20–0.43 [a]	Not indicated	Not indicated	0.17–0.40 [a]	0.49 [a]	Not indicated	0.14 [b]	0.43 [a]
Simple compression test	No. of specimens		Not indicated	4	5	…	15	9	5	13
	Dimensions (m) [c]	H:	Not indicated	0.58	0.43	…	Not indicated	0.53	1.26	0.28
		W:		0.38	0.25			0.29	1.26	0.30
		t:		0.24	0.25			0.20	0.36	0.45
	Aspect ratio (H/t)		…	2.4	1.7	…	…	2.7	3.5	0.62

(continued)

Table 4.6 (continued)

References			Meli [59]	San Bartolomé and Pehovaz [83]	Torrealva and Acero [94]	Liberatore et al. [54]	Yamín et al. (2007)	Wu et al. [97]	Silveira et al. [92]	Illampas et al. [45]
	Aspect ratio factor ('k_a') [d]		…	0.83	0.77	…	…	0.84	0.90	0.55
	Comp. strength ('f_c') (MPa)		1.32	0.86	0.85	…	1.10	0.94	0.33	0.88 / 1.73 [f]
	Unconfined comp. strength ($k_a f_c$) (MPa)		…	0.71	0.65	…	…	0.79	0.30	0.48 / 0.95 [f]
	Modulus of elasticity (MPa)		245	Not indicated	432	…	98	34 [e]	757	15.5 / 21.8 [f]
Diagonal compression test	No. of specimens		Not indicated	4	3	1	10	…	5	1
	Dimensions (m) [c]	H:	Not indicated	0.80	0.50	0.90	0.75–1.00	…	1.26	0.60
		t:		0.24	0.25	0.20	0.15–0.40	…	0.36	0.30
	Shear strength ('f_v') (MPa)		0.14	0.11	0.07	0.02	0.03	…	0.03	0.06
	f_v/f_c (%)		10	13	8	…	3	…	8	3
	$f_v/(k_a f_c)$ (%)		…	15	10	…	…	…	9	6
	Modulus of rigidity (MPa)		Not indicated	Not indicated	Not indicated	Not indicated	27	…	413	3

[a] Flexural tensile strength
[b] Splitting tensile strength
[c] H—Height; W—Width; t—Thickness
[d] Calculated according to NZS 4298 [82]
[e] Initial tangent modulus of elasticity

bricks, wherever applicable the codes and standards of fired clay bricks shall be extended to the CEB's and extruded blocks [64]. In this review, only laboratory tests were considered.

4.2.3.1 Sampling

The sampling size is one of the important criteria in the quality control of any production. Similarly for adobe production, the normative documents, codes/standards and the Australian handbook, recommends to test a minimum of 5 units per lot. The lot size recommendation varies from 2500 to 25,000 bricks for different standards (see Table 4.8 for more details).

For CEB's and Extruded blocks, the recommendation of the British/European standards for fired clay masonry brick [15] to test a minimum of six samples in each series (batch), shall be taken. This number of samples is widely backed by the researchers, also other established materials such as concrete.

4.2.3.2 Specimen Dimension and Aspect Ratio

Since earth based blocks such as adobe, CEB and extruded blocks are extremely locally driven, the blocks are available in wide range of varying dimensions. The block dimension is highly dependent on the local need, the manufacturer, and the mould dimension available. Therefore it is difficult have one standard dimension of earth blocks. Having one standard dimension would strengthen the possibility of developing one standard/code to test earth blocks and correlate the strength parameters easily. Nevertheless, some standards such as NZS 4298 [82], the Australian handbook [96] and British standards [15], have proposed an alternative to normalise the compressive strength by using aspect ratio correction factors. Since, the existing aspect ratio correction factors are developed for adobe or fired clay bricks, there is lack of accuracy when it is used in the case of CEB's and extruded bricks [64]. This leads to questions such as, (i) should aspect ratio correction factor for each block type be developed, or (ii) increase the block aspect ratio to eliminate the error in calculating compressive strength, or (iii) should the above two combinations be considered.

For the determination of the compressive strength of adobe, most normative documents recommend the testing of adobe blocks or cube specimens. The Australian Handbook [96] also refers to the possibility of testing cylindrical specimens. The dimension of the block to be tested is only mentioned in [68] and [66]. There is definitely lack of standardisation of specimen dimension for compressive, flexural and tensile strength test. Only NZS 4298 [82] and The Australian Handbook [96] include aspect ratio correction factors for the calculation of the unconfined compressive strength. Whereas British standards [15] for fired clay brick recommend to calculate the normalised compressive strength using shape factor. This normalised compressive strength would help in masonry design applications.

4.2.3.3 Specimen Conditioning and Capping

The moisture ingress is an inherent property of earth blocks. Studies show that the moisture present in the earth block has a direct relationship with their mechanical parameters [22, 57]. The moisture content of earth blocks varies with respect to the surrounding temperature and relative humidity. Therefore conditioning the earth block specimens prior to test and measuring their moisture content at test is absolutely necessary. For adobe, only [70] specifies the specimen conditioning criteria (see Table 4.8). The Australian handbook also suggests the conditions at which the test shall take place, but the conditioning protocols are not detailed. For CEB and extruded blocks, the recommendation of British [15] shall be applied to condition the specimens. The British standard [15] recommends four conditioning possibilities, (i) air dry, (ii) oven dry, (iii) 6% moisture and (iv) immersed in water for 15 h (saturated). Depending on the user requirement, the suitable conditioning shall be applied. Since the compressive strength and moisture at test is linearly related (within the residual suction range), large number of research studies adopt two extreme conditions, (i) oven dry and (ii) immersed in water for 15 h. The oven dry condition provides the highest compressive strength, whereas the saturated condition provides the minimum. The immersion in water (saturated) is possible only for earth blocks which are stabilised or fired. In the case of unfired earth blocks, the immersion in water will result in disintegration of soil in the water. Hence, 6% moisture condition or maximum possible moisture ingress condition shall be applicable. Representing mechanical parameters of blocks at minimum two moisture contents will help in predicting the mechanical parameters at any other moisture content.

The earth block specimens which are subjected to loading shall have a plane and smooth surface to establish a good contact with the loading plates to ensure uniform loading. The manufacturing technique of Adobe itself will leave uneven surfaces on the dried block. It would be difficult to uniformly load the specimen without a proper capping. Only [70] and the Australian handbook details the capping of the specimens. Thanks to manufacturing mechanism of CEB and extruded blocks, these blocks will relatively have a plane and clean surface. Yet these blocks should also be carefully examined for the unevenness in the surface before subjecting to loading. Some of the blocks prepared for masonry work may also have frogs. In the event of any unevenness or rough surface, the block surface shall be smoothened until the plane surface is achieved. Alternately, the surfaces of the blocks which come in contact with the loading plate shall be applied with cement mortar or plaster to achieve plane surface. In case of blocks with frogs, the frogs shall be filled with cement mortar or plaster. The cement mortar or plaster used shall have minimum compressive strength that is equal to the block compressive strength or 30 MPa, whichever is lesser [15].

Apart from preparing the specimen surface, since the loading plates that are in contact with earth blocks have higher stiffness in comparison with blocks, there will be a frictional resistance affecting the accuracy of strength calculated. Therefore some materials such as ply wood sheet, Teflon sheet, or neoprene layer is placed in between the specimen surfaces and loading plates of the press [15, 64]. This will help in reducing errors due to frictional component in calculating the compressive strength.

4.2.3.4 Specimen Positioning and Loading

The earth block compressive strength is generally measured by lying the block in the same direction as positioned in the masonry wall (Flat). Most of the normative documents/standards/codes recommend to load the block in flat position. Due to manufacturing technique of CEB with single ram, the density within the block is not uniformly distributed. The density within the block of CEB close to the moving compaction plate is high in comparison to the opposite end static plate. Therefore, the surface close to higher density of the CEB should face the bottom of the masonry course and similarly it should be positioned in the loading press [63, 64]. There are instances, when the blocks should be tested in other directions, depending on the user requirement. In that cases, direction in which the block is positioned in the masonry wall shall be tested and appropriate aspect ratio shall be applied. Since the aspect ratio of the block in flat position is less than 1, the over estimation of the block compressive strength is a concern [11]. Therefore, the correction factor established for aspect ratio in question shall be multiplied to the block compressive strength. Alternatively, a standardised block dimension shall be recommended as in case of modern industrialised building material. Also, there are some studies [12, 79] suggesting new methods to test the blocks in compression.

The rate of loading is another important criterion that will dictate the precision of the test result obtained. In general, the compressive strength test on block is carried out in load controlled method. From normative documents, RLD [75] recommends to test the block at 3.45 MPa/min (load control), while [70] recommends to test in strain control at less than 1.3 mm/min, the Australian handbook provides an option to choose among load or strain control. The British standard [15] recommends to initially load the specimen at any convenient rates, when half the expected maximum load had reached, adjust the rate such that the maximum load of the block is not reached less than approximately 1 min. For a general guide, the loading rate recommended for fired clay masonry unit by British standard [15] is presented in Table 4.7. For flexural test, all the normative documents/standards/codes recommend to carryout test in load control mode. To obtain more precise compressive strength of

Table 4.7 Loading rate recommended by BS EN 772–1:2011 + A1:2015 [15]

Expected compressive strength (MPa)	Loading rate (MPa/s)
< 10	0.05
11 to 20	0.15
21 to 40	0.3
41 to 80	0.6
> 80	1.0

the material and understand the failure pattern, displacement control mode would be ideal.

4.2.3.5 Compression Test Procedures

The compressive strength of earth blocks defines their value and suitability. The universal presentation of compressive strength of earth blocks has been discussed and debated for a long period. There are still questions pertaining to how the compressive strength of block should be universally presented, and how the blocks should be universally tested. Some of the questions that are hindering in developing the universal procedure are:

(i) What is the acceptable aspect ratio of the block to be tested? And how to incorporate the dimensional effects?
(ii) What correction factor for aspect ratio should be used?
(iii) Should there be a standard dimension and geometry of the specimen? Or shall the dimension factor be used to normalise the strength results?
(iv) How to incorporate the platen restraint effect on the compressive strength obtained? Or should there be defined platen condition?
(v) How should the compressive strength of the block be measured experimentally (which test and what aspect ratio)?

From literature it was found that the following tests were developed to study the compressive strength of earth blocks.

4.2.3.6 Direct Compression Test

Direct compression test is an established test procedure used for most of the masonry units such as fired clay brick, or concrete blocks (solid or hollow). This test procedure is internationally accepted and test results are well accepted. Direct compression test is carried out on the block as it is manufactured, or on the cubes or cylinders extracted from blocks. Figure 4.11 shows a schematic representation of direct compression test on a block.

In direct compression test, the block is laid flat in the similar direction as placed in the masonry course. Apart from preparing the specimen surface, it is highly recommended to use capping between the specimen surface and the platen to ensure a close fit with the platen. Generally ply wood, or Teflon sheet is used as a capping material. The compressive strength calculated is expressed either (i) directly without any correction, or (ii) by correcting for aspect ratio. Typically, compressive strength calculated is an average of 5 to 10 samples. In the case of British standard [15], the calculation of normalised compressive strength is recommended. Whereas in Australia they use aspect ratio to eliminate the platen restraint effect in compressive strength [64]. The aspect ratio correction factor or shape factors are all developed for

Table 4.8 Standard recommendations for the mechanical testing of Adobe

Mechanical tests for adobe specimens						
	Document	Sampling	Specimens	Test rate	Other indications	Strength limit
Compression test	14.7.4 NMAC [75]	5 units per 25,000 bricks [a]	Adobe blocks Length ≥ twice the width Smooth surfaces	3.45 MPa/min	Specimens tested in the flat position	Mean strength ≥ 2.07 MPa No individual unit may have a strength of less than 1.72 MPa
	[68]	6 units	100 mm cube specimens	…	…	Mean strength of 4 best specimens (out of 6) ≥ 1 MPa
	NZS 4298 [b] [82]	Prior to work start: 5 or more units During construction: 5 units for every 5000 bricks or part thereof	Aspect ratio between 0.4 and 5.0 200 mm cube specimens recommended	…	Specimens loaded in the same direction as in the wall	Least of the individual results in the set > 1.30 MPa (for samples with aspect ratio of 1.0; for other ratios the required result shall be 1.30 × 0.7/ka [c])

(continued)

Table 4.8 (continued)

Mechanical tests for adobe specimens

	Document	Sampling	Specimens	Test rate	Other indications	Strength limit
	Pima County Standard [70]	5 units	Adobe blocks Dried to constant weight (at 20 ± 9°C, relative humidity ≤ 50%) May be capped with calcined gypsum mortar or the bearing surfaces may be rubbed smooth and true	≤ 1.3 mm/min	Specimens loaded in the same direction as in the wall Use of 3.2–6.4 mm felt pad at the top and bottom of specimens	Mean strength ≥ 2.07 MPa No individual unit may have a strength of less than 1.72 MPa

(continued)

Table 4.8 (continued)

Mechanical tests for adobe specimens						
	Document	Sampling	Specimens	Test rate	Other indications	Strength limit
	Australian Handbook [96]	5 units per 2500–10,000 bricks [d]	Adobe blocks [e] or cylindrical specimens Oven-dry or saturated surface dry condition or at other measured moisture content Recesses filled with mortar, uneven surfaces regularized with mortar or dental plaster	1–5 mm/min or 9–42 MPa/min	Use of 4–6 mm thick plywood at the top and bottom of specimens Unconfined strength obtained by applying an aspect ratio correction factor to the measured values	…
Flexural test	14.7.4 NMAC [75]	5 units per 25,000 bricks [a]	Adobe blocks	3.45 MPa/min	Support bars: 51 mm diameter, 51 mm from each end of unit Loading bar: 51 mm diameter	Mean strength ≥ 0.34 MPa
	NZS 4298 [b] [82]	Prior to work start: 5 or more units During construction: 5 units for every 5000 bricks or part thereof	Adobe blocks	…	Support bars: 50 mm diameter Loading bar: 50 mm diameter	Least of the individual results in the set > 0.25 MPa

(continued)

Table 4.8 (continued)

Mechanical tests for adobe specimens						
	Document	Sampling	Specimens	Test rate	Other indications	Strength limit
	Pima County Standard [70]	5 units	Adobe blocks	2.22kN/min	Support bars: 51 mm diameter, 51 mm from each end of unit Loading bar: 51 mm diameter	Mean strength ≥ 0.34 MPa No individual unit shall have a strength of less than 0.24 MPa
	Australian Handbook [96]	5 units per 2500–10,000 bricks [d]	Adobe blocks Oven-dry or saturated surface dry condition or at other measured moisture content	2–6kN/min	Support bars: 25 ± 2 mm diameter Test span: specimen length less 50 ± 10 mm Loading bar: 25 ± 2 mm diameter	…
Splitting test [f]	[68]	6 units	Cylindrical specimens (height: 30.48 cm; diam.: 15.24 cm) Initial moisture: 20–25%	…	…	Mean strength of 4 best specimens (out of 6) ≥ 0.08 MPa

[a]Or at the discretion of the building official
[b]Indications for 'standard grade earth construction'
[c]k_a is the aspect ratio factor
[d]Or as defined by changes in constituent materials. The lot size should also take into consideration the method of production and experience of the manufacturer and builder
[e]A whole specimen should be used; in cases where specimen size cannot be accommodated, representative part-units cut from the whole may be used
[f]Also known as Brazilian test

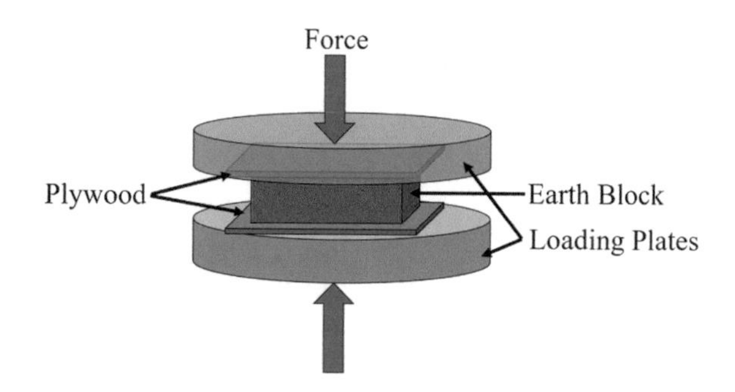

Fig. 4.11 Schematic representation of direct compression test on earth block

fired clay blocks rather than unfired clay blocks. Therefore there is lack of reliability on these factors application in unfired clay bricks [64].

Alternatively, to negotiate the aspect ratio, instead of blocks, cubes or cylindrical samples can be prepared from the same material and tested for their compressive strength. The earlier studies suggest a poor correlation of the cube or cylinder compressive strength with respect to block compressive strength [64]. This may be due to the change in manufacturing process and dimension of cube or cylinder. Champiré et al. [22] studied the mechanical properties of cylindrical samples having aspect ratio of 2 extracted from CEB's. Though the study's main agenda was not to study or compare the block strength with the extracted cylindrical sample strength, it suggests an alternative approach that is to cut the cube or cylinders from the block and test them. However, the impact of extraction process on compressive strength needs further investigation.

Three Points Bending

Three points bending test is a less accurate, indirect, quick and economical test method used to study the compressive strength of the blocks especially at field. Though this method underestimates the compressive strength of the block, it is widely accepted to be sufficient to have a first insight of this value [64]. The compressive strength is calculated from the flexural stress obtained in pure bending, based on the traction/bending stress theory. Morel and Pkla [63] highlight the shortcoming in calculation of compressive strength from traction/bending model and propose a 'compression strength model for the 3 points bending test'. This model assumes arch behaviour of two beams and calculates the compressive strength of the block with the help of failure load from 3 point bending test. The accuracy of compression strength model for 3 point bending test has been validated for limited samples [63]. Further investigation on their reliability and accuracy in calculating compressive strength needs to be carried out.

The main advantages of three points bending test are:

(i) The failure load required is 80–150 times lower than that is required in direct compression test.

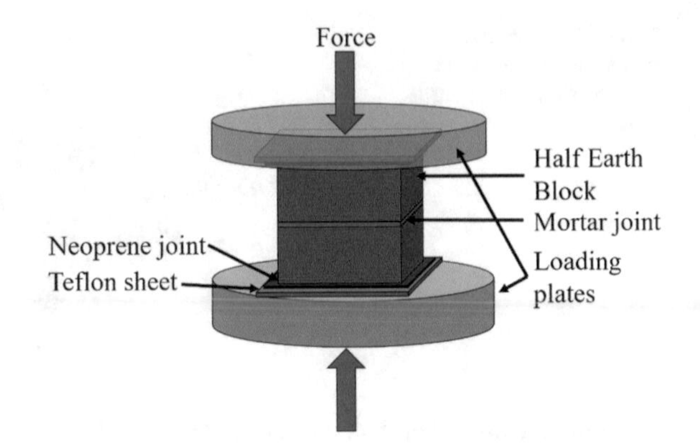

Fig. 4.12 Schematic representation of RILEM test

(ii) The test does not require to prepare the specimen surface or need of capping.
(iii) This experiment can be carried out in the field with minimal requirements.

Limitations are:

(i) The test method does not account for susceptibility of defects in blocks such as shrinkage cracks.
(ii) The accuracy of compressive strength calculated using these models are reduced, but it is accepted to be sufficient to enable a lower bound estimation.

RILEM Test

To counter the aspect ratio problem faced in direct compression test, RILEM technical committee 164-EBM developed in 1997 a test which doubles the aspect ratio of the specimen. The loading and testing of specimen is similar to direct compression test, but the specimen preparation is the key variation. Figure 4.12 shows the schematic representation of RILEM test set up. The specimen is either prepared by breaking the block into two perfect halves either by mechanical tool or by subjecting it to three points bending test. The two half blocks are stacked one above the other by providing a mortar joint. This will directly double the aspect ratio of the block. The stacked specimen typically replicates a single bed masonry prism. The mortar used to stack the blocks shall be of the same material used to manufacture the blocks. To enable the even distribution of load between the platen and block, the specimens are capped with a layer of neoprene. A sheet of Teflon is also placed between the platen and the specimen at each end to minimise the friction effect.

Morel et al. [64] carried out a comparative study between the RILEM test and direct compression test. The corrected direct compression test results were higher than the RILEM test results. This reduction in strength of RILEM can be explained in two ways: (i) RILEM test is not a masonry unit test instead it replicates masonry prism test with mortar bed joint, which has different stiffness compared to block, even if it is made of same raw material. (ii) The aspect ratio correction factor used to correct the compressive strength obtained from the direct compression test is inaccurate. In

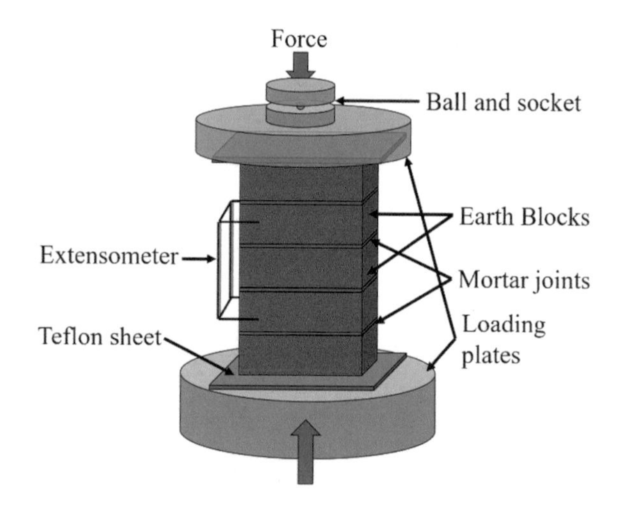

Fig. 4.13 Schematic representation of prism test

order to establish a good correlation of RILEM test and direct compression test, the accuracy of aspect ratio correction factor and a parameter to eliminate the effect of mortar joint influence should be well developed.

Prism Test

The prism test is also called a masonry prism test, in which the blocks are stacked in layers with mortar bed joints, as it is shown in Fig. 4.13. Generally, prism test will consist of 4 or 5 block stacked with 10–12 mm thick mortar joints. The mortar shall be similar to what is used for masonry construction. The prism test will help in understanding the compressive strength parameter of the masonry wall. The advantage of prism test is that samples would have higher aspect ratio of more than 5, which will eliminate the platen effect and the compressive strength obtained is more accurate for the masonry. Also the stress strain properties of the masonry can be extracted by attaching extensometers to the prism. The disadvantage is that, this test would consume more material for testing and the compressive strength parameters obtained are of the combined masonry rather than a single masonry unit (brick). A minimum of 5 samples should be tested for obtaining an accurate average of compressive strength. The stacked prism should be accurately constructed without any eccentricity.

Direct Compression Test—Block Placed Perpendicular to the Direction of Bedding

The block compressive strength is always measured lying in the direction similar to masonry course. But to study the stress strain characteristics of the masonry units, the aspect ratio of the block should be ≥ 2. Therefore the blocks can be positioned

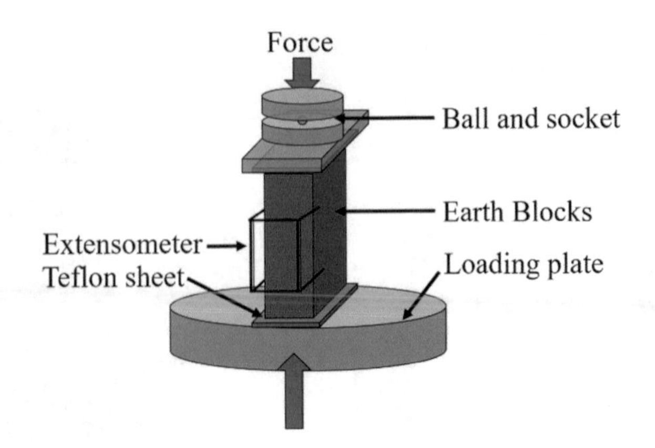

Fig. 4.14 Schematic representation of brick compression test in vertical direction

in the direction perpendicular to the direction how they are laid in the masonry, as it is shown in Fig. 4.14. The test procedure is similar to the direct compression test. The only difference is the attachment of extensometer on the blocks to record axial strain corresponding to compressive stress. Since the direction of masonry course and testing is not same, the compressive strength cannot be considered as the masonry unit compressive strength, only the stress strain characteristics shall be used for design considerations.

4.2.3.7 Compressive Strength Recommendations

Apart from recommending the test procedure, another main role of any standard/code is to recommend the minimum strength criteria of the material. Similarly, for Adobe, except the Australian handbook, all other standards mentioned in Table 4.8 recommend the minimum compressive strength. The minimum compressive strength of 1 MPa is recommended by [68], and [75] and [70] recommend 2.07 MPa as the minimum compressive strength of the Adobe. The minimum compressive strength should always be associated with the dry density of the block and the moisture content at the test. Without mentioning the dry density and moisture content, it would be difficult to understand the criteria.

4.2.3.8 Flexural and Tensile Strength Recommendations

For the determination of the tensile strength of adobe, most normative documents indicate flexural tests on Adobe blocks (flexural tensile strength, e.g. as discussed in Sect. 4.1.4.4), while [68] indicates splitting tests on cylindrical specimens (splitting tensile strength, e.g. as discussed in Sect. 4.1.4.2). Similar to the recommendation of minimum compressive strength, the minimum flexural strength and tensile strength of the Adobe is mentioned in the normative documents/standards/code. The [82], recommends a minimum flexural strength of 0.25 MPa, whereas [75] and [70] recommend 0.34 MPa as minimum flexural strength. Only [68] recommends the minimum tensile strength of 0.08 MPa for Adobe blocks.

4.2.3.9 Masonry

For the determination of the compressive strength of adobe masonry (i.e. specimens comprising multiple adobe blocks), the normative documents consulted indicate the testing of adobe prisms with different aspect ratios (varying between 2 and 5). The Australian Handbook [96] is the only document including aspect ratio correction factors for the calculation of the unconfined compressive strength of prism specimens. [68] also includes procedures for the determination of the shear strength of adobe masonry, recommending diagonal compression tests on small square walls. It was observed that the indications of most of the normative documents consulted are not very detailed. Also, the correction of the compressive strength with regards to the aspect ratio is fundamental, but is only addressed by two of the aforementioned documents. It is thus important to further develop the existing indications for the mechanical testing of adobe and adobe masonry and to move towards a uniform standardization of procedures, taking into account the various existing technical documents (Table 4.9).

4.3 Cob

4.3.1 Introduction

Cob is attested since the early Neolithic in the Middle East [85]. Cob built heritage can be found in Africa, Asia and Europe. This technique consists in mixing earth in a plastic state, with or without plant fibres, which is implemented wet, in order to build a monolithic and load-bearing or freestanding wall. Cob is more a family of very different techniques. It has been estimated that, at least, hundreds of variations exists for this process [42]. Some of them are close to Wattle and Daub, others to Adobe and others to Rammed Earth. Among load-bearing earth construction techniques, cob is the least studied one. This explains the little literature available for this technique.

Table 4.9 Standard recommendation for the mechanical testing of Adobe masonry

Mechanical tests for adobe masonry specimens						
	Document	Sampling	Specimens	Test rate	Other indications	Strength limit
Compression test	[68]	6 specimens	Prisms with aspect ratio of approximately 3	…	…	Mean strength of 4 best specimens (out of 6) $\geq$ 0.60 MPa
	NZS 4298[a] [82]	…	Prisms with an even number of bricks and aspect ratio of between 3 and 5	…	The tests shall be as per Appendix 2B of NZS 4210 (SNZ 2001)	…
	Australian Handbook [96]	5 specimens	Prisms with aspect ratio between 2 and 5, but not less than 3 courses high	0.5–1.0 MPa/min	Use of 4 mm thick plywood at the top and bottom of specimens Unconfined strength obtained by applying an aspect ratio correction factor to the measured values	…
Diagonal comp. test	[68]	6 specimens	Walls with dimensions of $0.65 \times 0.65 m^2$ (approx..)	…	…	Mean strength of 4 best samples (out of 6) $\geq$ 0.025 MPa

[a]Indications for 'standard grade earth construction'

4.3.2 Mechanical Behaviour

4.3.2.1 Mechanical Behaviour of Cob Mixture

Cob mixture clods are piled at plastic state to build lifts [42]. Slenderness ratio of lifts depends on the mechanical resistance of cob mixture: the higher the mechanical resistance, the higher the lift and the quicker the construction process. Past builders aimed at increasing lifts height in order to save time and reduce costs. Mechanical performance of cob mixture is therefore of great interest for the economical optimization of cob process. Typical UCS of cob mixture, when wet, is 0.05 MPa [43, 86], which is enough to build a lift of about 2 m high [74, 86].

4.3.2.2 Fibres

Fibres, when added, are thought to play a major role in the mechanical behaviour of cob mixture at plastic state by enhancing its cohesion and allowing the mixture to be implemented without the use of shuttering [86]. Hence, [86] defined an optimal straw and water content domain (Fig. 4.15) for which cob mixture is not too plastic, in order to ease cob mixture implementation, but not too dry, in order to facilitate fibre and earth mix. Fibres contribute to the distribution of drying shrinkage cracks of clayey earth, limiting disorders caused by shrinkage, but also creating fragilities that weaken the structure [52, 86]. This is the reason why optimal fibre contents are proposed in the literature.

Although fibres were not necessarily incorporated into cob mixture, no data is available on unfibered cob mechanical behaviour. This can be attributed to two different reasons: firstly, for the majority of authors, cob refers necessarily to

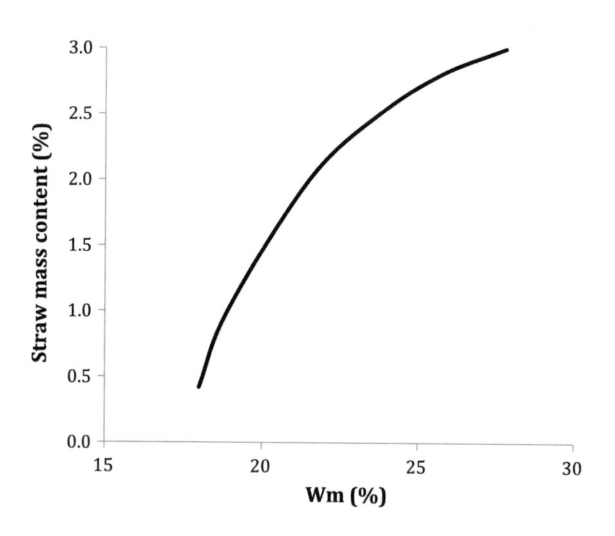

Fig. 4.15 Optimal cob mixture with regard to water content of manufacturing stage (W_m) and straw content, after [86]

fibre addition and secondly fibres are assumed to have a beneficial effect on the mechanical behaviour of cob before, during and after drying. As a consequence, all bibliographical results presented here concern only fibred cob.

The presence of fibres in cob material results in a ductile behaviour of simple compression test specimens, with large vertical strain [8, 27, 52, 61, 86, 95]. Crack patterns after the tests are almost random and even after breaking the fibres hold together the different parts of the broken specimens [61, 86].

4.3.2.3 Stiffness

The presence of fibres also induces a low stiffness of cob specimens, ranging from 170 to 335 MPa [99]. As a consequence, the contrast of stiffness with rigid construction material introduced vertically inside cob walls is responsible for damage mechanisms [81]. Same damage mechanism has been highlighted for cement-based plasters, too stiff, with regard to earthen walls. The architectural design of cob buildings should avoid the use of stiff materials placed vertically against cob walls. This more specifically concerns anti-seismic design.

4.3.2.4 Mechanical Behaviour of Cob Walls

Load

Ranges of cob compressive strength found in the literature are summarized in Fig. 4.16. Compressive strength typically range from 0.6 to 1.3 MPa [27, 52, 61, 72, 81 86, 95, 100]. The highest compressive stresses in cob buildings are likely to be at the base of gable end walls, and under roof trusses, where wall plates help to distribute the load [86]. Keefe [52] estimates that the compressive strength at plinth level in a traditional two-storey cob house with 550 to 600 mm thick walls, ranges from 0.08 to 0.10 MPa. Considering minimum compressive strengths found in the literature (Fig. 4.16), for cob heritage, cob compressive strength is at least five times higher than maximum compressive strength borne by cob walls [43, 52, 86].

Texture and Density

Several factors govern the mechanical behaviour of cob wall. A well-graded earth allows to maximise particles contact and enhance its mechanical strength [52]. Strength is also higher in clay-rich earths [51], but generate a more important shrinkage and can be responsible for shrinkage cracks that weaken the structure. The more the earth is compacted during implementation, the more the density and the more the strength are [52, 61]. Several authors agreed with a cob dry bulk density close to 1500 kg/m^3 [25, 61, 81, 95, 100], measured a cob bulk density close to 1600 kg m^{-3}.

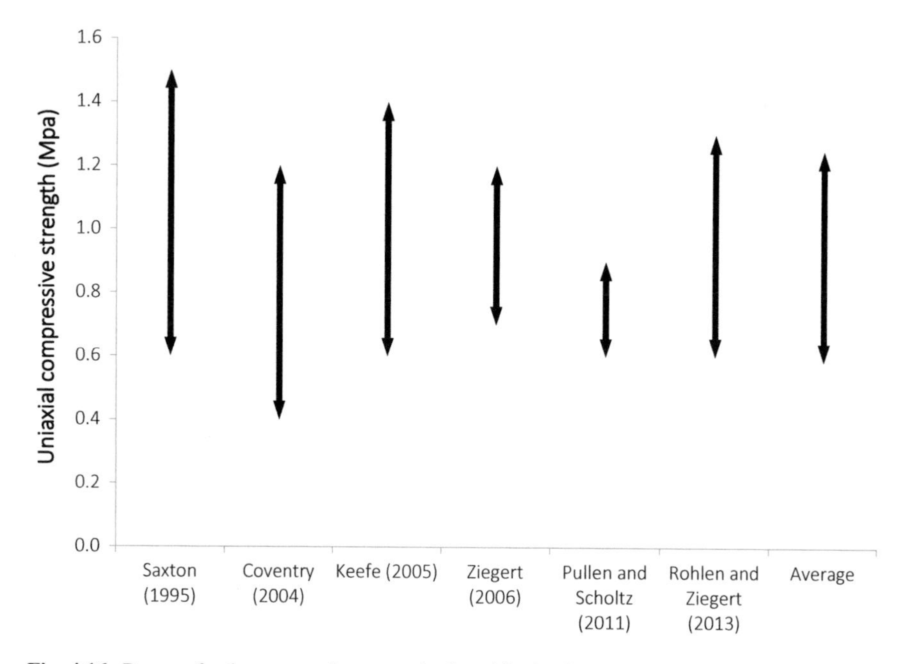

Fig. 4.16 Range of cob compressive strengths found in the literature [27, 52, 61, 72, 81, 86, 95, 100] and average values

Moisture

Cob material is traditionally not stabilised with a hydraulic binder. Therefore, cob walls are sensitive to water content variations. When too much water enters a cob wall, the clay particles which bind it together are forced apart, and the cob is first reduced to a plastic, then to a liquid state, with consequent structural failure. There is a coupling between mechanical strength and water content [43, 52, 61, 86] (Fig. 4.17). Saxton [86] considered that above 10% water content there is a major damage risk.

Nevertheless, Keefe [52] stated that intact soils samples removed from earth walls immediately following their collapse have shown a moisture content as low as 7% in some cases. Keefe [52] defines a Critical Moisture Content (CMC). According to him, this CMC is just below the Plastic Limit (PL). Indeed, the plastic limit of a sample is measured using the fine fraction of the soil (< 425 μm). It is therefore only relevant to soils composed entirely of fine material. As a consequence, plastic limit value has to be weighted by the 425 μm Passing ($P_{425\mu m}$) of the materials using the formula:

$$CMC = LP \times P_{425\mu m} \tag{4.1}$$

With: CMC: Critical Moisture Content (%), LP: Plastic Limit (%) and $P_{425\mu m}$: 425 μm Passing (%).

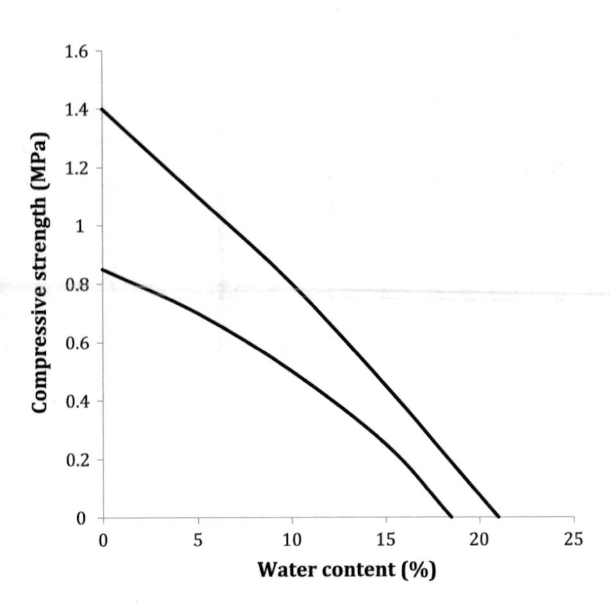

Fig. 4.17 Variation in compressive strength against water content, after [86]

Usually, equilibrium water content in a cob wall ranges from 3 to 4% [86], which is dry enough to provide sufficient strength to the material. Since they were built without damp proof courses, moisture increase more specifically concerns cob heritage buildings. Nonetheless, majority of failures in cob walls can be attributed to either neglect, inappropriate maintenance and repair [51, 65]. If well maintained by skilled craftsmen, moisture is self-regulated by the wall and cob does not pose any particular damage risk [33].

In modern cob walls, damp proof courses are generally introduced. Modern cob buildings are therefore less concerned by rising damp from the ground. If moisture issue is less significant for modern cob walls than for cob heritage walls, other moisture sources might be considered for this highly hygroscopic material [58]. As a consequence, architectural design for modern cob buildings should avoid the use of waterproof covering against cob walls.

4.3.3 Experimental Procedures

4.3.3.1 Compression Test Procedures

Cylinder Tests

Little testing of cob materials is stated in the literature, this is why only compressive strength tests are considered. Uniaxial Compressive Strength (UCS) and Young's modulus (E) determination are the most commonly cited tests for cob mechanical behaviour characterisation. Specimens' fabrication procedures for the determination

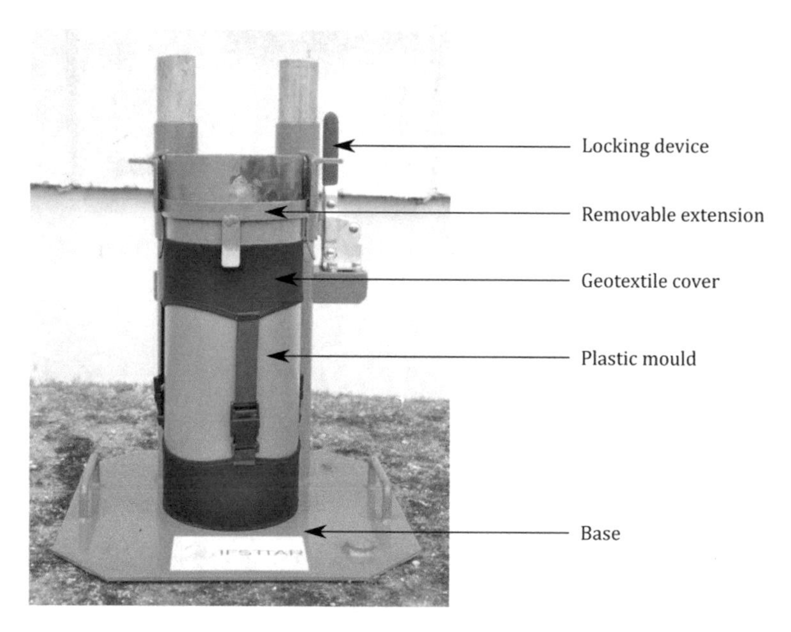

Fig. 4.18 Device for cob specimen production

of UCS and E refer to cylindrical test specimen with a slenderness ratio of 2:1 in order to accommodate the frictional forces due to the confinement caused by testing machine plates [8, 27, 43, 52, 72, 86, 95]. Specimens' sizes are either 100 × 200 mm or 150 × 300 mm since smaller size specimens are judged unrepresentative and larger ones are too heavy to be handled conveniently [27, 43, 52, 86, 95]. Cob mixture was placed inside cylinder moulds in several layers and compacted: (1) under dynamic load thanks to a Proctor compaction device [27, 43, 52, 86, 95] or (2) placed by hand [72]. Cob mixture was also placed inside cylinder moulds by compaction under a mechanically applied static load [8]. The aim during compaction was to produce specimens that did not contain noticeable air voids. As the cob mixture is in a soft state, over compaction would have little additional effect on the density [86].

As highlighted by [95] the use of concrete cardboard cylinders proposed by [86] is easy to use, but presents several limitations: (1) the side wall effect is strong and compaction is not efficient enough; (2) steel moulds used for concrete casting are impervious, the drying is therefore not homogeneous and drying times are very long; (3) earth material stick to the inner face of the mould and generates a specimen surface state of poor quality. A new sample production protocol was proposed by [95]. A flexible geotextile cover (a flexible synthetic fabric) is placed on the inner face of a cylinder mould wall (diameter 150 mm, height 300 mm) to reduce the friction and the adherence with soil (Fig. 4.18). Cob mixture is then compacted in 8 layers as suggested by [86].

This sample production protocol presents several advantages: (1) the flexibility of the geotextile cover reduces the side wall effect of the cylinder and reduces the

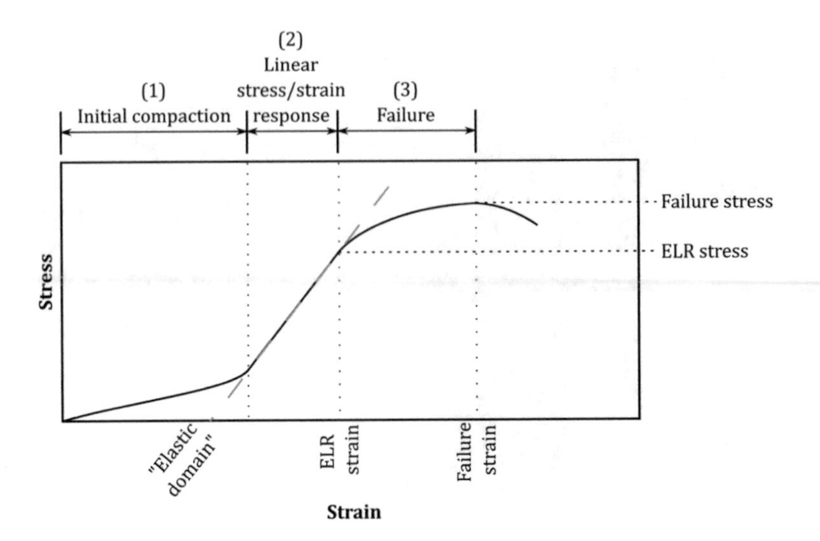

Fig. 4.19 Typical cob Uniaxial Compressive Strength test (ELR = End of Linear stress/strain Response), after [8]

adherence of earth to the inner face of the cylinder; (2) it provides specimens with a good surface state; (3) thanks to horizontal shrinkage, specimens can be unmoulded after a 24 h drying only. Specimens made with this protocol were compared to cob wall elements and provided satisfactory results [95].

Specimens are air-dried [52, 95], dried in a humidity and temperature controlled chamber (25 °C and 75% relative humidity) [27, 43, 86] or oven-dried (40 °C in [95] and 75 °C in [72]. Oven drying at higher temperatures can affect suction and therefore lower the mechanical strength of earth materials [24]. Moreover, Natural fibres are subjected to temperature decay. Usual recommendations impose a drying temperature lower than 60 °C. [95] have compared air-dried and oven-dried (40 °C) specimens and did not highlight any noticeable differences on their mechanical behaviour. The mechanical strength of earth materials depends on the water content of the specimens: the higher the moisture content, the lower the compressive strength [22, 36, 86]. For repeatability reasons, specimens should be conditioned in a humidity and temperature controlled chamber until weight stabilisation prior to testing. If conditioning is not possible, the water content during the mechanical test should be recorded.

Mechanical tests were controlled in displacement at a speed of 1 mm min^{-1} [27, 95]. A typical cob mechanical test description is proposed by [8]. Three phases are identified (Fig. 4.19) (1) initial compaction of the sample, (2) linear stress/strain response of the sample and (3) increasing crack growth within the sample and failure of the sample. Miccoli et al. [61] described the typical course of cob wallets testing as an elastic range with a low shear modulus followed by a plastic-type deformation of the specimen. In both cases, a linear stress/strain response is followed by a ductile failure, induced by fibres [27, 61].

Table 4.10 Corresponding value between rebound hammer display values and compressive strength, after Röhlen and Ziegert [80]

Rebound hammer value	35	40	45
Compressive strength (MPa)	0.7	1.0	1.3

Wallette Tests

Direct compression test have been conducted by Miccoli et al. [61] on 500 mm side wallets, and by Vinceslas et al. [95] on 600 mm side wallets. Handling and testing samples of this size is however quite complex. To avoid this issue, Röhlen and Ziegert [81] propose a non-destructive test that uses a rebound hammer and whose correspondace with compressive strength is shown in the Table 4.10.

Elastic Properties

Tangent Young's moduli values proposed in the literature are calculated according to total strain of test specimens, i.e. thanks to the displacement of testing machine plates [8, 27],Pullen and Scholz [72]. However, earth materials have an elasto-plastic behaviour [22]. As a consequence, in order to measure the elastic contribution only, secant moduli of repeated loading cycles should be considered. Furthermore, it has been demonstrated for rammed earth that frictional forces caused by confinement of testing machine plates have a high impact on E values [21]. Strain measurement for E calculation should only concern the central third part of cylindrical specimens. [95] determined the Young's moduli of their cob specimens by 3 cyclic loadings measured in the central third of the specimens using extensometers. The average secant moduli of their study range from 250 to 350 MPa.

Little literature exists concerning cob mechanical behaviour, and each author has developed its own testing protocol. This makes difficult data comparison and highlights a unification need for mechanical testing procedure (Jiménez Delgado and Guerrero [50, 61].

References

1. ASTM E 519/E519M-15 (2015) Standard test method for diagonal tension (shear) in masonry assemblages. ASTM International, West Conshohocken, PA
2. ASTM C78/C78M-18 (2018) Standard test method for flexural strength of concrete (using simple beam with third-point loading). ASTM International, West Conshohocken, PA
3. ASTM D3080/D3080M-11 (2011) D3080/D3080M-11. Standard test method for direct shear test of soils under consolidated drained conditions. ASTM International, West Conshohocken, PA
4. ASTM D4767-11 (2011) Standard test method for consolidated undrained triaxial compression test for cohesive soils. ASTM International, West Conshohocken, PA,

5. Abhilash HN, Morel J-C, Champiré F, Fabbri A (2019) A novel experimental study to investigate the interface properties of rammed earth. Proc Inst Civ Eng—Constr Mater. https://doi.org/10.1680/jcoma.18.00095
6. Abhilash HN, Morel J-C (2019) Stress–strain characteristics of unstabilised rammed earth. In: Reddy BVV, Mani M, Walker P (eds) Earthen dwellings and structures: current status in their adoption. Springer Singapore, Bangalore, India, pp 203–214
7. Achenza M, Fenu L (2007) On earth stabilization with natural polymers for earth masonry construction. Mater Struct 39:21–27. https://doi.org/10.1617/s11527-005-9000-0
8. Addison Greer MJ (1996) The effect of moisture content and composition on the compressive strength and rigidity of cob made from soil of the Breccia Measures near Teignmouth, Devon, PhD., Plymouth School of Architecture
9. Al Rim K, Ledhem A, Douzane O et al (1999) Influence of the proportion of wood on the thermal and mechanical performances of clay-cement-wood composites. Cem Concr Compos 21:269–276. https://doi.org/10.1016/S0958-9465(99)00008-6
10. Araki H, Koseki J, Sato T (2011) Mechanical properties of geo-materials used for constructing earthen walls in Japan. Builletin ERS, Inst Ind Sci Univ Tokyo, pp 101–112
11. Aubert JE, Fabbri a., Morel JC, Maillard P (2013) An earth block with a compressive strength higher than 45MPa! Constr Build Mater 47:366–369.https://doi.org/10.1016/j.conbuildmat.2013.05.068
12. Aubert JE, Maillard P, Morel JC, Al Rafii M (2015) Towards a simple compressive strength test for earth bricks ? Mater Struct 49:1641–1654. https://doi.org/10.13140/RG.2.1.4641.4242
13. BS 1377-7:1990 (1990) Soils for civil engineering purposes—Part 7: Shear strength tests (total stress). British Standards Institution
14. BS EN 1998-1:2004+A1:2013 (2013) Eurocode 8: design of structures for earthquake resistance—Part 1: General rules, seismic actions and rules for buildings. British Standards Institution
15. BS EN 772-1:2011+A1:2015 (2015) Methods of test for masonry units Part 1 : Determination of compressive strength
16. BS EN ISO 17892-9:2018 (2018) Geotechnical investigation and testing—Latoratory testing of soil Part 9 : Consolidated triaxial compression tests on water saturated soils. British Standards Institution
17. Baglioni E, Fratini F, Rovero L (2010) The materials utilised in the earthen buildings sited in Thedrâa Valley (Morocco): mineralogical and mechanicalcharacteristics. In: 6th seminar of earthen architecture in Portugal and 9th Ibero-American seminar on earthen architecture and construction, Coimbra, Portugal
18. Beckett CTS, Augarde CE, Easton D, Easton T (2017) Strength characterisation of soil-based construction materials. Géotechnique 68:400–409. https://doi.org/10.1680/jgeot.16.P.288
19. Bouguerra A, Ledhem A, de Barquin F et al (1998) Effect of microstructure on the mechanical and thermal properties of lightweight concrete prepared from clay, cement, and wood aggregates. Cem Concr Res 28:1179–1190. https://doi.org/10.1016/S0008-8846(98)00075-1
20. Bui Q, Morel J, Hans S, Walker P (2014) Effect of moisture content on the mechanical characteristics of rammed earth. Constr Build Mater 54:163–169
21. Bui Q-B (2008) Stabilité des structures en pisé : durabilité, caractéristiques mécaniques. Institut National des Sciences Appliquées de Lyon
22. Champiré F, Fabbri A, Morel J et al (2016) Impact of relative humidity on the mechanical behavior of compacted earth as a building material. Constr Build Mater 110:70–78. https://doi.org/10.1016/j.conbuildmat.2016.01.027
23. Cheah JSJ, Walker P, Heath A, Morgan TKKB (2012) Evaluating shear test methods for stabilised rammed earth. Proc Inst Civ Eng Constr Mater 165:325–334. https://doi.org/10.1680/coma.10.00061
24. Ciancio D, Jaquin P, Walker P (2013) Advances on the assessment of soil suitability for rammed earth. Constr Build Mater 42:40–47. https://doi.org/10.1016/j.conbuildmat.2012.12.049

25. Collet F, Bart M, Serres L, Miriel J (2010) Porous structure and hydric properties of cob. J Porous Media 13:111–124
26. Corbin A, Augarde C (2015) Investigation into the shear behaviour of rammed earth using shear box tests. In: Amziane S, Sonebi M, Charlet K (eds) First International conference on bio-based building materials. RILEM Proceedings, Clemont-Ferrand, France, France, pp 93–98
27. Coventry KA (2004) Specification development for the use of Devon cob in earthen construction, Ph.D., University of Plymouth—Faculty of Science
28. Demir I (2006) An investigation on the production of construction brick with processed waste tea. Build Environ 41:1274–1278. https://doi.org/10.1016/j.buildenv.2005.05.004
29. Dołżyk-Szypcio K (2019) Direct shear test for coarse granular soil. Int J Civ Eng 17:1871–1878. https://doi.org/10.1007/s40999-019-00417-2
30. El-Nabouch R, Bui QB, Plé O, Perrotin P (2018) Characterizing the shear parameters of rammed earth material by using a full-scale direct shear box. Constr Build Mater 171:414–420. https://doi.org/10.1016/j.conbuildmat.2018.03.142
31. El-Nabouch R, Bui Q-B, Perrotin P, Plé O (2017) Experimental and numerical studies on cohesion and friction angle of rammed earth material. In: Poromechanics 2017—proceedings of the 6th biot conference on poromechanics. ASCE, Paris, France
32. Eslami A, Ronagh HR, Mahini SS, Morshed R (2012) Experimental investigation and nonlinear FE analysis of historical masonry buildings—a case study. Constr Build Mater 35:251–260. https://doi.org/10.1016/j.conbuildmat.2012.04.002
33. Fabbri A, Morel JC (2016) Earthen materials and constructions. In: Harries KA, Sharma B (eds) Nonconventional and vernacular construction materials. Elsevier, Duxford (UK), pp 273–299
34. François B, Palazon L, Gerard P (2017) Structural behaviour of unstabilized rammed earth constructions submitted to hygroscopic conditions. Constr Build Mater 155:164–175. https://doi.org/10.1016/j.conbuildmat.2017.08.012
35. Fratini F, Pecchioni E, Rovero L, Tonietti U (2011) The earth in the architecture of the historical centre of Lamezia Terme (Italy): characterization for restoration. Appl Clay Sci 53:509–516. https://doi.org/10.1016/j.clay.2010.11.007
36. Gallipoli D, Bruno A., Perlot C, Salmon N (2014) Raw earth construction: Is there a role for unsaturated soil mechanics? In: Khalili N, Russel A, Khoshghalb A (eds) Unsaturated soils: research & applications. Taylor & Francis Group, London, UK, pp 55–62
37. Gavrilovic P, Sendova V, Ginell WS, et al (1998) Behaviour of adobe structures during shaking table tests and earthquakes. In: European conference on earthquake engineering. A A Balkema, p 172
38. Gerard P, Mahdad M, Robert McCormack A, François B (2015) A unified failure criterion for unstabilized rammed earth materials upon varying relative humidity conditions. Constr Build Mater 95:437–447. https://doi.org/10.1016/j.conbuildmat.2015.07.100
39. Gomes MI, Gonçalves TD, Faria P (2014) Unstabilized rammed earth: characterization of material collected from old constructions in South Portugal and comparison to normative requirements. Int J Archit Herit 8:185–212. https://doi.org/10.1080/15583058.2012.683133
40. Gooding DEM (1994) Improved processes for the production of soil-cement building blocks. University of Warwick
41. Hall M, Djerbib Y (2004) Rammed earth sample production: context, recommendations and consistency. Constr Build Mater 18:281–286. https://doi.org/10.1016/j.conbuildmat.2003.11.001
42. Hamard E, Cazacliu B, Razakamanantsoa A, Morel J-C (2016) Cob, a vernacular earth construction process in the context of modern sustainable building. Build Environ 106:103–119. https://doi.org/10.1016/j.buildenv.2016.06.009
43. Harries R, Saxton B, Coventry K (1995) The geological and geotechnical properties of earth material from central Devon in relation to its suitability for building in "Cob". In: Geoscience in South-West England. pp 441–444

44. Illampas R, Ioannou I, Charmpis DC (2016) Experimental assessment of adobe masonry assemblages under monotonic and loading–unloading compression. Mater Struct 50:79. https://doi.org/10.1617/s11527-016-0952-z
45. Illampas R (2014) Experimental and computational investigation of the structural response of adobe structures. University of Cyprus
46. Illampas R, Ioannou I, Charmpis DC (2014) Adobe bricks under compression: Experimental investigation and derivation of stress–strain equation. Constr Build Mater 53:83–90. https://doi.org/10.1016/j.conbuildmat.2013.11.103
47. Illampas R, Loizou Vasilios G, Ioannou I (2020) Effect of straw fiber reinforcement on the mechanical properties of adobe bricks. Poromechanics VI:1331–1338
48. Jaquin PA, Augarde CE, Gallipoli D, Toll DG (2009) The strength of unstabilised rammed earth materials. Géotechnique 59:487–490. https://doi.org/10.1680/geot.2007.00129
49. Jaquin PA, Augarde CE, Gerrard CM (2008) Chronological description of the spatial development of rammed Earth techniques. Int J Archit Herit 2:377–400. https://doi.org/10.1080/15583050801958826
50. Jiménez Delgado MC, Guerrero IC (2007) The selection of soils for unstabilised earth building: a normative review. Constr Build Mater 21:237–251. https://doi.org/10.1016/j.conbuildmat.2005.08.006
51. Keefe L (1993) The cob building of Devon 2—repair and maintenance
52. Keefe L (2005) Earth building: methods and materials, repair and conservation. Routledge
53. Ledhem A, Dheilly RM, Benmalek ML, Quéneudec M (2000) Properties of wood-based composites formulated with aggregate industry waste. Constr Build Mater 14:341–350. https://doi.org/10.1016/S0950-0618(00)00037-4
54. Liberatore D, Giuseppe S, Mucciarelli M, et al (2006) Typological and experimental investigation on the adobe buildings of Aliano (Basilicata, Italy). In: Lourenço PB, Roca P, Modena C, Agrawal S (eds) Structural analysis of historical constructions. New Delhi, pp 851–858
55. Liu K, Wang M, Wang Y (2015) Seismic retrofitting of rural rammed earth buildings using externally bonded fibers. Constr Build Mater 100:91–101. https://doi.org/10.1016/j.conbuildmat.2015.09.048
56. Maniatidis V, Walker P (2003) A review of rammed earth construction
57. Maskell D, Heath A, Walker P (2013) Laboratory scale testing of extruded earth masonry units. J Mater Des 45:359–364. https://doi.org/10.1016/j.matdes.2012.09.008
58. McGregor F, Heath A, Maskell D et al (2016) A review on the buffering capacity of earth building materials. Proc Inst Civ Eng—Constr Materhttps://doi.org/10.1680/jcoma.15.00035
59. Meli R (2005) Experiencias en México sobre reducción de vulnerabilidad sísmica de construcciones de adobe. In: SismoAdobe2005: International seminar on architecture, construction and conservation of earthen buildings in seismic areas, Lima, Peru
60. Miccoli L, Drougkas A, Müller U (2016) In-plane behaviour of rammed earth under cyclic loading: experimental testing and finite element modelling. Eng Struct 125:144–152. https://doi.org/10.1016/j.engstruct.2016.07.010
61. Miccoli L, Müller U, Fontana P (2014) Mechanical behaviour of earthen materials: a comparison between earth block masonry, rammed earth and cob. Constr Build Mater 61:327–339. https://doi.org/10.1016/j.conbuildmat.2014.03.009
62. Miccoli L, Oliveira DV, Silva RA et al (2015) Static behaviour of rammed earth : experimental testing and finite element modelling. Mater Struct 48:3443–3456. https://doi.org/10.1617/s11527-014-0411-7
63. Morel J-C, Pkla A (2002) A model to measure compressive strength of compressed earth blocks with the ' 3 points bending test.' Constr Build Mater 16:303–310
64. Morel J, Pkla A, Walker P (2007) Compressive strength testing of compressed earth blocks. Constr Build Mater 21:303–309. https://doi.org/10.1016/j.conbuildmat.2005.08.021
65. Morris W (1992) The cob building of Devon 1—history, building methods and conservation. Historic Building Trust, London (UK)
66. NZS-4297 (1998) Engineering design of earth buildings. Standard New Zealand

67. Nabouch R, Bui QB, Plé O et al (2016) Seismic assessment of rammed earth walls using pushover tests. Procedia Eng 145:1185–1192. https://doi.org/10.1016/j.proeng.2016.04.153
68. Norma E.080 (2017) Diseño y construcción con tierra reforzada, Ministerio de Vivienda, Construcción y Saneamiento (MVCS), Lima
69. Obonyo E, Exelbirt J, Baskaran M (2010) Durability of compressed earth bricks: assessing erosion resistance using the modified spray testing. Sustainability 2
70. PCDS (2012) Standard for earthen IRC structures, Pima County Development Services (PCDS), Tucson
71. Pavan GS, Ullas SN, Nanjunda Rao KS (2020) Shear behavior of cement stabilized rammed earth assemblages. J Build Eng 27:100966. https://doi.org/10.1016/j.jobe.2019.100966
72. Pullen QM, Scholz TV (2011) Index and engineering properties of Oregon cob. J Green Build 6:88–106. https://doi.org/10.3992/jgb.6.2.88
73. Quagliarini E, Lenci S (2010) The influence of natural stabilizers and natural fibres on the mechanical properties of ancient Roman adobe bricks. J Cult Herit 11:309–314. https://doi.org/10.1016/j.culher.2009.11.012
74. Quagliarini E, Stazi A, Pasqualini E, Fratalocchi E (2010) Cob construction in Italy: some lessons from the past. Sustainability 2:3291–3308. https://doi.org/10.3390/su2103291
75. RLD (2015) New Mexico earthen building materials code." New Mexico Administrative Code, Construction Industries Division of the Regulation and Licensing Department (RLD), New Mexico
76. Raju L, Venkatarama Reddy BV (2018) Influence of layer thickness and plasticizers on the characteristics of cement-stabilized rammed earth. J Mater Civ Eng 30:04018314. https://doi.org/10.1061/(asce)mt.1943-5533.0002539
77. Reddy BVV, Lal R, Rao KSN (2007) Enhancing bond strength and characteristics of soil-cement block masonry. J Mater Civ Eng 19:164–172
78. Rivera Torres JC, Muñoz Díaz EE (2005) Caracterización estructural de materiales de sistemas constructivos en tierra : El adobe / Structural characterization of materials used in construction systems with soil material : The adobe. Rev Int Desastr Nat Accid e Infraestruct Civ 5:135–148
79. Rodríguez-Mariscal JD, Solís M, Cifuentes H (2018) Methodological issues for the mechanical characterization of unfired earth bricks. Constr Build Mater 175:804–814. https://doi.org/10.1016/j.conbuildmat.2018.04.118
80. Rojat F, Hamard E, Fabbri A, Carnus B, McGregor F (2020) Towards an easy decision tool to assess soil suitability for earth building. Constr Build Mater 257. https://doi.org/10.1016/j.conbuildmat.2020.119544
81. Röhlen U, Ziegert C (2013) Construire en terre crue - Construction - Rénovation - Finition. Le Moniteur, Paris
82. SNZ (1998) NZS 4298:1998 Materials and workmanship for earth buildings. Standards New Zealand (SNZ), Wellington
83. San Bartolomé A, Pehovaz R (2005) Comportamiento a carga lateral cíclica de muros de adobe confinados. In: Ponencias del XV Congreso Nacional de Ingeniería Civil, Colegio de Ingenieros del Perú, Ayacucho, Peru (in Spanish), pp 209–214
84. Sarangapani G, Reddy BVV, Jagadish KS (2005) Brick-mortar bond and masonry compressive strength. J Mater Civ Eng 17:229–237. https://doi.org/10.1061/(ASCE)0899-1561(2005)17:2(229)
85. Sauvage M (2009) Les débuts de l'architecture de terre au Proche-Orient. In: Achenza M, Correia M, Guillaud H (eds) Mediterra 2009, 1st Mediterranean conference on earth architecture. EdicomEdizioni, Cagliari (Italy), pp 189–198
86. Saxton R (1995) The performance of cob as a building material. Struct Eng 73:111–115
87. Silva RA, Oliveira D V, Miccoli L, Schueremans L (2014a) Modelling of rammed earth under shear loading. In: Peña F, Chávez M (eds) SAHC2014–9th international conference on structural analysis of historical constructions. Mexico City, Mexico
88. Silva RA, Oliveira D V, Miranda T et al (2013) Rammed earth construction with granitic residual soils: the case study of northern Portugal. Constr Build Mater 47:181–191. https://doi.org/10.1016/j.conbuildmat.2013.05.047

89. Silva RA, Olliveira DV, Schueremans L et al (2014b) Shear behaviour of rammed earth walls repaired by means of grouting. In: 9th international masonry conference. International Masonry Society, Guimaraes, Portugal
90. Silveira D, Varum H, Costa A et al (2012) Mechanical properties of adobe bricks in ancient constructions. Constr Build Mater 28:36–44. https://doi.org/10.1016/j.conbuildmat.2011.08.046
91. Silveira D, Varum H, Costa A (2013) Influence of the testing procedures in the mechanical characterization of adobe bricks. Constr Build Mater 40:719–728. https://doi.org/10.1016/j.conbuildmat.2012.11.058
92. Silveira D, Varum H, Costa A, Carvalho J (2015) Mechanical properties and behavior of traditional adobe wall panels of the Aveiro District. J Mater Civ Eng 27:4014253. https://doi.org/10.1061/(ASCE)MT.1943-5533.0001194
93. Soudani L, Fabbri A, Morel J claude, et al (2015) A coupled hygrothermal model for earthen materials. Energy Build
94. Torrealva D, Acero J (2005) Reinforcing adobe buildings with exterior compatible Mesh: the final solution against the seismic vulnerability? SismoAdobe. Lima, Peru
95. Vinceslas T, Hamard E, Razakamanantsoa A, Bendahmane F (2018) Further development of a laboratory procedure to assess the mechanical performance of cob. Environ Geotech 0:1–8.https://doi.org/10.1680/jenge.17.00056
96. Walker P (2002) The Australian earth building handbook, HB 195–2002. Australia, Sydney
97. Wu F, Li G, Li H-N, Jia JQ (2013) Strength and stress-strain characteristics of traditional adobe block and masonry. Mater Struct 46(9):1449–1457
98. Xu L, Wong KK, Fabbri A et al (2018) Loading-unloading shear behavior of rammed earth upon varying clay content and relative humidity conditions. Soils Found 58:1001–1015. https://doi.org/10.1016/j.sandf.2018.05.005
99. Ziegert C (2008) Lehmwellerbau - Konstruktion, Schäden und Sanierung. Technical University of Berlin, Berichte aus dem Konstruktiven Ingenieurbau
100. Ziegert C (2006) Historical cob buildings in Germany—construction, damage and repairs. In: Patte E, Streiff F (eds) L'architecture en Bauge en Europe. Parc Naturel Régional des marais du Cotentin et du Bessin, Isigny-sur-Mer, pp 233–246

Chapter 5
Seismic Assessment of Earthen Structures

Quoc-Bao Bui, Ranime El-Nabouch, Lorenzo Miccoli, Jean-Claude Morel, Daniel V. Oliveira, Rui A. Silva, Dora Silveira, Humberto Varum, and Florent Vieux-Champagne

Abstract Earth is an ancient construction material that is attracting numerous scientific investigations due to the sustainable characteristics of this material and the significant heritage of existing earth constructions. Several studies have recently been conducted to investigate the seismic performance of earth buildings. This chapter presents the state of the art related to the seismic performance of earth structures. Since the seismic performance of a structure depends both on the dynamic and static characteristics of the material and the structure, the chapter starts with a summary of the static characteristics, followed by the results on the dynamic characteristics. Then the existing methods for the earthquake performance evaluation are presented. Finally, the techniques of seismic strengthening are cited and discussed. This chapter is a useful repertoire for further studies on the seismic design of earthen structures.

Keywords Earth structures · Seismic performance · Rammed earth · Adobes · Compressed earth blocks · Cob

---

Q.-B. Bui (✉)
Faculty of Civil Engineering, Ton Duc Thang University, Ho Chi Minh City, Vietnam
e-mail: buiquocbao@tdtu.edu.vn

R. El-Nabouch
LOCIE, CNRS, University Savoie Mont Blanc, Chambéry, France

L. Miccoli
Bundesanstalt für Materialforschung und –prüfung (BAM), Division Building Materials, Unter den Eichen 87, 12205 Berlin, Germany

J.-C. Morel
Coventry University, Coventry, UK

D. V. Oliveira · R. A. Silva
ISISE & IB-S, University of Minho, Guimarães, Portugal

D. Silveira
ADAI-LAETA, Itecons, Coimbra, Portugal

H. Varum
CONSTRUCT-LESE, Faculty of Engineering, University of Porto, Porto, Portugal

F. Vieux-Champagne
Université Grenoble Alpes, CNRS, Grenoble INP, 3SR, 38000 Grenoble, France

A. Fabbri et al. (eds.), *Testing and Characterisation of Earth-based Building Materials and Elements*, RILEM State-of-the-Art Reports 35,
https://doi.org/10.1007/978-3-030-83297-1_5

5.1 Introduction

Earth is one of the most common materials used in the past for construction. As a local material, earth has had a long and continuous history in many regions throughout the world, with different construction techniques: rammed earth (RE), cob, adobes, compressed earth blocks (CEB).

Rammed earth walls are made by compacting earth between vertical formworks. The earth is compacted into layers (approximately 10–15 cm thick) using a manual or pneumatic rammer. Today, the pneumatic rammer is more currently used. Due to the manufacturing mode, the rammed earth wall is the superposition of different horizontal earthen layers.

Cob is a mixture of earth and plant fibres, thus walls made of cob can be regarded as fibre-reinforced structural elements with a monolithic appearance [112]. It is made from a clay-based soil mixed with water and straw, sometimes with crushed flint or sand added. The manufacturing water content is about 25–30% in weight [131]. The current thickness is about 50–90 cm.

Adobes are unburnt bricks made from clayed soils, at plastic state (about 20–25% of water content, depending on the soil type [78]) and compressed by hand in moulds and then traditionally dried at the sun. CEB is a more recent construction technique—compared to other earth materials—where CEBs are also unburnt bricks; the soil is compressed within a mould by a machine and the water content is closed to that of a Proctor Optimum (about 12%). Earthen walls are constructed from adobes or CEB in a similar manner to a classical masonry wall: the presence of the horizontal and vertical joints by mortar.

With the development of other conventional materials (such as concrete), able to be produced in an industrial manner, earth materials have been less used during the last decades, especially in developed countries. However, in the context of sustainable development and preserving the heritage of earth buildings, earth materials are the subject of numerous scientific researches. Indeed, earth material has low embodied energy [117] and presents an interesting hygro-thermal behaviour [140]. Numerous studies have recently investigated different aspects of this material: durability [23], mechanical characteristics [24, 25, 27, 74, 96], hygro-thermal behaviour and living comfort [150], the new binder for stabilization [30, 99]. The number of studies on the seismic performance of earthen buildings has increased recently but still been moderate when compared to studies on earthquake behaviour of conventional materials. This paper aims to summarize the state of the art of different studies existing in the literature which will be useful for future studies on the seismic performance of earthen constructions.

The seismic performance of a structure depends both on the dynamic and static characteristics of the material and the structure. The structure has a satisfying seismic behaviour when the intrinsic resistances of the structure and the material are higher than the effects caused by the external actions (for example earthquake). The dynamic characteristics (natural frequencies, mode shapes, damping) are the key parameters to determine the seismic actions applied to the structure. For the static characteristics, some of them influence the dynamic characteristics (for example: Young's modulus, Poisson's ratio, density, structure geometry), other determine the performance of the structures under a seismic excitation (for example: compressive strength, tensile strength, shear strength). That is why this chapter starts with a summary about the static characteristics of earthen materials; then the dynamic characteristics will be presented. Finally, different seismic evaluation techniques and the corresponding results will be presented and discussed.

Earthen structures will be divided into two main categories: monolithic walls with high thickness (rammed earth and cob) and earthen masonry (earthen blocks + mortar).

5.2 Earth Materials—A Synthesis of Mechanical Characteristics

5.2.1 Compressive Strength

5.2.1.1 Rammed Earth

The compressive strength of rammed earth material was obtained in the range of 0.5–4 MPa in different studies for stabilised rammed earth [24, 43, 47, 48, 55, 74, 96, 97]. The variation of the values is caused by different parameters such as: the specimen's form (cylinder or parallelepiped) and specimen's dimensions (standard specimen or full-scale specimen), type of earth used, the compaction energy and the testing moisture content and the ambient relative humidity. The most current values of compressive strength for unstabilised rammed earth is about 1.5–2 MPa. When a high amount of hydraulic binders is used ($> 8\%$ in weight), the compressive strength of cement stabilised rammed earth can increase until 6–7 MPa and in some cases until more than 10 MPa [7, 47, 71].

5.2.1.2 Earthen Masonry (Adobe/CEB)

The mechanical characteristics of earthen masonries such as adobe masonry or CEB masonry depends both on the mechanical properties of the units (adobe/CEB) and the mortar.

A high variation of the results was noted in the literature, which is in function of the materials tested: type of soil used, unstabilised or stabilised, especially the amount of hydraulic binder used. The compressive strength of adobe blocks and CEB can vary from 0.3 to 5 MPa [11, 69, 82, 101, 107, 136, 146], 163, [118], [78] but in some exceptional cases, the compressive strength can be over than 10 MPa [8]. However, the most usual values found were in the rang 1–2 MPa.

The compressive strength of mortars used in earthen masonries (adobe/CEB) varies also in function of the type of mortar. The mortar compressive strength can vary from 0.3 to 6 MPa, but the usual values were in the range of 0.5–1.5 MPa [9].

The compressive strength of the earthen masonry walls (assembly of adobe/CEB and mortar) is a function of the strengths of the units (adobe/CEB) and the mortar. It was noted in several studies that the compressive strength of the overall masonry wall could be about 70% of the compressive strength of the units and mortar used to build the walls [135, 147].

5.2.2 *Tensile Strength*

Like other geomaterials, earthen material has a low tensile strength. The tensile strength of rammed earth material was found to be about 5–12% of its compressive strength for unstabilised rammed earth [5, 6, 27] and about 15–20% of the compressive strength for lime stabilised rammed earth [6].

For adobe/CEB units, their tensile strength varies between 0.11 and 0.43 MPa [11, 69, 82, 101, 118, 136, 145, 146, 163]. The usual tensile strength values were about of 0.2 MPa. It was noted that the tensile strength of adobe/CEB units was about 10–20% of the compressive strength for the same type of specimens [147].

5.2.3 *Shear Strength*

The shear strength of the earthen masonry (adobe/CEB) walls was ranged from 0.022 MPa to 0.032 MPa [82, 101, 139, 147, 153, 171]. In addition, the initial shear strength of earthen masonry triplets (dry and pre-wetted) measured by Fontana et al. [63] varies between 0.10 and 0.16 MPa. Moreover, it was observed that the shear strength of earthen masonries was about 8% of their compressive strength [147].

For rammed earth, Miccoli et al. [105] presented the mean shear strength of RE wallets tested under diagonal compression tests which was 0.70 MPa for a rammed earth with compressive strength of 3.73 MPa. Another study by Silva et al. [142] has found a shear strength of 0.18 MPa for a rammed earth with compressive strength of 1.1 MPa. By testing a rammed earth having a compressive strength of 1.90 MPa under standard direct shear tests, El-Nabouch et al. [57] obtained the shear strengths of 0.26 MPa and 0.16 MPa respectively for the upper and middle parts of an interlayer. Indeed, the upper part of a rammed earth layer is denser than the lower parts (middle and bottom of a layer) because the lower parts receive less compaction energy during the manufacturing than the upper parts. So it is suggested that the shear strength of rammed earth material is about 7–16% of its compressive strength, depending on the position within intralayers. In the study of El-Nabouch et al. [57], when a numerical model using homogenised intralayers (with interfaces between intralayers), a shear strength of 8% of the compressive strength could provide satisfying results when compared to experimental results.

5.2.4 *Young's Modulus*

The Young's modulus values for earthen materials has a high dispersion which is related to several factors: the inhomogeneity of the specimens, the type of the soil used, the method of measurement (on the whole height of the specimen or on the central part; by the displacement sensors or by DIC technique [55]), the testing moisture content [27, 74], the method of calculation (tangent modulus or secant modulus). It is assumed that the Young's modulus of earthen material should be determined from the displacements measured at the central part of the specimen and the Young's modulus correspond to the slope determined between 0 and 20% of the maximum stress [27].

For rammed earth material, the moduli obtained varied from 500 to 2000 MPa for the rammed earth material having compressive strength from 0.5 to 3 MPa.

For the units (adobe/CEB) of earthen masonry, it was noted in the literature a variation from 80 to 13 000 MPa, but the usual values were in the range of about 1000–2000 MPa [60, 64, 69, 130, 145]. The elastic modulus of earthen masonry walls is generally lower than that of the units, the usual elastic modulus values for earthen masonry walls were about 100–800 MPa.

5.2.5 *Shear Modulus*

The shear modulus of earthen masonry walls varied but the usual values between 300 and 500 MPa can be noted [147].

For rammed earth materials, Miccoli et al. [105], Silva et al. [142] obtained the shear moduli of rammed earth from 640 and 2300 MPa, respectively for rammed earth materials having compressive strength from 1.1 to 3.7 MPa.

5.2.6 *Poisson's Ratio*

For earthen masonries (adobe/CEB), few results about the Poisson's ratio were presented in the literature. A Poisson's ratio of approximately 0.10 was obtained from the test on one cubic lime stabilised adobe specimen in [145]. Given that this value results from a single test, it was only a first indication of the Poisson's ratio of the adobes. In that same study, the Poisson's ratio obtained for the adobe walls (adobe + mortar) ranged from 0.04 to 0.29, with a mean of 0.16 and coefficient of variation of 63%. A value of 0.37 was provided by Miccoli et al. [107] for unstabilised adobe masonry.

The Poisson's ratio of rammed earth varies from 0.2 to 0.35, in function of the moisture content, from quasi-dry state to humid state—after the manufacturing [27, 43].

5.2.7 Density

The density of the adobes is in the range of 16–18.5 kN/m^3 [107, 146]. For CEB, their density is in the range of 18–18.5 kN/m^3 [24, 78].

The density of rammed earth material varies from 18.5 to 22 kN/m^3 in function of the compaction energy during the manufacturing (manual or pneumatic) [23, 24, 27, 43, 149].

5.2.8 Friction Angle

The friction angle of rammed earth was investigated in El-Nabouch et al. [56] by using direct shear tests. A full-scale shear box (0.5 m width × 0.5 m length × 0.45 m height) was specifically developed to determine these shear parameters at two positions: inside the layers (called "intralayer") and at the interface between the layers. The values of friction angle obtained were of 40° for the interlayers and 37.3° for the interfaces. The interfaces had therefore the friction about of 94% of the intralayer values.

Standard "small" shear boxes were also used to test the specimens extracted from different positions inside an intralayer [57]. The results showed similar values of friction angles for the upper and middle parts of an intralayer.

5.2.9 Influences of Moisture Content

It was shown in several studies that stabilised earth material (by lime or cement) is less sensitive to moisture content than unstabilised earth. The compressive strength at the saturated state of lime stabilised earth adobe and rammed earth decreased about 20% when compared to the quasi-dry state (equilibrium moisture) [27, 98], while for unstabilised specimens, this decrease was of 80%.

The synthesis of the above values is presented in Table 5.1.

Table 5.1 Summary of the current values of static characteristics

Compressive strength, f_c (MPa)	Tensile strength, f_t	Shear strength	Young's modulus	Shear modulus	Poisson's ratio	Friction angle (°)
0.5–7	(10–20%)f_c	(7–15%)f_c	(500–1000)f_c	(300–600)f_c	0.15–0.35	37–40

5.3 Dynamic Characteristics of Earthen Structures

5.3.1 Typical Geometries of Earthen Buildings

Although there are some especial earthen buildings raising until 7 or 10 storeys [165], in the most cases, earthen buildings without particular reinforcement are built for 1 or 2 storeys. Three main types are usually identified:

- One-story buildings with earth walls
- Two-stories buildings with earth walls on both 2 stories
- Two-stories buildings with earth walls only for the ground floor and the 2nd floor in wooden structures. This type is more current for the new buildings where the principles of bio-climatic architecture are taken into account.

The typology of a building directly influences on its dynamic characteristics (natural frequencies, mode-shapes and damping). Indeed, the most important factors are the height and the masses (of the whole building and at each floor). Another important factor is the connection between different elements which determined the dynamic behavior of a building (likely a shear-beam model or a concentrated mass model) and the damping.

5.3.2 Natural Frequencies and Mode Shapes

As mentioned above, dynamic characteristics are the key parameters determining the seismic forces applied on a structure. The seismic force applied on a building depends on several different elements: the building's dynamic characteristics but also seismic zone and the soil where the building situates. The in-situ measurements are necessary to observe the influences of these different factors. However, in our knowledge, few investigations were carried out on this topic. Bui et al. [26] performed in-situ dynamic measurements by using accelerometers on the in-situ rammed earth buildings. Four rammed earth buildings were measured and their dynamic characteristics (natural frequencies, mode shapes, and damping) were identified.

These authors showed that the first mode was predominant and rammed earth houses had mainly a shear behaviour in dynamic. The study also showed that the empirical formula proposed in Eurocode 8 to determine the first natural period T_1 can be applied for rammed earth buildings.

$$T_1 = C_t \cdot h_n^{3/4}$$

where h_n is the building height calculated from natural ground; and C_t is the coefficient depending on the structural type, $C_t = 0.0488$ for RE buildings.

5.3.3 *Damping*

Several dynamic measurements were conducted on in-situ rammed earth walls and houses, by using accelerometers:

- Four houses from one to two storeys [26],
- three 22 years old walls of 0.4 m × 1 m × 1.1 m height [28],
- three new walls during the construction of a house, which had L cross-section and 2.6 m height [24].
- one cement stabilised RE scaled house [120].

The results show that the damping ratio for the measured were of 3.1–3.6% and 2.5–4.0% respectively for rammed earth walls and rammed earth houses. It is worth noting that for rammed earth houses, the damping does not only depend on the walls but also on other structural and non-structural elements. So for simple applications in practice, it is suggested that a damping ratio of 3% can be taken for rammed earth material.

5.4 Analytical and Numerical Modelling

5.4.1 *Simplified Modelling*

The use of simplified models to assess the structural safety of buildings is a useful approach for designers, since in general this type of models allows to perform safety verification in a short period of time. In these models, simplification is in general performed at the level of the structural geometry, material behavior and loading representation. Nevertheless, each type of simplification adopted must be compatible with the expected behavior of the structure. For instance, lumped mass models usually provide a good representation of the dynamic behavior of reinforced concrete framed structures, but may be inaccurate for earthen buildings due to the non-negligible mass distribution along the vertical direction. Thus, simplifications based on invalid assumptions may lead to misleading conclusions.

In masonry structures, simplification can be adequately achieved by using a macro-block approach, where the structure is divided into large rigid blocks, while the joints formed by them represent potential cracks [124]. The definition of the blocks is obtained by basing on observed damages or typical failure mechanisms [79], resulting in models with a reduced number of elements and degrees of freedom. For instance, these models can be analyzed using the limit analysis approach which bases on the static and kinematic theorems [77]. In alternative, the macro-element discretization based on the frame-equivalent idealization of masonry walls with openings was used to develop recent software for simplified analysis of masonry structures, such as SAM II [93] and TREMURI [127], which allow to perform nonlinear static or dynamic analyses.

For the case of earthen structures, the simplified modelling of adobe/CEB masonry can be dealt in the same manner as that of other masonry structures. For instance, adobe arches and vaults can be analyzed by means simplified models using limit analysis [114], while the in-plane and out-of-plane behavior of adobe walls can be analyzed by means macro-block modeling with limit analysis [148]. Monolithic earthen structures, such as rammed earth and cob can also apply the same simplified modelling approaches used for masonry, nevertheless their field of application needs further validation. For instance, Ciancio and Augarde [46] proposed the use of simplified models based on limit analysis to evaluate the capacity of rammed earth walls subjected to lateral wind forces. Another application of simplified modelling is reported in Maniatidis and Walker [96], where the capacity of RE columns loaded eccentrically is evaluated by using a basic pin-pin model. Bui et al. [26] also used a simplified model to analyze the dynamic behavior of rammed earth buildings, where the dynamic behavior was evaluated by means of shear beam models.

5.4.2 Finite Element Modelling

The finite element method (FEM) allows to represent accurately the geometry of structures, making it specially indicated for the analysis of complex structures. The FEM modelling of masonry structures can be carried out following either a micro-modelling or a macro-modelling strategy. In the former case, the model allows to adopt differentiated mechanical behaviors for units, mortar and interfaces [90], meaning that this strategy is expected to lead to an accurate simulation of the structural behavior, namely at the local level. Nevertheless, this level of discretization demands high refined meshes, whose implementation is time demanding, and the computation requires high computational time and effort. Furthermore, the constitutive laws of the materials and interfaces need to be fed with complex parameters (e.g. [87]), whose estimation may bring high uncertainty to the analyses. All these aspects make the micro-modelling strategy unfeasible for large structures, meaning that a macro-modelling strategy should be used instead [89]. In this case, masonry is assumed and modelled as a continuous and homogeneous material, whose constitutive laws can be derived from homogenization techniques or obtained directly from tests on masonry specimens [91]. The macro-modelling of a structure can also include other idealizations to reduce the time demand, namely by considering a two-dimensional model instead of tri-dimensional and by substituting structural elements of the structure by simpler FEM elements. For instance, truss, beam, panel, plate or shell elements can be used to represent columns, piers, arches and vaults, respectively [89].

Despite the generalized use of FEM modelling for masonry structures, this approach is not yet sufficiently established for the assessment of earthen structures. References on the numerical modelling of cob constructions are limited to two studies simulating the static behaviour of cob walls [112, 129]. The numerical studies performed on adobe structures deal in their majority with the numerical simulation of tests performed on specimens with different scales, namely of single units [22], adobe masonry wallets or walls [2, 38, 95, 108, 134, 151] and adobe masonry scaled buildings [73]. The constitutive laws adopted in these studies assume the non-linear behavior of the adobe masonry based on smeared crack models, typically used for other types of masonry. Here, the macro-modelling is the main modelling strategy used, whereas the micro-modelling strategy, even that in its simplified approach, is rarely used (e.g. [134]). Furthermore, the FEM modelling of CEB masonry is almost absent from the bibliography [116].

The numerical studies on FEM modelling of rammed earth also address the modeling of wallets and walls tested in laboratory [27, 33, 35, 76, 85, 105, 109, 121, 141]. In these cases, the selection of the material constitutive laws follows two main approaches, namely based on plasticity models and smeared crack models. In terms of modelling strategies, macro-modelling is in general preferred assuming a homogenous isotropic behavior of the material, while the micro-modelling strategy is used to evaluate the influence of the interfaces between layers on the structural behavior (e.g. [109]). FEM modelling has also been used to evaluate the structural behavior of rammed earth buildings [13, 70, 75, 125].

5.4.3 Discrete Element Modelling

The discrete element method (DEM) was originally developed to model jointed and fractured rock masses as a set of rigid blocks [51], which are prone to experience large displacements [10]. Nowadays, the use of DEM is more widespread, as it is being increasingly adopted in the numerical modeling of systems composed of multiple bodies, blocks or particles. DEM modeling of masonry structures is one of these recent applications and it follows a discontinuous approach, where the material is viewed as an assembly of distinct bodies, interacting along their boundaries. The original DEM formulations assume these bodies as presenting rigid behavior, but the latest advances allow to assume deformable behavior by, for instance, allowing the incorporation of internal FEM meshes [81]. Nevertheless, DEM modelling presents as main features the capacity of the model in simulating the interaction between bodies as loading progresses (e.g. contact, sliding and separation), and the capacity of simulating large displacements between blocks. These features allow to obtain a clear picture of the collapse mechanisms, which can constitute useful information for comprehend damage on existing structures, for applying simplified analysis methods and for interpreting experimental results. On the other hand, DEM modelling presents as disadvantage the general requirement for high computational effort due to the use of high number of bodies to represent the structure.

DEM modelling has also been used to simulate experimental tests on specimens representing earthen materials. Bui et al. [37] used DEM to simulate the interaction between layers of tested rammed earth wallets and walls, where the rammed earth layers were assumed as presenting deformable behaviour. It should be noted that this study is the first done on the subject, nevertheless DEM modelling did not show a relevant advantage towards the use of FEM for modelling rammed earth. In the case of earthen masonry structures, DEM modelling has been shown to be more relevant, as stated by some studies on the modelling of the structural behaviour of adobe masonry buildings [12, 65, 173]. In these cases, the use of DEM allowed a broad comprehension of the failure mechanisms due to seismic action, namely it provided an observation of the failure process evolution.

5.5 Seismic Assessment

5.5.1 Linear Static Analysis

Linear static analysis is among the methods included in Eurocode 8 [42] for seismic assessment of structures. This method presents the highest degree of simplification, since the material is considered to present linear-elastic behavior [123] and the seismic action is simulated by means of static horizontal forces proportional to the first mode shape of the structure or linearly increasing with the height. This method is not suitable for structures constituted by materials exhibiting low tensile strength, since cracks are generated at very low stress levels, thus inelastic deformations are not adequately estimated by linear static analyses. On the other hand, this method only requires the input of simple material parameters and is not demanding in terms of computational effort.

In the case of earthen structures, its mechanical behavior is highly non-linear due to the very low tensile strength of the material, meaning that a seismic assessment of these structures based on linear static analysis is unlikely to be adequate. Nevertheless, this type of analysis can be useful in an early stage of the assessment, for instance, to validate a model to be analyzed using more advanced analysis methods.

5.5.2 Non-linear Static Analysis

The seismic assessment of structures through non-linear static analyses, also known as pushover analyses, is also preconized in Eurocode 8 [42]. Although the seismic action is simulated by means of increasing static horizontal forces, the method takes into account the non-linear behavior, which can be addressed to physical (material non-linear behavior), geometrical and contact non-linearities (e.g. [103]). A non-linear static analysis allows to plot the capacity curve of the structure, which is roughly assumed as the envelope of the responses coming from dynamic analyses. Nevertheless, the reliability of this assumption strongly depends on the correct choice of the loading pattern and on the incremental control criterion of the analysis [66]. For the loading pattern, two main distributions are typically adopted, namely modal and mass proportional. In the first case, the load distribution is proportional to the first modal shape of the structure in order represent the structural dynamic amplification [45]. In the second case, the seismic load is directly proportional to the translational masses of the structure [88]. The aforementioned types of distribution are typically assumed as bounds for seismic analyses of regular buildings, where the results from dynamic analyses are usually assumed to be within these two solutions. For irregular buildings, where the contribution of higher modes may not be negligible, the two types of loading patterns abovementioned may lead to over- or under-prediction of the actual seismic response. To overcome these difficulties, multi-modal and adaptive pushover analyses have been proposed (e.g. [39, 45]), where the loading pattern is updated in order to account for the contribution of higher modes and variations in the modal shapes of the structure, resulting from its inelastic behavior.

As for earthen structures, pushover analysis has not yet been conveniently explored, as it requires a deep characterization of the material non-linear behavior and advanced knowledge from the users. Nevertheless, some works on this topic are worthy to be reported. Bui et al. [29] used pushover analyses with DEM to assess the shear capacity of RE walls, where the horizontal load was applied at the top of the walls. Zanotti [173] used mass-proportional pushover analysis of two-dimensional DEM models to evaluate the out-of-plane seismic capacity of an adobe church. The seismic performance of rammed earth buildings was also evaluated by means of mass proportional pushover analyses of FEM models [4, 13, 83, 125].

El-Nabouch et al. [55] performed experimental in-plane horizontal loading tests on rammed earth walls and used the theory of the pushover method to assess the seismic performance of rammed earth buildings at different seismicity zones. The walls of dimensions 1.5 m-width × 1.5 m-height × 0.25 m-thickness and 1.5 m-width × 1.0 m-height × 0.25 m-thickness were subjected to horizontal loading tests to obtain the nonlinear shear force–displacement curves. The performance points were determined by transposing these shear force–displacement curves to an acceleration–displacement system and using the standard spectra presented in Eurocode 8 [42]. The assessment showed satisfying performance on seismicity zones ranging from "very low" to "medium" with type-A soil (very good soil). For type-B soil, acceptable results were only found for seismicity zones from "very low" to "moderate". As for the failure patterns of walls, no brittle behaviour was observed; horizontal cracks between the interfaces started to appear from the horizontal load corresponding to 85% of the maximum horizontal load.

5.5.3 Response-Spectrum Modal Analysis

Response-spectrum modal analysis is a method also preconized in Eurocode 8 [42], where the behaviour of the structure is assumed as linear-elastic and the earthquake load is represented by means of a response spectrum. The contribution of all relevant modes for the global response of the structure can be considered. Gomes et al. [70] performed response-spectrum modal analyses to evaluate the seismic performance of FEM models of a new rammed earth house, where different structural solutions were tested under the local seismic hazard.

The major drawback of the response spectrum analysis is that it does not take into account the non-linear behavior on the seismic response of the structure. As mode shapes and natural frequencies depend on the stiffness of the structure [92], which decreases with damage progression, this method is also unlikely to be adequate for seismic assessment of earthen structures, given their dominant non-linear behavior.

5.5.4 Non-linear Dynamic Analysis

Also preconized in Eurocode 8 [42], non-linear dynamic analysis is the most comprehensive method for seismic assessment of structures, as it considers both the non-linear behavior and the time varying nature of earthquakes. Here, the seismic action is simulated by natural, synthetic or artificial accelerograms, which requires a very high computational effort and advanced expertize from the user. Furthermore, incremental dynamic analysis, based on the consideration of a series of non-linear time-history analyses with increasing intensity, can also be performed [155].

Also in the case of earthen structures, non-linear dynamic analysis is considered as the best analytical seismic assessment method, since it is able to take into account the expressive non-linear behavior of these structures and its severe influence on the dynamic response. Despite the time demand limitations of non-linear dynamic analysis, some works have been already carried out on this topic. Bakeer and Jäger [12] and Furukawa and Ohta [65] performed non-linear dynamic analyses of DEM models of adobe buildings. Librici [83] performed time-history dynamic analyses of a FEM model representing a typical RE dwelling from Southern Portugal, where the results are compared with mass-proportional pushover analyses. Bui et al. [31] used the dynamic analysis with discrete element method to investigate the in-plane seismic performance of rammed earth walls. In those cases, non-linear dynamic analysis enabled to assess the dynamic behaviour of earthen structures under dynamic excitations and compared with the results obtained by using the non-linear static method (pushover).

5.5.5 Experimental Tests with Shaking Table

Although the number of studies using shaking tables to test earthen structure is still limited, some studies can be cited here. Ruiz et al. [138] used a shaking table to test the performance of scaled models strengthened with boundary wooden elements. Nanjunda Rao et al. [120] performed "shocking table tests" on a cement stabilised rammed earth house model to investigate its dynamic behaviour. From the dynamic responses, the damping values of the cement stabilised RE house model were obtained. However, the amplitude of the shock seemed to be much higher than that of the usual earthquakes, so it was difficult to evaluate the seismic performance of the model tested.

Wang et al. [168] performed the tests on two full-scale RE single-floor one-room models with dimensions of 2.4 m-width × 2.6 m-length × 2.1 m-height and the thickness of rammed earth walls were of 0.4 m dwellings. One model was without reinforcement and the other was reinforced by horizontally bonding double layers of tarpaulin strips around the outside and inside surfaces of the walls. The results showed that the seismic performance of the reinforced model was improved with an increase in the load-bearing capability from 0.4 to over 0.95 g, and in the structural stiffness (improved by a factor of 2.33). These results confirmed the effectiveness of the reinforcement technique proposed.

Rafi and Lodi [133] performed shaking table tests of shear wall panels made of cob retrofitted out with vertical bamboo sticks and horizontal layers of plastic coated steel wire mesh. The retrofit design proposed was effective in improving the in-plane response of the cob shear wall and can be used for both the existing and new cob structures.

5.6 Seismic Strengthening

5.6.1 Adobes Structures

5.6.1.1 Typical Seismic Damages on Adobe Structures

Adobe masonry is characterized by low tensile and shear strengths and brittle behaviour [19]. Thus, adobe constructions may perform poorly when under seismic loading. In fact, there are many examples of recent earthquakes in which adobe buildings were severely affected (e.g. [17, 59, 68, 94]). The 2010 Chile earthquake and the following tsunami, for example, damaged nearly 370,000 buildings, of which about 37% were adobe buildings [59]. In the 2015 Nepal earthquake, about 95% of the damage occurred in unreinforced masonry and adobe constructions [68].

The typical seismic damages in adobe constructions include: (a) diagonal in-plane cracks; (b) cracks at openings; (c) vertical cracks at wall intersections; (d) separation of walls in the corners; (e) disintegration of the upper portion of the walls; (f) out-of-plane damage or collapse of gable-end and other walls; (g) separation between roof and walls; (h) flexural damage at mid-height of the walls [170]. The predominant failure mode consists in the out-of-plane movement of the walls after the formation of vertical cracks at the intersection of walls [102]. After the collapse of the walls, the roof may lose support and collapse as well.

5.6.1.2 Main Strategies of Effective Seismic Strengthening for Adobe Structures

The main aim of effective seismic strengthening for adobe structures should be to prolong the stability of buildings during earthquakes [102], namely by preventing the out-of-plane overturning of walls. The objective is not to avoid cracking but to prevent the widening of cracks, i.e. to limit the relative displacement of the elements separated by cracks and thus avoid collapse and consequent injuries and loss of lives [169]. Hence, it is necessary to ensure a proper connection between structural elements and to provide deformation capacity without reaching collapse.

Many different solutions have been used successfully to strengthen adobe constructions. One of the simplest and most essential strengthening solutions is a ring beam placed at the top of the walls, tying them together and supporting the roof (Fig. 5.1) [19]. This beam must be continuous and well connected to the walls and roof. This solution, however, should be complemented with other strategies. Many of the strengthening strategies tested and used successfully consist of a system of vertical and horizontal ductile tension-resistant elements, applied internally or externally to the walls (Fig. 5.2, [21, 44, 52, 62, 113, 152]. These strengthening elements are usually connected together and must also be adequately connected to the other structural elements (foundation, floor and roof structures), to improve structural integrity [19]. The relative displacement between structural elements can also be limited using local ties [169]. The use of horizontal transversal bracing elements at the level of the roof and floor structures, adequately connected to the walls, also

Fig. 5.1 Horizontal ring beam [19]

Fig. 5.2 Adobe walls Reinforced with polymer mesh [19]

helps to prevent the out-of-plane collapse of walls [101, 102]. Another complementary strategy consists in the use of pilasters and buttresses in critical points of the structure (e.g. at corners and at intermediate points in long walls), contributing in this way to prevent the overturning of walls [19].

5.6.1.3 Research on the Development of Seismic Strengthening Solutions for Adobe Structures

Important research on effective strengthening solutions for adobe construction has been conducted worldwide. Some examples of work developed by different institutions are described in the following paragraphs.

At the Pontifical Catholic University of Peru (PUCP), research on this topic has been carried out in the last 45 years [20]. In the 1970s, static tests using a tilt table were conducted on adobe specimens, and different strengthening materials (e.g. wood, cane, and wire) were tested [49]. From the 1980s until the present, various strengthening solutions have been assessed, resorting to full-scale shaking table tests [20]. At the beginning of the twenty-first century, the research focused on creating strengthening solutions with industrial materials, more economical and easily available. Cyclic tests were conducted on full-scale walls and seismic tests on full-scale house models and vaulted elements [16, 18, 154]. In these tests, the solution using polymer mesh applied to the surfaces of walls and covered with earth mortar was particularly successful. Recently, PUCP developed another effective strengthening solution made with a mesh of synthetic ropes [21].

The Getty Conservation Institute (GCI) has also given great attention to the seismic retrofit of adobe construction. In the 1990s, the GCI created the Getty Seismic Adobe Project, which aimed to study the seismic behaviour of historic adobe buildings and develop low-impact retrofit solutions. Shaking table tests of reduced-scale models of adobe walls and buildings were conducted, and seismic strengthening solutions, such as vertical and horizontal straps, vertical centre-core rods, and bond beams, were tested with success [152]. More recently, the GCI, with the collaboration of other institutions, launched the Seismic Retrofitting Project, aiming to develop retrofit solutions that can be easily implemented, using locally available materials and know-how [40].

At the Autonomous University of Mexico State, the use of synthetic meshes (geogrid) to strengthen adobe constructions was also studied [159]. This solution proved successful in in-plane cyclic tests on full-scale adobe walls. At the National Autonomous University of Mexico, rural house models, strengthened with different solutions, were subjected to shaking table tests [101]. The most effective solution was a steel mesh used in both surfaces of the walls.

At the University of the Andes, in Colombia, steel mesh and wood reinforcement solutions were tested in in-plane cyclic tests, out-of-plane monotonic tests, and shaking table tests [172]. The wood confining elements showed better seismic performance than the steel mesh solution.

Research for the development of low-cost, low-tech strengthening strategies for adobe construction has also been conducted at the University of Technology, in Australia [52]. A system made with stiff external vertical reinforcement (e.g. bamboo), external horizontal reinforcement (e.g. bamboo or wire), and a wooden ring beam led to a significant enhancement in the seismic behaviour of adobe specimens tested on the shaking table.

At the University of Aveiro, in Portugal, the structural response of full-scale adobe elements, with and without strengthening, has also been investigated in full-scale horizontal cyclic tests [62, 156]. A strengthening solution made with a polymer mesh applied to the surfaces of walls and embedded in earth mortar proved very effective.

Researchers from the Victoria University of Wellington and PUCP also carried out work to explore alternative low-cost strengthening techniques. An external strengthening solution with straps cut from used car tires was used to strengthen an adobe model that was tested in the shaking table [44]. The results obtained showed a significant improvement in the behaviour of the model.

It can thus be concluded from the results of the research conducted worldwide that the reinforcement solutions developed can significantly improve the seismic behaviour of adobe constructions by ensuring better structural continuity and confinement.

5.6.2 Rammed Earth Structures

The number of experimental campaigns on strengthening systems for rammed earth is still limited when compared with the number of studies carried out on adobe masonry [15]. Liu et al. [84] tested the effectiveness of a strengthening system using externally bonded fibres to increase the energy dissipation of rammed earth walls. The findings confirmed that the reinforced walls shown an increase of ductility and horizontal-load capacity. Bernat-Maso et al. [14] tested the behaviour of RE samples reinforced with textile grids. The use of flexible textile grids with large spacing between fibre tows increased the flexural strength and the flexural toughness of the prismatic samples tested.

Wang et al. [167] performed shaking table tests to validate a strengthening system based on externally bonded fibres. Shaking table tests on lab scale models strengthened with boundary wooden elements were performed by Ruiz et al. [138].

A few more studies are available on stabilised rammed earth walls strengthened with bamboo canes [67], steel rods [166] and post tensioned steel bars [32, 72] tested under in-plane cyclic loading (Fig. 5.3). Miccoli et al. [110, 111] tested in-plane cyclic loading unstabilised rammed walls strengthened with polyester fabric strips. Although the walls showed a limited ductile behaviour, the strengthening limits the diagonal cracks spread providing an increase of horizontal load and displacement capacity.

RE arches reinforced made with jute fabric and earthen matrix proposed by Fagone et al. [61] showed a significant increase of bearing capacity and kinematical ductility.

Metallic anchors are still commonly applied to improve the wall-to-wall and wall-to-floor connections and to enhance the seismic response of buildings [34, 164]. Gomes et al. [70] simulated the effectiveness of steel cables to restore the box-like behaviour of a rammed earth structure. Pull-out tests on rammed earth walls were carried out [106] to test the behaviour grouted anchors constituted by stainless steel rods (with or without nuts) and lime-based grout [119]. Another approach to repair

Fig. 5.3 Reinforcement by vertical rods [32]

cracks caused by seismic events is based on injecting earthen [142, 143] or lime-based grout [119] on the damaged walls. Results show that injections contribute to recovering a satisfactory amount of shear strength, even if the recovery of initial shear stiffness is less effective.

5.6.3 *Timber Frame Structure with Infill and Timber Laced Masonry*

Timber frame structures with infill (TFSI) earth based material (wattle and daub, adobe masonry, stone masonry with an earth mortar) are traditional building found worldwide (Fig. 5.4) [160]. Since recent years, several studies bring scientific knowledge relative to the structural behaviour (mainly seismic performance) of these structures [1, 3, 41, 100, 128, 132, 161].

Hybrid structures that can be nearest from the half-timbered masonry structures, such as Dhajji-Dewari [3], Taq systems [80] or corresponding more to reinforced masonry structures (Borbone system) [137], timber laced masonry [162] can be mentioned.

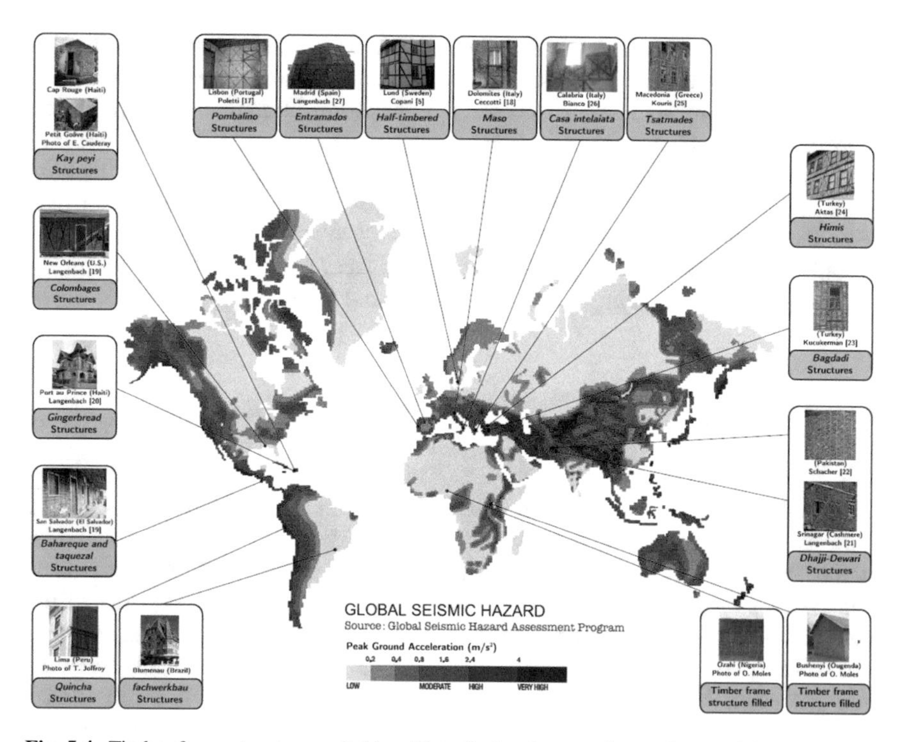

Fig. 5.4 Timber frame structures suitable with to the local constraints and potential

In the case of the wood frame building with infill, the bearing structure is very often the timber frame. Its behaviour is mainly governed by the connections between the wood elements (tenon-mortise, metal fasteners, etc.); nevertheless the infill plays a significant role that is more or less important depending of the type of the building typology: wood structure framework and infill composition [1, 3, 128, 132, 137, 160]. The infill can enhance the stiffness, ductility and energy dissipation of the walls, influence the vertical and the lateral resistance of the wall due to the confinement effect; prevent from large deformations in the connections and therefore allows the wood structure to behave as a lattice i.e. correctly transmit loads to the ground; modify the failure modes. For these reasons, the infill can be assimilated as a reinforcement of the wood structure.

Regarding timber laced masonry where on contrary the wood can be assimilated as a reinforcement of the masonry, Vintzileou et al. [162] showed that timber reinforcement provides confinement to rubble stone masonry leading to a moderate enhancement of its compressive strength and to a significant enhancement of the vertical deformation. The timber mays also allow the masonry wall to sustain higher shear load and undergo large shear cracks without disintegration.

TFSI are more often traditional ancient and less often new buildings (e.g. post-earthquake reconstruction in Haiti or Nepal). Therefore reinforcement and retrofitting are not distinguished below. Several types of reinforcement can be identify (Fig. 5.5):

- wood and/or connection reinforcement: since the timber frame is the bearing structure, the retrofitting solutions aim to reinforce the timber at the critical points (connections) by applying: sheets of Fibre Reinforced Polymer—FRP (Fig. 5.5a [50], Fig. 5.5b [158], Fig. 5.5c [53]), steel plate (Fig. 5.5d [126]); bars (steel or glass FRP, Fig. 5.5e [50, 126]), bolted (Fig. 5.5f [157]); repairing wood elements and/or connections: new nails, removed damaged wood part and glued a new one, replace entire wood piece [157, 161].
- infill: build a fibre reinforced earth mortar (e.g. sisal fibre [160, 161]); retrofit the infill by mean of applying a cement-based mortar [157].

5.7 Conclusions

This chapter presents the state of the art of the existing studies in the literature on the seismic performance of earth structures. Earthen structures were divided into two main categories: earthen masonry including earthen blocks (adobe/CEB/extruded) + mortar, and monolithic earthen walls (rammed earth/cob). Since the seismic performance of a structure depends both on the dynamic and static characteristics of the material and the structure, the paper starts first with a summary of the static characteristics. Then the studies on the dynamic characteristics of earthen structures were presented (natural frequencies, damping ratio) and the relevant values or relationship were recommended. The different existing methods for earthquake performance evaluation were presented, from simple to complex approaches. Analyses about the robustness of each method and its relevancy (or not) for earthen structure

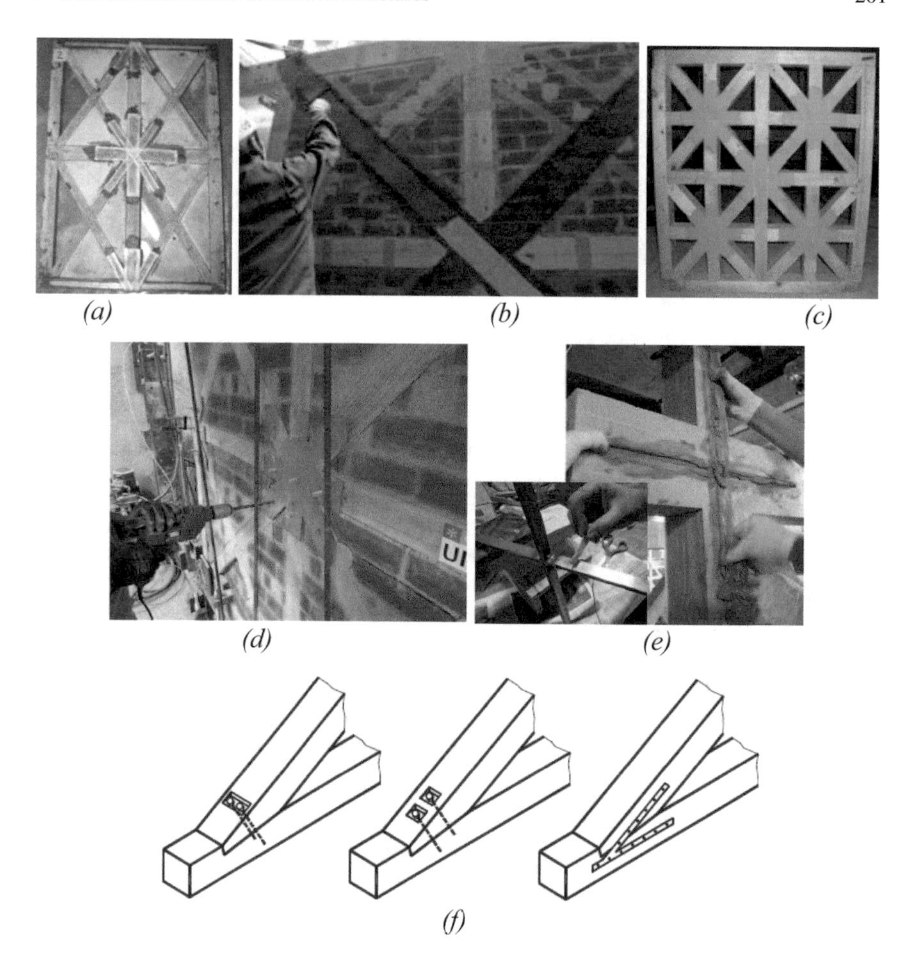

Fig. 5.5 Retrofitting technics: **a** glass FRP sheets and bars; **b** aramid FRP; **c** glass FRP; **d** custom sheet plate; **e** steel plate bar; **f** steel rods and plate

were discussed. Finally, the existing techniques for seismic strengthening of earthen structures were cited which is a useful repertoire for further studies on the seismic reinforcements.

Acknowledgements This chapter is prepared in the case of the RILEM Technical Committee TC 279-TCE (Testing and characterisation of earth-based building materials and elements).

The authors wish to thank the French National Research Agency (ANR) for funding the PRIMATERRE project (ANR-12-Villes et Bâtiments Durables). Several results presented in this paper are the outcome of this project.

The funding from the ReBuMAT project (German-Vietnamese Collaborative Project on Resource-efficient Construction using Sustainable Building Materials) is acknowledged.

Part of this work was also carried out within the framework of project POCI-01-0145-FEDER-016737 (PTDC/ECM-EST/2777/2014), financed by FEDER funds through the Competitivity Factors Operational Programme—COMPETE and by FCT—Foundation for Science and Technology. The funding provided is kindly acknowledged.

This work was partly financed by FEDER funds through the Competitivity Factors Operational Programme—COMPETE and by national funds through FCT—Foundation for Science and Technology within the scope of project SafEarth (PTDC/ECM-EST/2777/2014).

References

1. Aktaş YD, Akyüz U, Türer A et al (2014) Seismic resistance evaluation of traditional Ottoman timber-frame Hımış houses: frame loadings and material tests. Earthq Spectra 30(4):1711–1732
2. Al Aqtash U, Bandini P, Cooper SL (2017) Numerical approach to model the effect of moisture in adobe masonry walls subjected to in-plane loading. Int J Archit Herit 11(6):805–815
3. Ali Q, Schacher T, Ashraf M et al (2012) In-plane behavior of the Dhajji-Dewari structural system (wooden braced frame with masonry infill). Earthq Spectra 28(3):835–858
4. Allahvirdizadeh R, Oliveira DV, Silva RA (2019) Numerical modeling of the seismic out-of-plane response of a plain and TRM-strengthened rammed earth subassembly. Eng Struct 193:43–56
5. Araki H, Koseki J, Sato T (2011) Mechanical Properties of geomaterials used for constructing earthen walls in Japan. Bull ERS Inst Ind Sci Univ Tokyo 44:101–111
6. Araki H, Koseki J, Sato T (2016) Tensile strength of compacted rammed earth materials. Soils Found 56(2):189–204
7. Arrigoni A, Beckett C, Ciancio D, Dotelli G (2017) Life cycle analysis of environmental impact vs. durability of stabilised rammed earth. Constr Build Mater 142:128–136
8. Aubert JE, Fabbri A, Morel JC, Maillard P (2013) An earth block with a compressive strength higher than 45 MPa! Constr Build Mater 47:366–369
9. Azeredo G, Morel JC, Barbosa NP (2007) Compressive strength of earth mortars. J Urban Environ Eng 1(1):26–35
10. Azevedo J, Sincraian G (2000) Modelling the seismic behaviour of monumental masonry structures. In: Proceedings of the Archii, vol 59, p 60
11. Baglioni E, Fratini F, Rovero L (2010) The materials utilised in the earthen buildings sited in the Drâa Valley (Morocco): mineralogical and mechanical characteristics. In: Proceedings of 6th seminar of earthen architecture in Portugal and 9th Ibero-American seminar on earthen architecture and construction, Center for Archaeological Studies at the Universities of Coimbra and Porto, Coimbra, Portugal
12. Bakeer T, Jäger W (2007) Collapse analysis of reinforced and unreinforced adobe masonry structures under earthquake actions–case study: Bam Citadel. In: 10th international conference, structural studies repairs and maintenance of heritage architecture. WIT transactions on the built environment, vol 95, pp 577–586
13. Barros RS, Costa A, Varum H, Rodrigues H, Lourenço PB, Vasconcelos G (2015) Seismic behaviour analysis and retrofitting of a row building. Seismic retrofitting: learning from vernacular architecture, pp 213–218
14. Bernat-Maso E, Gil L, Escrig C (2016) Textile-reinforced rammed earth: Experimental characterisation of flexural strength and thoughness. Constr Build Mater 106:470–479
15. Bhattacharya S, Nayak S, Dutta S (2014) A critical review of retrofitting methods for unreinforced masonry structures. Int J Disaster Risk Reduct 7:51–67
16. Blondet M, Torrealva D, Villa García G, Ginocchio F, Madueño I (2005) Using industrial materials for the construction of safe adobe houses in seismic areas. In: Proceedings of

EarthBuild2005: international earth building conference, Faculty of Design, Architecture and Building, University of Technology Sydney, Australia
17. Blondet M (2008) Behavior of earthen buildings during the Pisco Earthquake of August 15, 2007. Earthquake Engineering Research Institute, Oakland
18. Blondet M, Torrealva D, Vargas J, Velasquez J, Tarque N (2006) Seismic reinforcement of adobe houses using external polymer mesh. In: First European conference on earthquake engineering and seismology, Swiss Society for Earthquake Engineering and Structural Dynamics, Geneva, Switzerland
19. Blondet M, Villa Garcia G, Brzev S, Rubiños A (2011) Earthquake-resistant construction of adobe buildings: a tutorial, 2nd edn. Earthquake Engineering Research Institute, Oakland
20. Blondet M, Vargas J, Tarque N, Iwaki C (2011) Construcción sismorresistente en tierra: la gran experiencia contemporánea de la Pontificia Universidad Católica del Perú. Inf Constr 63(523):41–50
21. Blondet M, Vargas J, Tarque N, Soto J, Sosa C, Sarmiento J (2015) Refuerzo sísmico de mallas de sogas sintéticas para construcciones de adobe. In: Proceedings of 15th SIACOT: Ibero-American seminar on earthen architecture and construction, Faculty of Architecture and Urbanism of the University of Cuenca, Cuenca, Ecuador
22. Bove A, Misseri G, Rovero L, Tonietti U (2016) Experimental and numerical analyses on the antiseismic effectiveness of fiber textile for earthen buildings. J Mater Environ Sci 7(10):3548–3557
23. Bui QB, Morel JC, Reddy BVV, Ghayad W (2009) Durability of rammed earth walls exposed for 20 years to natural weathering. Build Environ 44(5):912–919
24. Bui QB, Morel JC, Hans S, Meunier N (2009) Compression behaviour of nonindustrial materials in civil engineering by three scale experiments: the case of rammed earth. Mater Struct 42(8):1101–1116
25. Bui QB, Morel JC (2009) Assessing the anisotropy of rammed earth. Constr Build Mater 23(9):3005–3011
26. Bui QB, Hans S, Morel JC, Do AP (2011) First exploratory study on dynamic characteristics of rammed earth buildings. Eng Struct 33(12):3690–3695
27. Bui QB, Morel JC, Hans S, Walker P (2014) Effect of moisture content on the mechanical characteristics of rammed earth. Constr Build Mater 54:163–169
28. Bui QB, Morel JC (2014) First exploratory study on the ageing of rammed earth material. Materials 8(1):1–15
29. Bui QB, Bui TT, Limam A (2016) Assessing the seismic performance of rammed earth walls by using discrete elements. Cogent Eng 3(1):1200835
30. Bui Q-B, Prud'homme E, Grillet A-C, Prime N (2017), An experimental study on earthen materials stabilized by geopolymer. Lecture notes in civil engineering. Springer, pp 319–328
31. Bui QB, Limam A, Bui TT (2018) Dynamic discrete element modelling for seismic assessment of rammed earth walls. Eng Struct 175:690–769
32. Bui QB, Bui TT, El-Nabouch R, Thai DK (2019) Vertical rods as a seismic reinforcement technique for rammed earth walls: an assessment. Adv Civ Eng Article ID 1285937:12 p
33. Bui QB, Bui TT, Tran MP, Bui TL, Le HA (2019) Assessing the seismic behavior of rammed earth walls with an L-form cross-section. Sustainability 11:1296
34. Bui QB, Bui TT, Jaffré M, Teytu L (2020) Steel nail embedded in rammed earth wall to support vertical loads: an investigation. Constr Build Mater 234:117836
35. Bui TL, Bui TT, Bui QB, Nguyen XH, Limam A (2020) Out-of-plane behavior of rammed earth walls under seismic loading: finite element simulation. Structures 24:191–208
36. Bui TT, Bui QB, Limam A, Maximilien S (2014) Failure of rammed earth walls: from observations to quantifications. Constr Build Mater 51:295–302
37. Bui TT, Bui QB, Limam A, Morel JC (2015) Modeling rammed earth wall using discrete element method. Continuum Mech Thermodyn 28(1–2):523–538
38. Calderón WR, Muñoz MR (2009) Calibración de modelos de elementos finitos de muros de adobe por optimización. Ingeniería e Investigación 29(2):10–19

39. Casarotti C, Pinho R (2007) An adaptive capacity spectrum method for assessment of bridges subjected to earthquake action. Bull Earthq Eng 5(3):377–390
40. Cancino C, Macdonald S, Lardinois S, D'Ayala D, Fonseca C, Torrealva D, Vicente E (2012) The Seismic Retrofitting Project: methodology for seismic retrofitting of historic earthen sites after the 2007 earthquake. In: Proceedings of Terra 2012: 11th international conference on the study and conservation of earthen architecture heritage, Pontifical Catholic University of Peru, Lima, Peru
41. Ceccotti A, Faccio P, Nart M, Sandhaas C, Simeone P (2006) Seismic behaviour of historic timber-frame buildings in the Italian Dolomites. In: International wood committee-15th international symposium, Istanbul and Rize, Turkey
42. CEN (2005) Eurocode 8—Design of structures for earthquake resistance. European Committee for Standardization, Brussels
43. Champiré F, Fabbri A, Morel JC et al (2016) Impact of relative humidity o the mechanical behavior of compacted earth as a building material. Constr Build Mater 110:70–78
44. Charleson A, Blondet M (2012) Seismic reinforcement for adobe houses with straps from used car tires. Earthq Spectra 28(2):511–530
45. Chopra A, Goel R (2002) A modal pushover analysis procedure for estimating seismic demands for buildings. Earthq Eng Struct Dynam 31(3):561–582
46. Ciancio D, Augarde C (2013) Capacity of unreinforced rammed earth walls subject to lateral wind force: elastic analysis versus ultimate strength analysis. Mater Struct 46(9):1569–1585
47. Ciancio D, Gibbings J (2012) Experimental investigation on the compressive strength of cored and molded cement-stabilized rammed earth specimens. Constr Build Mater 28(1):294–304
48. Ciancio D, Jaquin P (2011) An overview of some current recommendations on the suitability of soils for rammed earth. In: Proceedings of international workshop on rammed earth materials and sustainable structures & Hakka Tulou forum 2011: structures of sustainability, pp 28–31
49. Corazao M, Blondet M (1973) Estudio experimental del comportameniento estructural de las construcciones de adobe frente a solicitaciones sísmicas, Banco Peruano de los Constructores, Lima
50. Cruz H, Moura JP, Machado JS (2001) The use of FRP in the strengthening of timber reinforced masonry load-bearing walls. Historical constructions. Guimarães, Portugal, p 847
51. Cundall PA (1971) The measurement and analysis of acceleration in rock slopes. Ph.D. Thesis, University of London, United Kingdom
52. Dowling D, Samali B (2009) Low-cost and low-tech reinforcement systems for improved earthquake resistance of mud brick buildings. In: Hardy M, Cancino C, Ostergren G (eds) Proceedings of Getty Seismic Adobe Project 2006 Colloquium, The Getty Conservation Institute, Los Angeles, USA, pp 23–33
53. Dutu A, Sakata H, Yamazaki Y (2017) Comparison between different types of connections and their influence on timber frames with masonry infill structures' seismic behavior. In: 16th world conference on earthquake engineering
54. El-Nabouch R, Bui QB, Perrotin P, Plé O, Plassiard JP (2015) Numerical modeling of rammed earth constructions: analysis and recommendations. In: Proceedings of 1st international conference on bio-based building materials, Clermont-Ferrand, France
55. El-Nabouch R, Bui QB, Plé O, Perrotin P (2017) Assessing the in-plane seismic performance of rammed earth walls by using horizontal loading tests. Eng Struct 145:153–161
56. El-Nabouch R, Bui QB, Plé O, Perrotin P (2018) Characterizing the shear parameters of rammed earth material by using a full-scale direct shear box. Constr Build Mater 171:414–420
57. El-Nabouch R, Bui Q B, Perrotin P, Plé O (2018) Shear parameters of rammed earth material: results from different approaches. Adv Mater Sci Eng
58. El-Nabouch R, Bui QB, Plé O, Perrotin P (2019) Rammed earth under horizontal loadings: proposition of limit states. Constr Build Mater 220:238–244
59. Elnashai A, Gencturk B, Kwon OS, Al-Qadi I, Hashash Y, Roesler J, Kim S, Jeong SH, Dukes J, Valdivia A (2010) The Maule (Chile) earthquake of February 27, 2010: consequence assessment and case studies. MAE Center

60. Eslami A, Ronagh HR, Mahini SS, Morshed R (2012) Experimental investigation and nonlinear FE analysis of historical masonry buildings. Constr Build Mater 35:251–260
61. Fagone M, Loccarini F, Ranocchiai G (2017) Strength evaluation of jute fabric for the reinforcement of rammed earth structures. Compos B Eng 113:1–13
62. Figueiredo A, Varum H, Costa A et al (2013) Seismic retrofitting solution of an adobe masonry wall. Mater Struct 46(1–2):203–219
63. Fontana P, Miccoli L, Grünberg U (2018) Experimental investigations on the initial shear strength of masonry with earth mortars. Int J Masonry Res Innov 3(1):34–49
64. Fratini F, Pecchioni E, Rovero L, Tonietti U (2011) The earth in the architecture of the historical centre of Lamezia Terme (Italy): characterization for restoration. Appl Clay Sci 53(3):509–516
65. Furukawa A, Ohta Y (2009) Failure process of masonry buildings during earthquake and associated casualty risk evaluation. Nat Hazards 49(1):25–51
66. Galasco A, Lagomarsino S, Penna A (2006) On the use of pushover analysis for existing masonry buildings. In: Proceedings of 1st European conference on earthquake engineering and seismology, Geneva, Switzerland
67. Gao ZN, Yang XD, Tao Z, Chen ZS, Jiao CJ (2009) Experimental study of rammed-earth wall with bamboo cane under monotonic horizontal-load. J Kunming Univ Sci Technol (Sci Technol) 2:015
68. Gautam D, Rodrigues H, Bhetwal K et al (2016) Common structural and construction deficiencies of Nepalese buildings. Innov Infrastruct Solut 1(1):1
69. Gavrilovic P, Sendova V, Ginell WS, Tolles L (1998) Behaviour of adobe structures during shaking table tests and earthquakes. In: Earthquake engineering proceeding of the 11th European conference, Balkema, Rotterdam, Netherlands
70. Gomes MI, Lopes M, de Brito J (2011) Seismic resistance of earth construction in Portugal. Eng Struct 33(3):932–941
71. Hall M, Lindsay R, Krayenhoff M (2012) Modern earth buildings—materials, engineering, constructions and applications. Woodhead Publishing, 800p. ISBN: 9780857090263
72. Hamilton H, McBride J, Grill J (2006) Cyclic testing of rammed-earth walls containing post-tensioned reinforcement. Earthq Spectra 22(4):937–959
73. Illampas R, Charmpis DC, Ioannou I (2014) Laboratory testing and finite element simulation of the structural response of an adobe masonry building under horizontal loading. Eng Struct 80:362–376
74. Jaquin PA, Augarde CE, Gallipoli D, Toll DG (2009) The strength of unstabilised rammed earth materials. Géotechnique 59(5):487–490
75. Jaquin PA, Augarde CE, Gerrard CM (2004) Analysis of Tapial structures for modern use and conservation. In: Proceedings of structural analysis of historical constructions, Padua, Italy, 2004
76. Jaquin PA, Augarde CE, Gerrard M (2006) Analysis of historic rammed earth construction. In: Proceeding of 5th international conference on structural analysis of historical constructions, New Delhi, India
77. Kamenjarzh J (1996) Limit analysis of solids and structures. CRC Press
78. Kouakou CH, Morel JC (2009) Strength and elasto-plastic properties of non-industrial building materials manufactured with clay as a natural binder. Appl Clay Sci 44:27–34
79. Lagomarsino S, Podesta S (2004) Seismic vulnerability of ancient churches: I. Damage assessment and emergency planning. Earthq Spectra 20(2):377–394
80. Langhenbach R (2009) Don't tear it down. Preserving the earthquake resistant vernacular architecture of Kashmir. United Nations Educational, Scientific and Cultural Organization
81. Lemos JV (2007) Discrete element modeling of masonry structures. Int J Archit Herit 1(2):190–213
82. Liberatore D, Spera G, Mucciarelli M, Gallipoli MR, Santarsiero D, Tancredi C et al (2006) Typological and experimental investigation on the adobe buildings of Aliano (Basilicata, Italy). In: Proceeding of the 5th international conference on structural analysis of historical constructions, Macmillan India, New Delhi, India, pp 851–858

83. Librici C (2016) Modelling of the seismic performance of a rammed earth building. MSc. Thesis, University of Minho, Guimarães
84. Liu K, Wang M, Wang Y (2015) Seismic retrofitting of rural rammed earth buildings using externally bonded fibers. Constr Build Mater 100:91–101
85. Liu K, Wang Y, Wang M (2014) Experimental and numerical study of enhancing the seismic behavior of rammed earth buildings. Adv Mater Res 919–921:925–931
86. Lombillo I, Villegas L, Fodde E, Thomas C (2014) In situ mechanical investigation of rammed earth: calibration of minor destructive testing. Constr Build Mater 51:451–460
87. Lourenço PB (1996) Computational strategies for masonry structures. Ph.D. Thesis, Technical University of Delft, Netherlands
88. Lourenço PB (2001) Analysis of historical constructions: from thrust-lines to advanced simulations. In: Proceeding of the historical constructions, Guimarães, Portugal
89. Lourenço PB (2002) Computations on historic masonry structures. Prog Struct Mat Eng 4(3):301–319
90. Lourenço PB, Rots JG (1997) Multisurface interface model for analysis of masonry structures. J Eng Mech 123(7):660–668
91. Lourenço PB, Rots JG, Blaauwendraad J (1998) Continuum model for masonry: parameter estimation and validation. J Struct Eng 124(6):642–652
92. Lubowiecka I, Armesto J, Arias P, Lorenzo H (2009) Historic bridge modelling using laser scanning, ground penetrating radar and finite element methods in the context of structural dynamics. Eng Struct 31(11):2667–2676
93. Magenes G, Della Fontana A (1998) Simplified non-linear seismic analysis of masonry buildings. In: Proceedings of Br. Masonry Soc. No. 8
94. Mahdi T (2005) Behavior of adobe buildings in the 2003 Bam earthquake. In: Proceedings of SismoAdobe2005: international seminar on architecture, construction and conservation of earthen buildings in seismic areas, Pontifical Catholic University of Peru, Lima, Peru
95. Mahini SS (2015) Smeared crack material modelling for the nonlinear analysis of CFRP-strengthened historical brick vaults with adobe piers. Constr Build Mater 74:201–218
96. Maniatidis V, Walker P (2008) Structural capacity of rammed earth in compression. J Mater Civ Eng 20(3):230–238
97. Maniatidis V, Walker P (2003) A review of rammed earth construction. University of Bath
98. Martins H (2015) Estudio de las propiedades de las fábricas históricas de adobe como soporte a intervenciones de rehabilitación. Ph.D. Thesis, Technical University of Madrid, Madrid
99. Maskell D, Heath A, Walker PJ (2014) Geopolymer stabilization of unfired earth masonry units. Key Eng Mater 600:175–185
100. Meireles H, Bento R, Cattari S, Lagomarsino S (2012) A hysteretic model for "frontal" walls in Pombalino buildings. Bull Earthq Eng 10(5):1481–1502
101. Meli R (2005) Experiencias en México sobre reducción de vulnerabilidad sísmica de construcciones de adobe. In: Proceedings of SismoAdobe2005: international seminar on architecture, construction and conservation of earthen buildings in seismic areas, Pontifical Catholic University of Peru, Lima, Peru
102. Memari A, Kauffman A (2005) Review of existing seismic retrofit methodologies for adobe dwellings and introduction of a new concept. In: Proceedings of SismoAdobe2005: international seminar on architecture, construction and conservation of earthen buildings in seismic areas, Pontifical Catholic University of Peru, Lima, Peru
103. Mendes N, Lourenço PB (2009) Seismic assessment of masonry "Gaioleiro" buildings in Lisbon, Portugal. J Earthq Eng 14(1):80–101
104. Mesbah A, Morel JC, Walker P, Ghavami Kh (2004) Development of a direct tensile test for compacted earth blocks reinforced with natural fibers. J Mater Civ Eng 16(1):95–98
105. Miccoli L, Drougkas A, Müller U (2016) In-plane behaviour of rammed earth under cyclic loading: experimental testing and finite element modelling. Eng Struct 125:144–152
106. Miccoli L, Fontana P (2014) Bond strength performances of anchor pins for earthen buildings. A comparison between earth block masonry, rammed earth and cob. In: Proceedings of 9th international conference on structural analysis of historical constructions, Mexico City, Mexico

107. Miccoli L, Müller U, Fontana P (2014) Mechanical behaviour of earthen materials: a comparison between earth block masonry, rammed earth and cob. Constr Build Mater 61:327–339
108. Miccoli L, Garofano A, Fontana P, Müller U (2015) Experimental testing and finite element modelling of earth block masonry. Eng Struct 104:80–94
109. Miccoli L, Oliveira DV, Silva RA et al (2015) Static behaviour of rammed earth: experimental testing and finite element modelling. Mater Struct 48(10):3443–3456
110. Miccoli L, Fontana P Müller U (2016) In-plane shear behaviour of earthen materials panels strengthened with polyester fabric strips. In: 10th international conference on structural analysis of historical constructions, pp 1099–1105
111. Miccoli L, Müller U, Pospíšil S (2017) Rammed earth walls strengthened with polyester fabric strips: experimental analysis under in-plane cyclic loading. Constr Build Mater 149:29–36
112. Miccoli L, Silva RA, Oliveira DV, Müller U (2019) Static behaviour of cob: experimental testing and finite-element modeling. J Mater Civ Eng 31(4):04019021
113. Michiels TLG (2015) Seismic retrofitting techniques for historic adobe buildings. Int J Archit Herit 9(8):1059–1068
114. Minke G (2006) Building with Earth: design and technology of a sustainable architecture. Birkhauser Basel, Berlin
115. Mostafa M, Uddin N (2016) Experimental analysis of Compressed Earth Block (CEB) with banana fibers resisting flexural and compression forces. Case studies in construction materials, vol 5, pp 53–63
116. Moreira Sturm T (2015) Experimental characterization of dry-stack interlocking compressed earth block masonry. Ph.D. Thesis, University of Minho, Portugal
117. Morel JC, Mesbah A, Oggero M, Walker P (2001) Building houses with local materials: means to drastically reduce the environmental impact of construction. Build Environ 36(10):1119–1126
118. Morel JC, Pkla A, Walker P (2007) Compressive strength testing of compressed earth blocks. Constr Build Mater 21(2):303–309
119. Müller U, Miccoli L, Fontana P (2016) Development of a lime based grout for cracks repair in earthen constructions. Constr Build Mater 110:323–332
120. Nanjunda Rao KS, Anitha M, Venkatarama Reddy BV (2015) Dynamic behavior of scaled Cement Stabilized Rammed Earth building models. In: Ciancio D, Beckett C (eds) Rammed earth construction cutting-edge research on traditional and modern rammed earth. Taylor & Francis
121. Nowamooz H, Chazallon C (2011) Finite element modelling of a rammed earth wall. Constr Build Mater 25(4):2112–2121
122. NZS 4298 (1998) Materials and workmanship for earth buildings. Standards New Zealand, Wellington
123. Oliveira DV (2003) Experimental and numerical analysis of blocky masonry structures under cyclic loading. Ph.D. Thesis, University of Minho, Portugal
124. Orduña A, Roeder G, Araiza JC (2006) Development of macro-block models for masonry walls subject to lateral loading. In: Proceedings of structural analysis of historic construction, New Delhi, India, pp 1075–1082
125. Ortega Heras J, Vasconcelos G, Lourenço PB, Correia M, Rodrigues H, Varum H (2015) Evaluation of seismic vulnerability assessment parameters for Portuguese vernacular constructions with nonlinear numerical analysis. In: 5th Proceeding ECCOMAS thematic conference on computational methods in structural dynamics and earthquake engineering, Crete Island, Greece
126. Parisi MA, Piazza M (2008) Seismic strengthening of traditional carpentry joints. In: Proceedings of 14th conference of earthquake engineering, Beijing
127. Penna A, Lagomarsino S, Galasco A (2014) A nonlinear macroelement model for the seismic analysis of masonry buildings. Earthq Eng Struct Dynam 43(2):159–179
128. Poletti E, Vasconcelos G (2015) Seismic behaviour of traditional timber frame walls: experimental results on unreinforced walls. Bull Earthq Eng 13(3):885–916

129. Quagliarini E, Maracchini G. (2018). Experimental and FEM investigation of cob walls under compression. Adv Civ Eng 2018
130. Quagliarini E, Lenci S (2010) The influence of natural stabilizers and natural fibres on the mechanical properties of ancient Roman adobe bricks. J Cult Herit 11(3):309–314
131. Quagliarini E, Stazi A, Pasqualini E, Fratalocchi E (2010) Cob construction in Italy: some lessons from the past. Sustainability 2:3291–3308
132. Quinn N, D'Ayala D (2014) In-plane experimental testing on historic Quincha walls. In: 9th international conference on structural analysis of historical constructions
133. Rafi MM, Lodi SH (2017) Comparison of dynamic behaviours of retrofitted and unretrofitted cob material walls. Bull Earthq Eng 15:3855–3869
134. Rafsanjani SH, Bakhshi A, Ghannad MA et al (2015) Predictive tri-linear benchmark curve for in-plane behavior of adobe walls. Int J Archit Herit
135. Reddy VBV, Gupta A (2006) Strength and elastic properties of stabilized mud block masonry using cement-soil mortars. J Mater Civ Eng 18(3):472
136. Rivera J, Muñoz E (2005) Caracterización estructural de materiales de sistemas constructivos en tierra: el adobe. Revista Internacional de Desastres Naturales, Accidentes e Infraestructura Civil 5(2):135–148
137. Ruggieri N, Tampone G, Zinno R (2015) In-plane versus out-of-plane "behavior" of an Italian timber framed system—the Borbone constructive system: historical analysis and experimental evaluation. Int J Archit Herit 9(6):696–711
138. Ruiz D, López C, Unigarro S, Domínguez M (2014) Seismic rehabilitation of sixteenth- and seventeenth-century rammed earth-built churches in the Andean Highlands: field and laboratory study. J Perform Constr Facil 29(6):04014144
139. San Bartolomé A, Pehovaz R (2005) Comportamiento a carga lateral cíclica de muros de adobe confinados. In: Proceedings of SismoAdobe2005: international seminar on architecture, construction and conservation of earthen buildings in seismic areas, Pontifical Catholic University of Peru, Lima, Peru
140. Soudani L, Fabbri A, Morel JC et al (2016) Assessment of the validity of some common assumptions in hygrothermal modeling of earth based materials. Energy Build 116:498–511
141. Silva R, Olliveira D, Schueremans L et al (2014) Shear behaviour of rammed earth walls repaired by means of grouting. In: 9th international masonry conference, pp 1–12
142. Silva RA, Oliveira DV, Schueremans L et al (2016) Effectiveness of the repair of unstabilised rammed earth with injection of mud grouts. Constr Build Mater 127:861–871
143. Silva RA, Schueremans L, Oliveira DV et al (2012) On the development of unmodified mud grouts for repairing earth constructions: rheology, strength and adhesion. Mater Struct 45(10):1497–1512
144. Silva RA, Oliveira DV, Schueremans L, Lourenço PB, Miranda T (2014) Modelling the structural behaviour of rammed earth components. In: Proceeding of the 12th international conference on computational structures technology
145. Silveira D, Varum H, Costa A (2013) Influence of the testing procedures in the mechanical characterization of adobe bricks. Constr Build Mater 40:719–728
146. Silveira D, Varum H, Costa A et al (2012) Mechanical properties of adobe bricks in ancient constructions. Constr Build Mater 28(1):36–44
147. Silveira D, Varum H, Costa A, Carvalho J (2015) Mechanical properties and behavior of traditional adobe wall panels of the Aveiro district. J Mater Civ Eng 27(9):04014253
148. Solís M, Torrealva D, Santillán P, Montoya G (2015) Análisis del comportamiento a flexión de muros de adobe reforzados con geomallas. Inf Constr 67(539):1–10
149. Standards Australia (2002) The Australian earth building handbook. Standards Australia, Sydney, Australia
150. Taylor P, Fuller RJ, Luther MB (2008) Energy use and thermal comfort in a rammed earth office building. Energy Build 40(5):793–800
151. Tarque N, Camata G, Varum H et al (2014) Numerical simulation of an adobe wall under in-plane loading. Earthq Struct 6(6):627–646

152. Tolles E (2009) Getty Seismic Adobe Project research and testing program. In: Proceedings of the Getty Seismic Adobe Project 2006 Colloquium, The Getty Conservation Institute, Los Angeles, USA, pp 34–41
153. Torrealva D Acero J (2005) Reinforcing adobe buildings with exterior compatible mesh. The final solution against the seismic vulnerability. In: Proceedings of the SismoAdobe2005: international seminar on architecture, construction and conservation of earthen buildings in seismic areas, Pontifical Catholic University of Peru, Lima, Peru
154. Torrealva D, Vargas J, Blondet M (2009) Earthquake resistant design criteria and testing of adobe buildings at Pontificia Universidad Católica del Perú. In: Proceedings of the Getty Seismic Adobe Project 2006 Colloquium. The Getty Conservation Institute, Los Angeles, USA, pp 3–10
155. Vamvatsikos D, Cornell CA (2002) Incremental dynamic analysis. Earthq Eng Struct Dynam 31(3):491–514
156. Varum H, Figueiredo A, Silveira D et al (2011) Outputs from the research developed at the University of Aveiro regarding the mechanical characterization of existing adobe constructions in Portugal. Inf Constr 63(523):127–142
157. Vasconcelos G, Poletti E (2015) Traditional timber frame walls: mechanical behavior analysis. Handbook of research on seismic assessment and rehabilitation of historic structures, pp 30–59
158. Vasconcelos G, Poletti E, Salavessa E et al (2013) In-plane shear behaviour of traditional timber walls. Eng Struct 56:1028–1048
159. Vera R, Miranda S (2005) Reparación de muros de adobe con el uso de mallas sintéticas. In: Proceedings of SismoAdobe2005: international seminar on architecture, construction and conservation of earthen buildings in seismic areas, Pontifical Catholic University of Peru, Lima, Peru
160. Vieux-Champagne F, Sieffert Y, Grange S et al (2014) Experimental analysis of seismic resistance of timber-framed structures with stones and earth infill. Eng Struct 69:102–115
161. Vieux-Champagne F, Sieffert Y, Grange S et al (2017) Experimental analysis of a shake table test of timber-framed structures with stone and earth infill. Earthq Spectra 33(3):1075–1100
162. Vintzileou E (2008) Effect of timber ties on the behavior of historic masonry. J Struct Eng 134(6):961–972
163. Walker P (1995) Strength, durability and shrinkage characteristics of cement stabilised soil blocks. Cement Concr Compos 17(4):301–310
164. Walker P, Dobson S (2001) Pullout tests on deformed and plain rebars in cement-stabilized rammed earth. J Mater Civ Eng 13(4):291–297
165. Walker P, Keable R, Martin J, Maniatidis V (2005) Rammed Earth—design and construction guidelines. BRE Bookshop
166. Walker R, Morris H (1998) Development of new performance based standards for earth building. In: Proceedings of the Australasian structural engineering conference, Auckland, pp 477–84
167. Wang Y, Wang M, Liu K et al (2016) Shaking table tests on seismic retrofitting of rammed-earth structures. Bull Earthq Eng 15(3):1037–1055
168. Wang Y, Wang M, Liu K, Wen P, Yang X (2017) Shaking table tests on seismic retrofitting of rammed-earth structures. Bull Earthq Eng 15:1037–1055
169. Webster F (2009) Application of stability-based retrofit measures on some historic and older adobe buildings in California. In: Proceeding of the Getty Seismic Adobe Project 2006 Colloquium, The Getty Conservation Institute, Los Angeles, USA, pp 147–158
170. Webster F, Tolles E (2000) Earthquake damage to historic and older adobe buildings during the 1994 Northridge, California Earthquake. In: Proceedings of the 12th world conference on earthquake engineering, New Zealand Society for Earthquake Engineering, Auckland, New Zealand

171. Wu F, Li G, Li HN, Jia JQ (2013) Strength and stress-strain characteristics of traditional adobe block and masonry. Mater Struct 46(9):1449–1457
172. Yamín L, Phillips C, Reyes J, Ruiz D (2007) Estudios de vulnerabilidad sísmica, rehabilitación y refuerzo de casas en adobe y tapia pisada. Apuntes 20(2):286–303
173. Zanotti S (2015) Seismic analysis of the church of Kuñoo Tambo (Peru). Msc. Thesis, University of Minho, Portugal

Chapter 6
Durability of Earth Materials: Weathering Agents, Testing Procedures and Stabilisation Methods

Domenico Gallipoli, Agostino W. Bruno, Quoc-Bao Bui, Antonin Fabbri, Paulina Faria, Daniel V. Oliveira, Claudiane Ouellet-Plamondon, and Rui A. Silva

Abstract This chapter reviews the potential impact of six environmental agents (water, ice, wind, fire, solar radiation and chemical attack) on the long-term stability of earth buildings together with some of the most common techniques for measuring and improving material durability. Liquid water appears the most detrimental of all environmental agents, not only because it can significantly reduce capillary cohesion inside the material but also because water can penetrate inside buildings through multiple routes, e.g. rainfall, foundation rise, ambient humidity and utilities leakage. Water can also be very damaging when it is present in solid form as the expansion of pore ice may induce cracking of the earth material. The high resistance of earth buildings to wind is instead proven by the good conditions of many historic structures in windy regions. Earth buildings also exhibit good resistance to fire as the exposure to very high temperatures may even improve material durability. Solar radiation

D. Gallipoli (✉)
Laboratoire SIAME, Fédération IPRA, Université de Pau et Des Pays de l'Adour, Anglet, France
e-mail: domenico.gallipoli@univ-pau.fr; domenico.gallipoli@unige.it

Dipartimento di Ingegneria Civile, Chimica e Ambientale, Università degli Studi di Genova, Genoa, Italy

A. W. Bruno
School of Engineering, University of Newcastle, Newcastle, UK

Q.-B. Bui
Faculty of Civil Engineering, Ton Duc Thang University, Ho Chi Minh City, Vietnam

A. Fabbri
LTDS-ENTPE, CNRS, University of Lyon, UMR 5513, Vaulx-en-Velin, France

P. Faria
CERIS and NOVA School of Science and Technology, NOVA University of Lisbon, Caparica, Portugal

D. V. Oliveira · R. A. Silva
ISISE & IB-S, University of Minho, Guimarães, Portugal

C. Ouellet-Plamondon
Département de Génie de La Construction, École de technologie supérieure, 1100, rue Notre-Dame Ouest, Montréal, Qc 3C 1K3, Canada

A. Fabbri et al. (eds.), *Testing and Characterisation of Earth-based Building Materials and Elements*, RILEM State-of-the-Art Reports 35,
https://doi.org/10.1007/978-3-030-83297-1_6

has, in general, a beneficial effect on the stability of earth buildings as it promotes water evaporation with a consequent increase of capillary cohesion. Solar radiation may, however, have a detrimental effect if the earth is stabilised by organic binders that are sensitive to photodegradation because, in this case, it may produce material damages ranging from a simple surface discoloration to a much more serious deterioration of the intergranular bonds. Unstabilized earth is generally inert and, hence, largely unaffected by chemicals though, in some instances, the precipitation of salt crystals inside the pore water can induce material cracking. Chemical degradation can instead be severe in both stabilised earth (due to the dissolution of intergranular bonds) and steel-reinforced earth (due to the corrosion of rebars). No international standard protocol exists to measure the durability of earth materials, which is currently assessed by multiple experimental procedures depending on which environmental agent is considered. Testing standards may, however, be devised in the future by differentiating between weathering protocols, which reproduce the effect of each agent on the earth sample, and durability protocols, which adopt a unique experimental procedure to measure a given material property regardless of weathering history.

Keywords Earth weathering · Earth ageing · Environmental impact · Durability testing · Durability improvement

6.1 Introduction

Raw earth is one of the oldest materials ever used for the construction of human dwellings. The first records of earth buildings date back to the Neolithic period around 8000 BC and have been found in Mesopotamia, a region roughly corresponding to modern Iraq [129]. Different civilizations along the southern Mediterranean coast have subsequently embraced the use of earth as a construction material. For example, Egyptians were familiar with adobe construction between 2000 and 1000 BC, as suggested by the Exodus Book of the Old Testament, which mentions that the Pharaoh commanded that Israelites should not be given straw to make bricks.

Earth construction evolved over the years and led to the manufacture of fired bricks, which produced a step change in the construction of masonry structures. The first fired earth bricks appeared in Mesopotamia during the Early Bronze period, i.e. around the 3rd millennium BC. However, their use became widespread much later, from the fourth century BC onwards, when the Greeks, and later the Romans, disseminated a viable earth firing technology across Europe and beyond. This led to the construction of masonry structures not too dissimilar from current ones. The oldest standing fired brick building in the world is probably the Theatre of Marcellus in Rome (Italy), which was built between the 13 and the 17 BC, during the last years of the Roman republic. In the sixteenth century, an extra floor was built on top of the ancient roman structure to host the apartments of the Orsini's, an influential family of the Italian Renaissance. The durability of the original building is demonstrated by the fact that some of these apartments are still inhabited nowadays.

The oldest standing raw earth buildings are instead found in the Taos Pueblo (USA), which is a complex of ancient adobe dwellings currently inhabited by about 150 people. This complex is much younger than the Theatre of Marcellus as it dates back to about one thousand years ago [62]. The buildings of the Taos Pueblo have been preserved in their current state thanks to regular maintenance consisting in the application, every year, of a sacrificial earth render on the external surface to protect the underneath structure from weathering.

As suggested by the above examples, the oldest standing buildings worldwide are made of fired earth despite this material is more recent than raw earth. This is because of the relatively good durability of fired earth, which is due to the treatment of the material at high temperatures between 900 and 1100 °C. This thermal treatment induces the transformation of the clay fraction into metamorphic rock and therefore increases the resilience to weathering [20]. Unfortunately, the thermal treatment also increases the embodied energy and manufacturing costs of the material while reducing moisture buffering capacity and hygrothermal inertia, which explains why fired brick structures are generally more expensive to build and operate compared to raw earth ones. Research has therefore been focusing on the development of novel stabilisation methods that improve the durability of raw earth while preserving its advantageous environmental prerogatives and low financial costs.

This chapter reviews the main weathering agents that affect the durability of earth buildings and describes the physical mechanisms through which each one of these agents weakens the material. The chapter also discusses some of the laboratory procedures that are currently used to estimate the long-term durability of raw earth when exposed to environmental actions. Finally, it examines the main stabilisation techniques that have been employed to increase the weathering resilience of earth materials and, therefore, to enhance their long-term durability.

6.2 Weathering Actions

This section describes the main environmental actions that affect the durability of raw earth and presents examples of the impact of these actions on earth buildings across the world.

6.2.1 Water

The strong affinity between soil and water is the cause of the large hygrothermal inertia of raw earth and explains the high energy efficiency of this material. Unfortunately, the affinity between soil and water is also the cause of the poor durability of earth buildings when exposed to rainfall or capillary rise from the foundation ground. An increase of water content in the earth pores produces a decrease of capillary tension, which in turn reduces both the stiffness and strength of the material [23,

46, 54, 67]. The mechanical deterioration of earth materials at high water contents has been experimentally observed both in the laboratory, at the scale of small samples, and in the field, at the scale of building walls.

In general, as long as the water content stays between 3 and 4%, the strength and durability of unstabilised earth remain relatively large. For example, Quagliarini et al. [106] found that the compressive strength of adobe walls ranged between 0.8 and 1.2 MPa when the water content was 2.45%, which corresponded to equilibrium conditions under an ambient humidity of 47%. Quagliarini et al. [106] also observed that this level of strength can ensure a relatively large margin of safety at the base of a two-storey earth building.

Nevertheless, an increase of water content can produce a noticeable reduction of both strength and stiffness. For example, Bui et al. [28] showed that an increase of water content from 2 to 12% reduces the strength and stiffness of compacted earth by a factor larger than four. The wetting of poorly compacted earth walls can also result in the collapse of buildings, as shown by Scarato et al. [113], who concluded that the main cause of pathologies in rammed earth buildings was the abnormal increase of pore moisture at the interface between walls and foundations. This increase of pore moisture may be caused, for example, by the accumulation of water in adjacent backfills or by the run-off from nearby slopes, which promote groundwater infiltration and favour capillary rise through the building foundations. Figure 6.1 shows the failure of an earth building in Lyon (France), which was caused by a large increase of water content at the wall base. The wetting-induced collapse of poorly compacted earth can also be explained by constitutive models that predict the stress–strain response of unsaturated soils, as proposed by geotechnical engineers over the past decades (e.g. [5, 53, 78]).

The occurrence of drying-wetting cycles, caused for example by fluctuations of indoor and outdoor humidity, can induce a periodic shrinkage-swelling of earth materials with a consequent deterioration of mechanical properties [19, 32]. Shrinkage-swelling cycles are associated to the progressive reorientation and reorganization of

Fig. 6.1 Collapse of an earth building in Lyon, France, due to the accumulation of humidity at the wall base

Fig. 6.2 Paderne castle in Algarve, Portugal, built in the twelfth century by using earth stabilised with air lime

earth particles [15, 96], which may in turn promote cracking and erosion. The impact of shrinkage-swelling cycles on the long-term performance of earth buildings has not yet been accurately quantified, but it depends on both the material type (i.e. mineralogy, grain size distribution) and the construction technique (i.e. compressed earth blocks, rammed earth, adobe, cob).

For centuries, the water durability of raw earth has been enhanced by chemical stabilisation with cement or lime (e.g. [101, 124]). An example of well-preserved ancient earth structure is Paderne Castle in Algarve (Portugal), which was built in the twelfth century by using earth stabilised with air lime (Fig. 6.2). Alternative stabilisers, including unusual options such as cow dung [91] or plant aggregates [77], have also been employed to reduce environmental impact [89].

In general, stabilisation improves water durability but the chemical interaction between the binder and the earth produces a modification of the porous structure. This generates a number of negative side effects including a reduction of moisture buffering capacity (and, hence, hygrothermal inertia), a faster deterioration of mechanical characteristics during fire or freeze–thaw cycles and greater difficulties to recycle the earth upon demolition.

6.2.2 Ice

The impact of freeze–thaw cycles on the durability of earth buildings has been scarcely studied in the literature probably because earth buildings are mostly located in temperate regions where frost is unlikely. This means that the suitability of raw earth as a building material in freezing climates remains to be ascertained.

Current evidence suggests that the periodic transition between liquid and solid states causes the volumetric variation of pore water and a consequent application of cyclic pressures on the granular skeleton, which may result in the deformation, spalling and erosion of earth walls. The magnitude of frost damages depends on both the amount of free water present inside the pores and the specific composition of the earth. Damages may range from the appearance of small defects on the wall surface to a deeper alteration of structural integrity [87]. Surface defects comprise cracks, flakes, blistering, peeling, loss of adhesion and boniness while structural defects consist in the occurrence of large settlements that are generally associated to foundation uplift and/or earth bulging at the wall base. In general, higher levels of porosity, especially if associated to the presence of relatively big voids, better accommodate the volumetric variation of pore water during phase changes and therefore increase frost durability. This explains why highly porous handmade loam adobes are more resistant to frost than dense extruded clay bricks.

The cohesive strength of unstabilised earth is mainly ensured by the bonding action of capillary water lenses at inter-granular contacts, which is stronger at higher clay contents. Because of this, finer materials tend to be more sensitive to freeze–thaw cycles than coarser ones as discussed by Minke [92], who suggested that a reduction of clay content below 16% significantly reduces the vulnerability of raw earth to frost erosion. Rammed earth walls tend to exhibit relatively low clay content, which explains why this particular construction technique is preferred in cold climates.

Aubert and Gasc-barbier [12] suggested that the loss of cohesion caused by freezing–thaw cycles during the cold season may generate micro-cracks inside the earth, which grow during the hot season as a consequence of desiccation [92]. This progressive opening of cracks augments water adsorption during the subsequent cold seasons and therefore accelerates the degradation of the material. Aubert and Gasc-barbier [12] also indicated that freezing–thaw cycles tend to harden the intact earth between cracks and therefore produce a local reduction of material porosity.

Earth buildings tend to be more vulnerable to frost when the amount of moisture inside the material pores is higher. This is, for example, the case immediately after construction when water content is uniform and generally high across earth walls. Delong [37] monitored buildings constructed during the winter season, when temperatures were less than zero degrees Celsius, and concluded that freeze–thaw cycles cause a progressive loosening of the granular structure in freshly posed earth. In the same way, earth foundations are not durable in freezing climates because of their high water content due to capillary rise from the underlying ground. The incorporation of thermal insulation inside perimeter walls should also be carefully considered in freezing climates as the presence of a heat barrier lowers the temperature of outermost part of the wall, thus facilitating water condensation and increasing vulnerability to frost.

The effects of freeze–thaw cycles on earth buildings are here demonstrated by the analysis of two cases from the United States. The first case is the Irving House (Fig. 6.3), which is located in the region of Geneva, New York. The National Centers for Environmental Information [95] indicates that, in this region, the average high temperature is 14 °C while the average low temperature is 4 °C with extreme values of 35 °C and −22 °C. The Irving House was built in 1846 with unfired earth bricks made of clay, sand, gravel, organic matter and water (with a clay content ranging between 8 and 22%). The blue frame in Fig. 6.3 indicates the part of Irving House

(a)

(b)

Fig. 6.3 **a** Irving House built in 1846 in Geneva, New York State, USA (the blue frame indicates the part built in adobe), **b** adobe bricks in the attic of Irving House [108]

that is made of adobe bricks. The external wall surface was protected with the typical render used in this region during the nineteenth century, which included one volume of clay, one volume of sand, one volume of lime, one volume of ash and half volume of beef blood together with fibres such as horsehairs or straw [108].

The second case is the Jackson-Einspahr House (Fig. 6.4), which is located in Holstein, Nebraska. The National Centers for Environmental Information [95] indicates that, in this region, the average high temperature is 17 °C while the average low temperature is 4 °C with extreme values of 39 °C and −25 °C. The Jackson-Einspahr House was built in 1881 with unfired earth bricks made of clayey soil covered with Prairies grass earth clump, a material locally known as "sod" ("motte de terre" in French), without any mortar [108]. The original roof was made of wooden boards covered with tarpaper.

The climates of the above two regions are relatively similar and are characterised by the widespread occurrence of frost in the cold season. Yet, comparison of Figs. 6.3 and 6.4 indicates that the Irving House is relatively well-preserved, apart from some localised defects on the inner wall surface, while the Jackson-Einspahr House shows extensive signs of weathering. This difference may be due to the regular application of a protective render on the external wall surface of the Irving House, which was not the case for the Jackson-Einspahr House. The application of this protective coating significantly reduced the impact of freeze–thaw cycles on the underlying earth structure. The above observations emphasize the importance of the regular inspection of exposed walls to detect the early signs of cracking or spalling and to ensure a timely remediation of the protective coating [73].

Earth buildings must comply with standard construction codes, which typically include norms to limit the damage caused by frost. To satisfy these requirements, the earth is usually stabilised with a combination of chemical binders, such as

Fig. 6.4 Jackson-Einspahr House built in 1881 in Holstein, Nebraska State, USA [108]

cement, lime or resins [58, 84], fibre reinforcement and mechanical densification [103]. The use of superplasticisers also allows reducing the amount of water inside earth concretes, which in turn increases the density and strength of the material while reducing the vulnerability to frost [100].

The durability of chemical stabilisers to freeze–thaw cycles is another important aspect to consider. Guettala et al. [58] showed that a clayey sand stabilised with 8% cement/resin exhibited a small mass loss of about 1.8% after undergoing twelve freeze–thaw cycles according to the American norm ASTM D560 [11]. On the contrary, the same material stabilised with 8% lime exhibited a more than twofold mass loss of 3.7%. Based on these results, Guettala et al. [58] recommended that compressed earth bricks should be compacted to a pressure of at least 10 MPa and stabilised with 5% cement to maximise frost resilience. A similar study by Shibi and Kamei [115] highlighted that a kaolinitic earth stabilised with 5% cement exhibited a 50% reduction of compressive strength after five freeze–thaw cycles. Shibi and Kamei [115] also observed that the addition of 5–20% of basanite, i.e. hemihydrate calcium sulfate $CaSO_4 \cdot 1/2\ H_2O$, and 10%–20% of coal ash noticeably improved the freeze–thawing resistance of the material. The vulnerability to frost depends on the characteristics of both the earth material and the stabiliser, thus it is advisable to test the chosen earth with different types of stabiliser prior to embarking on a construction project.

6.2.3 Wind

Wind is another environmental action that undermines the durability of earth structures by causing either immediate or progressive damages [127]. Immediate damages are relatively uncommon and are the result of exceptionally strong winds (e.g. hurricanes, cyclones or typhoons), whose turbulence generates high pressure differentials across structural elements and may therefore lead to failure [72]. On the contrary, progressive damages are much more common and are the result of weak to moderate winds, which drive slow erosive processes that endanger the long-term durability of unmaintained buildings [98]. Progressive damages are associated to the formation of small air eddies and vortices that attack the surface of earth buildings by continuously removing loosely bonded particles. The intensity of this erosive process mainly depends on the momentum of the wind impacting the building walls [127].

Progressive erosion occurs according to different mechanisms in arid and wet environments. In arid environments, moisture shortage augments the availability of loose particles that are lifted by the moving air, thus increasing the kinetic energy of the wind [80]. A wind of sufficiently high speed may carry suspended particles from dry lands such as deserts [33], which hit the building either directly or after bouncing on the ground, thus contributing to the progressive erosion of the wall surface. As erosion progresses, additional particles are detached from the building and become available to be lifted by the wind, thus enhancing the abrasion of the wall surface. Conversely, in wet environments, moisture-laden winds can penetrate the earth walls [64] and therefore weaken the capillary bonds between grains [67]. A sequence of dry and wet winds can also induce cyclic variations of moisture content inside the exposed earth which, in the presence of expansive clays, causes the periodic shrinkage-swelling of the building walls [126]. Swelling is produced by an increase of moisture content inside the earth and may cause the detachment of protective coatings [127] while shrinkage is produced by a decrease of moisture content and may cause cracking. The periodic variation of moisture content can also induce a migration of salts from the core of the wall to the surface, which can lead to the appearance of efflorescences and subflorescences causing the detachment of protective coatings [99].

Progressive damage advances very slowly and it is therefore mostly visible in historical buildings that have been exposed to the action of wind for centuries. Good examples of the progressive damages caused by wind can be observed in some sections of the Great Wall of China [80], the Alhambra palace in Spain [52] and the Paderne Castle in Portugal [35]. Figure 6.5 shows the progressive erosion caused by the wind at the base of a rammed earth wall of the Paderne Castle. Due to the slow progression of these damages, serious consequences can generally be avoided by applying a protective render on the exposed walls and/or by periodically replacing the eroded material [55].

6.2.4 Fire

Fire is an accidental action which can have devastating consequences on the performance of buildings and the life of occupants. During a blaze, the combustion of materials generates toxic smoke and gas, whose inhalation is the main cause of

Fig. 6.5 Erosion at the base of a rammed earth wall in Paderne castle, Portugal

death. Combustible and hazardous materials include textile fabrics, furniture, wall paper and paint, carpeting, insulation and plastic household materials [70]. Moreover, the combustion of timber beams and columns, which are frequent in the roofs and floors of earth buildings, not only produces harmful smoke and gas, but also reduces the cross section of structural elements and may ultimately lead to the collapse of the building [25].

Conventional construction materials, such as concrete [4], steel [79] and masonry bricks [111], are non-combustible, though the exposure to high temperatures may eventually produce a degradation of their structural performance. Earth materials with densities higher than 1700 kg/m^3 are also deemed to be non-combustible according to the German norm DIN 4102 [39] and are expected to withstand the actions of fire and high temperatures [18, 92, 114, 117]. Earth materials with densities lower than 1700 kg/m^3 often incorporate reinforcing fibres and their resistance to fire therefore depends on the amount of these combustible components.

During prolonged exposure to fire, unstabilised fine earth tends to shrink and crack while unstabilised coarse earth experiences a loss of cohesion that is the consequence of the evaporation of the capillary water lenses bonding particles together [63]. In stabilised earth, high temperatures may instead promote chemical reactions that undermine inter-particle cementation and therefore lead to a disaggregation of the material. Spalling can also occur in low porosity earth because of the build-up of pore vapour pressures at high temperatures [75].

Unfortunately, very little experimental investigation is available about the response of earth structures to fire. This lack of experimental data, together with the large diversity of earth materials and manufacturing methods, does not allow a simple definition of the mechanisms through which fire affects earth buildings. It may even be possible that the exposure of unstabilised earth to fire improves, rather than degrading, the mechanical performance of the material by transforming the clay fraction into metamorphic rock. This hypothesis is consistent with the production of fired earth bricks, which rely on the lithification of clay minerals at high temperatures to increase strength and stiffness [36]. The chemical, physical and mechanical modifications undergone by the clay fraction during fire depend, however, on the mineralogy of the earth and the modality of exposure to high temperatures [50, 76]. At this stage, it is therefore difficult to speculate whether the exposure of a building to fire will produce an improvement or degradation of the mechanical properties of the earth material. Finally, with reference to the hygroscopic properties, the mineralogical transformation of the clay at high temperatures entails a reduction of the moisture buffering capacity of the earth and, hence, a decrease of the hygro-thermal inertia of the building walls [20].

6.2.5 Solar Radiation

Solar radiation propagates in the atmosphere as electromagnetic waves, which are classified as UV-A, UV-B and UV-C radiations according to their wavelength and photon energy (Table 6.1).

Generally, UV radiation has a beneficial effect on the mechanical characteristics of freshly posed unstabilised earth as it increases the material temperature and therefore promotes the evaporation of excess pore water. This facilitates the development of capillary pore water tensions, which are the main source of inter-particle bonding, thus increasing strength and stiffness [16, 21, 28, 67]. For the same reason, UV radiation may negatively affect the cure of freshly posed stabilised earth, especially in the presence of mineral binders that require moisture for hydration (e.g. hydraulic limes) or carbon dioxide for carbonation (e.g. air limes).

Exposure to solar radiation can also reduce the durability of "photosensitive" binders that react with the photon energy of UV rays [6]. This is, for example, the case of polymeric binders, either synthetic (e.g. polyvinyl acetate, acrylic or latex emulsions) or natural (e.g. resins, waxes, oils, fats). These binders are effective in improving the mechanical properties of the earth (e.g. [42, 71]) but they contain chromophoric groups, such as carbon–carbon (C=C) and carbonyl (C=O), that absorb the photon energy of the UV radiation. The exposure to UV radiations can therefore cause photoreactions that induce a gradual degradation of the chemical bonds inside the binding phase (e.g. [13, 90, 130]). This degradation starts with a yellow discoloration

Table 6.1 UV radiations present in sunlight

Solar radiation	Wavelength (nm)	Photon energy (eV)
UV-A (long-wave)	315–400	3.10–3.94
UV-B (medium-wave)	280–315	3.94–4.43
UV-C (short-wave)	100–280	4.43–12.4

of the material and then evolves into the chain scission of chromophoric groups, which generates the embrittlement of the binder and the formation of micro-cracks. The transition from a simple discoloration to mechanical degradation is gradual and is controlled by the exposure time [13]. This detrimental effect of UV radiation is even more critical in warmer climates as high temperatures accelerate the deterioration of polymeric binders and produce a faster weakening of the stabilised earth [120]. Degradation is also faster in wetter climates as rainwater continuously washes the weathered material surface and exposes a fresh undamaged layer of stabilised earth to UV radiation. Finally, ozone depletion increases the intensity of UV-B radiation and may also accelerate the deterioration of polymeric stabilisers.

6.2.6 Chemical Agents

The durability of earth materials is also undermined by chemical agents that are routinely found in the environment [29, 57, 86]. The impact of these agents is most significant if the earth is stabilised with chemical binders or steel reinforced [27, 34]. Steel reinforcement is currently employed in stabilised rammed earth but is not recommended in unstabilised earth due to the lack of mechanical anchorage. Rebar schedules are usually selected according to the design codes of reinforced concrete but the construction processe is very different from that of reinforced concrete as rammed earth is compacted rather than poured in place. A careful compaction of the earth around rebars is crucial to allow a good adhesion of the material to the embedded reinforcement. To ensure this, vertical rebars are installed before the earth is compacted around them while horizontal rebars are placed between subsequent compaction lifts, making sure that they properly sit on the underlying earth.

High concentrations of chloride ions can cause the corrosion of steel bars and can therefore adversely affect the performance of reinforced earth [86]. Chlorides may be either inherent to the earth or may be artificially introduced by mixing the earth with contaminated water (e.g. salt water). Moisture laden winds can also carry chlorides that penetrate the building surface and then migrate towards the core of the walls [57]. Sulphate salts are also inherent to soils and exist in cement, groundwater, seawater, industrial waste and acid rain. Sulphate hydration products attack the inter-particle bonds of cement stabilised earth and provoke the progressive weakening of the material. For this reason, sulphate-resisting cements should always be used to stabilise earth materials that contain notable amounts of sulphate salts. Note that both chlorides and sulphates are commonly present in earth materials that are contaminated with animal excrements.

Chlorides and sulphates are highly hygroscopic salts and are easily dissolved inside pore water. In the presence of a hydraulic gradient, the dissolved salts are then transported to the wall surface where they produce a local increase of concentration. As pore water evaporates, the salts precipitate causing efflorescences on the wall surface or subflorescences behind the wall surface (this is, for example, the case when a vapor barrier is present). Subflorescences are particularly damaging because they are associated to a precipitation of salt crystals inside the material pores, which causes the development of swelling pressures and, hence, the formation of cracks. The above degradation mechanisms also affect unstabilised earth walls, which are normally classified as resilient to chemical attacks. In reinforced earth, the growth of crystals and the formation of cracks may expose steel elements to weathering, which in turn provokes corrosion.

Carbonation is another chemical process that can undermine the durability of reinforced earth walls. This process consists in the slow reaction of the carbon dioxide from the atmosphere with the calcium hydroxide produced by the hydration of Portland cement inside stabilised earth [86]. This reaction produces calcium carbonate, which increases material strength and stiffness but also causes the acidification of the pore water and, hence, promotes the corrosion of steel reinforcements. Moreover, Portland cement is vulnerable to acids and a decrease of the pH of pore water can facilitate the dissolution of the binding fraction. Note also that earth acidification can be produced not only by exposure to carbon dioxide but also to sulphur dioxide and other types of industrial waste.

A highly alkaline environment can equally undermine the long term durability of earth buildings. High pH levels may transform the naturally present silica inside soils into gels, which absorb water and expand in volume, thus causing the formation of cracks inside earth walls. This transformation takes place if alkaline cements, or alkali-reactive aggregates, are present at high moisture contents. Low alkaline binders, such as blast furnace slag or pulverised fuel ash, should therefore be preferably used for the construction of basements and foundations, which are directly exposed to water due to capillary rise from the underlying ground.

C–S–H hydrates are the principal component of the binding fraction of cement-stabilised earth materials [86]. These hydrates cause a viscoplastic response of the material when loaded, which is due to the rearrangement of C–S–H particles at nanoscale level. This phenomenon, which may be explained by the free-volume dynamics theory of granular physics [109], is one of the main sources of creep in cement-stabilised earth materials [26]. C–S–H hydrates are instead absent in unstabilised earth where the occurrence of creep is mainly the consequence of the propagation of micro-cracks [26]. Loaded earth may experience the formation of micro-cracks, which produce a hygric imbalance within the material and the consequent generation of pore water pressure gradients. This further helps the propagation of micro-cracks and the amplification of creep, which is associated to a progressive decrease of stiffness at the structural scale.

6.3 Field Measurement of Durability

Bui et al. [29] published one of the first long-term experimental study of the durability of full scale earth walls exposed to environmental actions. In this study, unstabilised and lime stabilised rammed earth walls with a thickness of 400 mm were exposed to the wet continental climate of Grenoble (France) for a period of 20 years while

Fig. 6.6 Reed-cob experimental cellule: **a** shortly after construction with unrendered walls and **b** 5 years after construction during the application of a protective render on the wall surface

surface erosion was measured by stereo-photogrammetry. The final erosion depth was about 2 mm (0.5% of the wall thickness) for the lime stabilised walls and about 6.4 mm (1.6% of the wall thickness) for the unstabilised walls. In a similar study, Faria et al. [49] investigated the durability of rammed earth walls exposed for a decade to the weather of Serpa (Portugal). Both studies from Bui et al. [29] and Faria et al. [49] concluded that lime stabilisation reduces surface degradation and erosion rate, though structural durability remained acceptable even for the unstabilised walls.

In 2014, a small earth building was erected in a semi-rural area 3 km from the Atlantic coast of Portugal (Fig. 6.6), which is characterised by the occurrence of strong rains and winds from the South. The walls of the building were made of lightweight cob stabilised with lime putty and pozzolan. The walls were manufactured in a sequence of 10 cm cob lifts separated by layers of reed fibres and were supported by a concrete foundation with a thermally insulated roof. A detailed description of the wall material and building technology is given in Val et al. [121] and Carneiro et al. [31]. The North and East walls of the building were left unprotected and directly exposed to the atmosphere, while the South and West walls were lightly protected by a lime wash (Fig. 6a). The building was continuously monitored since construction and no sign of significant erosion were detected after 5 years, at which time the walls were rendered with an air lime stabilised earth mortar (Fig. 6b).

6.4 Laboratory Measurement of Durability

Field studies provide the most accurate assessment of the durability of earth buildings but they are rare because of the high financial costs and lengthy execution times. Laboratory tests are much more common but they are less representative of the actual performance of earth buildings [125]. Distinct types of laboratory tests have been proposed to assess the durability of earth materials exposed to different environmental actions. This has contributed to the proliferation of experimental protocols, which has in turn impeded the formulation of widely accepted standards. In this section some of the existing testing methods are reviewed and a possible way to simplify and unify current approaches is proposed.

6.4.1 Water

Several laboratory tests have been devised to assess the durability of earthen materials exposed to liquid water. These tests can be grouped in three main families depending on the modality of application of the water action and the particular aspect of material behaviour under investigation. The first family of tests assesses the resilience of the earth to the erosion caused by water impact and includes the Accelerated Erosion Test (AET), the Geelong Drip Test and the Swinburne Accelerated Erosion Test (SAET), which are described in the Australian earth building handbook [125]. Due to the important role played by the kinetic energy of the impacting water, these tests are better suited to assess the durability of earth materials under the combined action of water and wind. Instead, the second family of tests assesses the durability of earth materials when exposed to standing, or slowly flowing, water. It includes the wet-dry appraisal test of the Australian earth building handbook [125], the contact test, the swelling/shrinkage test of the French standard [1] and the suction and dip tests of the German standard [40]. Finally, the third family of tests quantifies the potential mechanical collapse of earth materials subjected to wetting and requires the performance of confined or unconfined compression tests on samples at different degrees of saturation.

6.4.2 Ice

Laboratory tests have also been devised to evaluate the durability of earth materials exposed to freeze–thaw cycles. Some of these tests have been adapted from geotechnical standards that assess the frost durability of road pavements or railway tracks. These standards are, however, unnecessarily harsh because geotechnical infrastructure generally experiences tougher mechanical actions than earth buildings. For example, some procedures accentuate material weathering by brushing the sample surface after each freeze–thaw cycle [11], which has been criticised by Shihata [116] as excessively severe for earth building applications.

Current test protocols generally measure the mass loss experienced by earth samples subjected to a number of freeze–thaw cycles. These procedures mainly differ for the modalities of preparation, curing and equalization of the samples as well as for the imposed temperature cycles [73]. Properties such as compressive strength [118], tensile strength [2] and ultrasound wave velocity [8] can also be measured after each freeze–thaw cycle to infer the durability of the material.

One important mechanism through which freeze–thaw cycles undermine the durability of earth buildings is the development of tensile stresses inside exposed walls, which are caused by the volumetric change of pore water during the transition between liquid and solid states. This mechanism is deliberately amplified during laboratory experiments by humidifying the samples after each freeze–thaw cycle to compensate for the water lost during the previous test stages. This humidification augments the damages caused by the expansion/contraction of pore water during phase changes and has been considered overly severe by some authors [12].

6.4.3 Wind

A number of authors have proposed experimental procedures to quantify the durability of earth materials to the action of wind. The peeling test [41] is a simple procedure that can be performed both in the laboratory or in-situ to assess the cohesion of the earth surface [47]. Similarly, the wearing test of the American standard ASTM D559 [10] quantifies the erosion produced by the stroke of a wire brush on the surface of an earth sample. The Accelerated Erosion Test (AET), the Geelong Drip Test and the Swinburne Accelerated Erosion Test (SAET) [125] are also good options to assess the durability of earth materials under the combined actions of wind and water, as previously discussed.

At a larger scale, [80] and [105] investigated the erosion of adobe buildings by driven sands inside a wind tunnel. A relatively large earth sample was placed at the end of the working section of a wind tunnel while the floor was covered with a bed of sand. During the test, the sand was lifted and carried by the wind, thus enhancing the erosion of the earth sample, which was quantified by measuring the mass loss at the end of the experiment. This is a relatively fast test as the sample is exposed to the wind for a relatively short time, between 10 and 60 min depending on the wind speed. Moreover, the speed of the wind and the size of the sample can be varied to account for different in-situ conditions. Han et al. [61] also proposed the use of portable wind tunnels, with a length of about 13 m, to perform an in-situ evaluation of wind erosion on real buildings.

Past studies have generally highlighted the importance of the abrasive action exerted by wind-driven particles, especially in arid climates. This is an important aspect that will have to be duly considered during the development of future experimental standards.

6.4.4 Solar Radiation

In the absence of standard experimental protocols, a realistic approach for assessing the effect of solar radiation on durability consists in the measurement of the photodegradation experienced by earth samples after natural in-situ weathering [88].

Photodegradation can be measured both at the macroscopic scale, by means of standard laboratory tests, and at the microscopic scale, by Fourier Transform Infrared or X-Ray diffraction tests [30]. The latter tests provide information about the changes of structure and mineralogical composition experienced by the material after exposure to UV radiations. Natural in-situ weathering, albeit inexpensive, is however lengthy and may prove inaccurate due to the simultaneous action of multiple environmental agents, which make impossible to isolate the effect of solar radiation. Because of this, a number of laboratory protocols have been devised to reproduce the weathering caused by UV radiations. These protocols have the advantage of being much faster than field weathering, though their representativeness of the actual degradation mechanisms remains debatable [17]. One of these protocols consists in exposing earth samples to the UV-A or UV-B radiations of electric lamps inside a degradation chamber equipped with a ventilation system to avoid excessive heating [30]. This replicates the action of short wavelength radiation by the sun, which is mostly responsible for the photodegradation of stabilised earth [69]. Another protocol makes use of filtered xenon arc lamps, which offer the advantage of producing a radiation that is similar to that of sunlight throughout the UV spectrum together with the possibility of generating a monochromatic radiation [69]. This latter feature allows quantifying the material degradation caused by different radiation wavelengths. Generally, both artificial and natural polymeric binders exhibit higher degradation rates when exposed to shorter radiation wavelengths [6].

6.4.5 Fire and Chemical Agents

Very little research has been undertaken to assess the impact of fire and chemical agents on the durability of earth materials. In the absence of suitable experimental standards, the durability of earth materials exposed to fire or chemical agents can be assessed by means of conventional laboratory tests that are employed to characterise the hydromechanical behaviour of other construction materials. These tests may be performed before and after material weathering to compare results and, hence, to assess durability. Similar to the case of solar radiation, this approach necessitates the definition of suitable weathering protocols that replicate the actual degradation experienced by the in-situ material during exposure to fire and chemical agents. These weathering protocols may, again, be similar to those already proposed for other construction materials exposed to fire and chemical agents. With reference to fire, a standard code of practice is offered by the recommendations of the RILEM technical committee TC 200-HTC for the characterisation of concrete behaviour at high temperatures [110]. Instead, with reference to chemical agents, suitable protocols may follow the guidelines of the European Committee for Standardization for evaluating the resistance of concrete structures to severe chemical attacks [43] or the recommendations of the RILEM technical committee TC 271-271-ASC for assessing the durability of construction materials exposed to salt crystallization [85]. Some of these weathering procedures require the direct contact between the material samples and the water, which is practicable for stabilised earth but requires a degree of adaptation for unstabilised earth.

6.4.6 Standardization of Durability Tests

The above short review has indicated the existence of a large number of experimental procedures to evaluate the durability of earth materials exposed to environmental actions. Some of these procedures are rather difficult to implement by practitioners in ordinary civil engineering laboratories. Moreover, the large number and relative complexity of experimental methods have so far hindered the formulation of widely adopted testing standards. A significant simplification may, however, be achieved by replacing the present multitude of protocols with a single standard durability test that is valid regardless of the particular environmental action under consideration. The specific impact of each individual environmental action could then be studied by subjecting the earth samples to distinct weathering processes, which reproduce the field effect of the chosen action. After weathering, all samples are therefore subjected to the same standard test to assess durability to the chosen environmental action. In this way, the choice of the durability test becomes independent of the environmental action under consideration.

The standard durability test could be a simple abrasion test, such as that described by Minke [92] or the ICONTEC NTC 5324 [65], consisting in the application of a scrubbing action on the sample surface for a given period of time by means of a metallic brush or sandpaper loaded by a fixed weight. The durability of the sample could then be directly related to the abrasion resistance of the material, which is measured by the mass lost during scrubbing. Alternatively, the durability could be evaluated by comparing standard material properties, such as strength, stiffness and hygrothermal inertia, before and after weathering.

The definition of a single standard test, or a small set of standard tests, would therefore offer the advantage of streamlining experimental procedures, making them accessible to a wide range of academic and industrial organizations.

6.5 Stabilisation Methods for Enhancing Earth Durability

The durability of earth materials may be improved by different stabilisation methods, whose choice depends on the particular construction technology under consideration [22, 49]. The choice of the stabilisation method is also affected by the mineralogical characteristics of the clay fraction and, in particular, by the relative proportions of illite, kaolinite and montmorillonite minerals.

In general, adobes, cob and wattle and daub are physically stabilised by augmenting the fine fraction of the earth and/or by incorporating natural fibres. Instead, compressed earth blocks and rammed earth walls are generally stabilised via the addition of chemical binders, such as cement or lime. Cement or lime stabilisation is particularly appropriate in areas where floods are frequent and when earth walls are not separated from the underlying ground by concrete or stone foundations [3].

In the following part of this section, the methods of stabilisation are classified as organic or inorganic. Organic stabilisation makes use of waterproofing additives and/or reinforcing fibres of organic origin while inorganic stabilisation relies on the addition of chemical binders and/or the modification of earth grading. Organic additives and inorganic binders can also be combined together to produce hybrid stabilisation methods. This is, for example, the case when quicklime is mixed with animal or vegetal fat to produce an air lime putty that exhibits waterproofing characteristics.

6.5.1 Organic Stabilisation

Organic stabilisation makes use of plant aggregates (e.g. corn cob particles, olive pit and cork particles), plant fibres (e.g. straw or husks), animal waste (e.g. cow dung or pigeon droppings) and natural polymers (e.g. starch, vegetal gels from cactus, agave, algae, cereal flour, natural resins) to improve the mechanical and durability characteristics of earth materials [74, 77].

Plant derivatives have historically been used for improving the strength and durability of earth buildings. Some notable examples include Nopal mucilage (Opuntia sp.), mauve leaves and stems (Sida rhombifolia) and gaucima bark (Guazuma ulmifolia), which have all been employed as earth stabilisers by ancient civilizations in Latin America and the Mediterranean [74]. Bitumen diluted in vegetal oil has also been used by pre-Columbian civilisations to improve the water resistance of earth buildings in humid tropical areas [74, 128]. Similar techniques were also adopted by the Babylonians [14]. More recently, bitumen water emulsions have been recommended by O´Connor [97] for earth stabilisation. Kita et al. [74] also studied the durability of adobe wallets with a protective render made of earth, mucilage and bitumen concluding that the presence of mucilage limit drying shrinkage while bitumen reduces rain erosion.

Linseed oil and karité butter are other fat products of vegetable origin that have been used to stabilise earth materials. Lima et al. [82] tested an illitic clayey earth mortar stabilised with 1–5% of linseed oil observing an increase of flexural, compressive and tensile strength, together with an augmentation of dry abrasion resistance, in the stabilised material compared to the unstabilised one. The addition of linseed oil reduced the vulnerability of the material to water but also produced a decrease of hygroscopicity and, hence, moisture buffering capacity [48]. In another work by Minke [93], earth mortars with 6% boiled linseed oil exhibited good resistance to water erosion while showing relatively low vapour permeability compared to unstabilised mortars.

Polysaccharides biopolymers are long-chained, well-structured sugar macromolecules, which have also been used for earth stabilisation. These biopolymers change the electrostatic charge of clay particles, which reduces the amount of water that is necessary to achieve a good workability of the earth. Subsequently, the progressive flocculation of clay particles compacts the material, thus increasing durability [42].

Lipids are another family of biopolymers originating from the fat of living organisms, which have been used for the stabilisation of earth materials. The terms oil, butter and cera designate the physical state of lipids, which ranges from liquid to solid depending on temperature [123]. Lipids are flexible and insoluble in water, which improves both workability and water resistance but reduces vapour permeability.

Proteins are an additional category of biopolymers that have been used for the stabilisation of earth materials. They are composed of a hydrophobic part, which is absorbed by clay platelets, and a hydrophilic part, which covers the clay surface. They can therefore create a film that prevents water infiltration [51] while maintaining high vapour permeability [42]. Proteins can bind clay particles together and hence improve the resistance of the material to water erosion. The most common proteins used for earth stabilisation are caseins and tannins. The latter ones are present in almost all plants and tend to form iron tannate, which is particularly effective in gluing earth particles together [123].

Recent research has also focused on the biomineralisation of earth materials by means of microbes or enzymes. Dhami and Mukherjee [38] tested calcium biomineralised earth blocks achieving a 40% reduction in water absorption, together with a notable decrease of linear expansion compared to the unstabilised material. Similarly, Mukherjee et al. [94] tested calcium biomineralised earth blocks achieving a 34% reduction of water content after immersion, together with a 10% increase of wet compressive strength. Ivanov et al. [66] compared calcium and iron biomineralisation of earth blocks showing that the latter method reduces water permeability at smaller financial costs, though it results in lower levels of compressive strength. Microbially induced iron-oxide precipitation (MIIP) was also employed to enhance the durability of earth mortars [122] with the stabilised mortar exhibiting lower flexural and compressive strength than the unstabilised one, which confirms the results of Ivanov et al. [66].

Organic stabilisation of earth materials has also been achieved via the addition of natural or synthetic fibres. Recycled synthetic fibres from industrial or household waste are generally preferred to reduce environmental impact and embodied energy.

Finally, the utilisation of eco-friendly bio-products has been recently considered as a possibility for improving the durability of earth plasters on exposed walls. For example, MIIP has been employed by Parracha et al. [102] for the stabilisation of earth plasters resulting in a reduction of both moisture absorption and damage after contact with water (Fig. 6.7).

Fig. 6.7 Water repellence of an earth mortar stabilized by microbially induced iron-oxide precipitation (MIIP) [48]

6.5.2 *Inorganic Stabilisation*

Cement and lime are the most common inorganic stabilisers for improving the mechanical and durability characteristics of earth materials. Between the two, cement produces the highest improvement of strength and durability but it also exhibits the highest environmental impact in terms of embodied energy and carbon footprint. Air lime (or simply lime) is older than cement and has been used for centuries to improve the mechanical characteristics of earth buildings. Lime neutralises the free clay cations and promotes the formation of neo-silicates and hydrated calcium aluminates [60]. The advantageous properties of lime are well known to geotechnical engineers, who commonly make use of this binder to enhance the mechanical characteristics of infrastructure embankments or soil subgrades during road construction.

Lime stabilised rammed earth, known as "taipa militar" in Portuguese [101] or "tapia la real" in Spanish [24], was used in the Iberian Peninsula during the Muslim period from the 8th to the fifteenth century to build military structures such as Paderne Castle, Silves Castle or Álcacer do Sal Castle. In these buildings, three different rammed earth techniques can be distinguished corresponding to different levels of durability: (a) ordinary unstabilised rammed earth, (b) "la real" or "military" rammed earth stabilised with around 10% of lime (see Guerrero Baca et al. [60]) and (c) "calicostrada" rammed earth where lime mortar is placed on the surface of the formworks before compaction of each lift. This last technique has the purpose of both increasing the durability of the wall to driving rain and enhancing the adhesion of the external render, if present.

Eires et al. [42] studied the behaviour of different kaolinitic earth samples stabilised with 4% of cement or quicklime, 4% of quicklime plus 1% of used soyabean oil and 4% of quicklime plus 0.1% of commercial sodium hydroxide. The behaviour of these materials was assessed by measuring dry and wet compressive strength, capillary absorption, spraying resistance and vapour permeability. Results show that quicklime stabilisation increases compressive strength in both dry and wet conditions while reducing surface erosion, especially if a small quantity of soya-bean oil or sodium hydroxide is added. All stabilised samples showed lower levels of vapour permeability compared to unstabilised ones.

Gomes et al. [56] tested a kaolinitic earth mortar stabilised with distinct percentages (i.e. 5, 10 and 15%) of four different binders, namely air lime, hydraulic lime, Portland cement and natural cement. The study showed a relatively small improvement of mechanical properties with increasing amount of binder. Conversely, the capillarity coefficient, that is the slope of water absorption against square root of time [44], markedly reduced with growing binder percentage. This was particularly evident for the samples stabilised with Portland cement, which exhibited a very slow drying rate after exposure to water. These results suggest that the application of stabilised mortars on the surface of unstabilised earth walls (as it is often the case during the restoration of archaeological buildings) may augment, rather than reducing, the risk of moisture-related pathologies. In general, protective renderings must be compatible with the underlying earth by ensuring that moisture can easily evaporate to the atmosphere. Moisture transfers across unstabilised earth walls are common (due, for example, to capillary rise from the foundation ground) and the application of a protective rendering that does not allow evaporation at the same rate of the incoming moisture flow may compromise, instead of enhancing, the durability of the structure. The addition of hemp fibres reduced the linear and volumetric shrinkage of the mortars, with the only exception of the mortar stabilised with hydraulic lime. The addition of fibres also increased the flexural and compressive strength of the stabilised mortars without, however, achieving the same levels of the unstabilised one. On a negative note, the addition of fibres produced undesirable biological growths, with the only exception of the mortar stabilised with air lime [112]. These biological growths were favoured by the relatively slow drying rate of the stabilised mortars, which means that good ventilation must be ensured during and immediately after construction.

Gypsum is an inorganic stabiliser that exhibits lower embodied energy than lime but is also more vulnerable to water. Vulnerability to water can be reduced by treating the gypsum at high temperatures, which however increases the energy and carbon footprint of the material. Lima et al. [81] tested an illitic earth mortar stabilised with 5, 10 and 20% of low fired hemihydrated gypsum. Results from these tests showed that an increase of gypsum content decreased drying shrinkage and increased both strength and surface cohesion with no significant reduction of the moisture sorption/desorption capacity.

Natural or artificial pozzolan is an inorganic stabiliser that hardens when mixed with water and calcium hydroxide, or with other materials that release calcium hydroxide such as Portland cement and quicklime. Natural pozzolan is meteorized volcano lava while artificial pozzolan is mainly the by-product of energy production by thermo-electric or biomass plants. Sometimes artificial pozzolans are sourced from the waste generated by the manufacture of red ceramics or the demolition of buildings. The thermal treatment of kaolin [104] also leads to the formation of metakaolin that can be used as a pozzolan. Most pozzolans have lower embodied energy than other inorganic stabilisers, which makes them an attractive option for reducing environmental impact. Guerrero Baca and Soria Lopez [59] investigated the behaviour of lightly compacted earth, stabilised with air lime and pozzolan, inside formworks. The resulting monolithic walls were lighter, faster to dry and more durable than conventional rammed earth walls. Moreover, the material did not present any sign of deterioration after six months of immersion in water.

Salts, including sodium chloride and sodium hydroxide, have also been used to improve the durability of earth materials. Sodium chloride limits clay flocculation and, hence, improves earth workability [7], which in turn reduces the amount of kneading water. The reduction of kneading water produces a decrease of porosity and, therefore, enhances the mechanical characteristics of the material. Sodium hydroxide is another salt capable of developing binder reactions that improve the mechanical characteristics of the earth. In particular, sodium hydroxide has been used to induce the geopolymerisation of aluminosilicates, inherently present in soils, into long mineral chains which bind earth grains together [22, 23, 42]. However, that the long-term durability of salt-stabilised earth is yet to be evaluated and salt stabilisers should therefore be used with caution. This is particularly true if sodium chloride is used for the stabilisation of earth incorporating steel rebars, which may then become vulnerable to corrosion.

6.6 Conclusions

The lack of a clear understanding about the mechanisms through which environmental agents affect the durability of earth buildings has so far hindered the dissemination of raw earth as a routine construction material. To overcome this gap of knowledge, the present chapter has reviewed the effects of water, ice, wind, fire, solar radiation and chemistry on the long-term durability of earth buildings. Current laboratory protocols for characterising the durability of earth exposed to weathering have been discussed in detail, together with the most common stabilisation techniques for improving the material resilience.

The most detrimental environmental agent is water, which is very pervasive in buildings due to meteoric precipitation, capillary rise from the foundation ground and ambient humidity. One of the main mechanisms through which water reduces material durability is the increase of moisture content inside the earth pores with the consequent decrease of capillary tension and inter-particle bonding.

Earth buildings are also permanently exposed to wind but, unlike water, wind produces relatively slow damages, which can be amended by means of a regular maintenance of the building envelope.

Ice and fire also have an effect on the durability of earth buildings, though the impact of these two actions is yet to be precisely quantified. The preponderance of earth buildings in areas characterized by temperate climates, where negative temperatures are rare, is one of the reasons why durability to freezing has been little investigated. In the case of fire, the exposure of unstabilised earth to high temperatures may even increase, instead of reducing, structural durability in the same way as firing improves the mechanical properties of the earth during the production of clay bricks. This hypothesis is also consistent with the existence of ancient earth ruins that have been exposed to fire during their lifetime and have remained reasonably well conserved until present age.

Solar radiation has generally a beneficial effect on earth durability as it promotes water evaporation and therefore favours the development of capillary tensions inside the material pores, which are the main source of inter-particle bonding in unstabilised earth. In the case of organically stabilised earth, however, solar radiation may have an unfavourable effect, especially if photosensitive binders, such as synthetic or natural polymers, are employed. The photodegradation of these polymeric binders progresses over time from unaesthetic discolorations to the destruction of chemical bonds between earth grains with the consequent weakening of the material.

Unstabilised earth is mostly unaffected by exposure to acid or alkaline environments and is often classified as chemically inert. Nevertheless, the dissolution of salts, such as chlorides and sulphates, inside the pore water can cause the localised precipitation of crystals, which may in turn generate swelling pressures and the appearance of cracks also in unstabilised earth. In general, salt crystallization can have either an adverse or beneficial effect on the earth material depending on factors such as the availability of pore water, the cyclic variation of moisture, the type of salts, the nature of the earth and the chosen building technique. Stabilised or steel-reinforced earth is more sensitive to chemical aggression than unstabilised earth as material degradation may also occur through other processes that are not dissimilar from those observed in conventional building materials.

The most common deterioration process affecting the durability of earth buildings is progressive cracking, which may lead to spalling and erosion. In the majority of cases, cracking is initiated by the cyclic swelling-shrinkage of the earth, which produces the delamination of the building surface. Earth swelling-shrinkage is often observed in the presence of expansive clay minerals and is caused by wetting–drying cycles due, for example, to seasonal variations of climate, sequences of moist and dry winds and fluctuations of the water table underneath building foundations. Swelling-shrinkage may also be the consequence of freeze–thaw cycles, which produce a volumetric change of pore water between liquid and solid states. The impact of freeze–thaw cycles is most severe if the material is in a wet state as it happens immediately after construction. Finally, swelling-shrinkage deformations might also be induced by the growth of crystals inside the earth pores, which is caused by a cyclic precipitation-dissolution of salts. If precipitation-dissolution cycles take place on the wall surface, the consequences are relatively minor and are limited to unaesthetic efflorescences that can be easily cleaned. However, if precipitation-dissolution cycles take place inside the wall, for example behind a vapour barrier, the consequences might be more serious as the confined growth of crystals produces stress concentrations that may affect structural integrity.

Few studies have investigated the long-term durability of full scale earth buildings exposed to in-situ weathering. These studies have unsurprisingly shown that stabilised earth structures exhibit lower erosion rates than unstabilised ones. They have also indicated that the cumulative erosion, measured over several years, is relatively small and may be deemed acceptable even for unstabilised earth buildings, if one takes into consideration the average service life of the structure.

Distinct laboratory tests have been developed to assess the durability of earth materials exposed to environmental actions, which has led to the multiplication of experimental protocols. Some of these protocols are complicated, time consuming and require equipment that is often unavailable in conventional laboratories. The proliferation of rather complex experimental procedures has also impeded the formulation of universally accepted testing standards for characterizing the durability of earth materials. A simplification may, however, be achieved by separating the durability protocols from the weathering protocols. In other words, one single normalised test may be used to measure the durability of the weathered material while distinct experimental protocols may be employed to reproduce the effects of the different environmental actions. In this way, the choice of the test assessing material durability becomes independent of the environmental action under consideration. A relatively simple durability test may be chosen to maximise accessibility by practitioners and facilitate the formulation of universally accepted laboratory standards.

Stabilisation has been employed since thousands of years to improve the engineering properties of earth materials. Some stabilisation techniques date back to ancient civilizations such as the Babylonians, who employed plant derivatives to enhance the durability of earth buildings exposed to weathering. In general, stabilisation improves durability but also reduces hygroscopicity, and hence the moisture buffering capacity of the earth, which in turn produces a decrease of hygrothermal inertia and vapour permeability. The specific physical and mineralogical characteristics of each earth should therefore be considered before selecting a suitable stabilisation method in order to achieve an optimum balance between durability, strength, environmental impact and financial costs.

References

1. AFNOR XP-P13-901 (2001) Compressed earth blocks for walls and partitions: definitions—specifications—test methods—delivery acceptance conditions
2. Akagawa S, Nishisato K (2009) Tensile strength of frozen soil in the temperature range of the frozen fringe. Cold Reg Sci Technol 57(1):13–22
3. Alam I, Naseer A, Shah AA (2015) Economical stabilization of clay for earth buildings construction in rainy and flood prone areas. Constr Build Mater 77:154–159
4. Ali F, Nadjai A, Silcock G, Abu-Tair A (2004) Outcomes of a major research on fire resistance of concrete columns. Fire Saf J 39(6):433–445
5. Alonso EE, Gens A, Josa A (1990) A constitutive model for partially saturated soils. Géotech 40(3):405–430
6. Andrady AL, Hamid SH, Hu X, Torikai A (1998) Effects of increased solar ultraviolet radiation on materials. J Photochem Photobiol, B 46(1):96–103
7. Anger R, Fontaine L, Houben H (2009) Influence of salt content and pH on earthen material workability. (in French). In: Mediterra 2009—1st Mediterranean conference on earth architecture, Edicom Edition, Cagliari (CD)

8. ASTM D2845-08 (2017) Standard test method for laboratory determination of pulse velocities and ultrasonic elastic constants of rock. ASTM international: West Conshohocken, PA, USA
9. ASTM D558-11 (2011) Standard test methods for moisture-density (unit weight) relations of soil-cement mixtures. ASTM International: West Conshohocken, PA, USA
10. ASTM D559/D559M-15 (2015) Standard Test Methods for Wetting and Drying Compacted Soil-Cement Mixtures. ASTM International: West Conshohocken, PA, USA
11. ASTM D560/D560M-16 (2016). Standard Test Methods for Freezing and Thawing Compacted Soil-Cement Mixtures. ASTM International: West Conshohocken, PA, USA
12. Aubert JE, Gasc-Barbier M (2012) Hardening of clayey soil blocks during freezing and thawing cycles. Appl Clay Sci 65:1–5
13. Azwa ZN, BF, Manalo AC, Karunasena W (2013) A review on the degradability of polymeric composites based on natural fibres. Mater. Des. 47, pp 424-442
14. Barton GA (1926) On binding-reeds, bitumen, and other commodities in ancient Babylonia. J Am Orient Soc 46:297–302
15. Basma AA, Al-Homoud AS, Malkawi AIH, Al-Bashabsheh MA (1996) Swelling-shrinkage behavior of natural expansive clays. Appl Clay Sci 11(2–4):211–227
16. Beckett CTS, Augarde CE (2012) The effect of humidity and temperature on the compressive strength of rammed earth. In: Proceedings of 2nd European conference on unsaturated soils (pp 287–292)
17. Beninia KCCC, Voorwald HJC, Cioffi MOH (2011) Mechanical properties of HIPS/sugarcane bagasse fiber composites after accelerated weathering. Procedia Eng 10:3246–3251
18. Bestraten Castells SC, Hormias Laperal E, Altemir Montaner A (2011) Construcción con tierra en el siglo XXI. Inf Constr 63(523):5–20
19. Bruno AW (2016) Hygro-mechanical characterisation of hypercompacted earth for building construction, PhD Thesis, Université de Pau et des Pays de l'Adour
20. Bruno AW, Gallipoli D, Perlot C, Mendes J (2019) Optimization of bricks production by earth hypercompaction prior to firing. J Clean Prod 214:475–482
21. Bruno AW, Gallipoli D, Perlot C, Mendes J (2017) Mechanical behaviour of hypercompacted earth for building construction. Mater Struct 50(2):160
22. Bruno AW, Gallipoli D, Perlot C, Mendes J (2017) Effect of stabilisation on mechanical properties, moisture buffering and water durability of hypercompacted earth. Constr Build Mater 149:733–740
23. Bruno AW, Perlot C, Mendes J, Gallipoli D (2018) A microstructural insight into the hygro-mechanical behaviour of a stabilised hypercompacted earth. Mater Struct 51(1):32
24. Bruno P (2005) Military rammed earth constructions: fortifications of the period of muslim domination. In: Arquitectura de terra em Portugal/Earth architecture in Portugal, Argumentum pp. p-39
25. Buchanan AH (2000) Fire performance of timber construction. Prog Struct Mater Eng 2(3):278–289
26. Bui QB, Morel JC (2015) First exploratory study on the ageing of rammed earth material. Mater 8:1–15
27. Bui QB, Bui TT, El-Nabouch R, Thai D-K (2019) Vertical rods as a seismic reinforcement technique for rammed earth walls: an assessment. Adv Civil Eng Article ID 1285937, 12 p.
28. Bui QB, Morel JC, Hans S, Walker P (2014) Effect of moisture content on the mechanical characteristics of rammed earth. Constr Build Mater 54:163–169
29. Bui QB, Morel JC, Reddy BV, Ghayad W (2009) Durability of rammed earth walls exposed for 20 years to natural weathering. Build Environ 44(5):912–919
30. Cadena C, Acosta D (2014) Effects of Solar UV radiation on materials used in agricultural industry in Salta, Argentina: study and characterization. J Mater Sci Chem Eng 2(04):1
31. Carneiro P, Jerónimo A, Silva V, Cartaxo F, Faria P (2016) Improving building technologies with a sustainable strategy. Procedia Social Behavioral Sci 216:829–840
32. Champiré F, Fabbri A, Morel JC, Wong H, McGregor F (2016) Impact of hygrometry on mechanical behavior of compacted earth for building constructions. Constr Build Mater 110:70–78

33. Chepil WS, Woodruff NP (1963) The physics of wind erosion and its control. Adv Agron 15:211–302
34. Ciancio D, Robinson S (2011) Use of the strut-and-tie model in the analysis of reinforced cement-stabilized rammed earth lintels. J Mater Civ Eng 23(5):587–596
35. Cóias V, Costa JP (2006) Terra Projectada: Um Novo Método de Reabilitação de Construções em Taipa. Houses and cities Built with earth: conservation, significance and urban quality, 59–61
36. Cultrone G, Rodriguez-Navarro C, Sebastian E, Cazalla O, De La Torre MJ (2001) Carbonate and silicate phase reactions during ceramic firing. Eur J Mineral 13(3):621–634
37. DeLong HH (1959) Rammed earth walls, in agricultural experiment station circulars. SDSU agricultural experiment station
38. Dhami NK, Mukherjee A (2015) Can we benefit from the microbes present in rammed earth?. In: Rammed earth construction: cutting-edge research on traditional and modern rammed earth, p 89
39. DIN 4102 (1998) Fire behaviour of buildings materials and buildings components-part 1: buildings materials, concepts, requirements and tests
40. DIN 18945 (2013) Earth blocks—Terms and definitions, requirements, test methods
41. Drdácký M, Lesák J, Niedoba K, Valach J (2014) Peeling tests for assessing the cohesion and consolidation characteristics of mortar and render surfaces. Mater Struct 45(6):1947–1963
42. Eires R, Camões A, Jalali S (2017) Enhancing water resistance of earthen buildings with quicklime and oil. J Clean Prod 142:3281–3292
43. EN 13529 (2003) Products and systems for the protection and repair of concrete structures. Test methods.In: Resistance to severe chemical attack. Brussels, CEN
44. EN 15801 (2009) Conservation of cultural property. Test methods.In: Determination of water absorption by capillarity. Brussels, CEN
45. Fabbri A, Champiré F, Soudani L, McGregor F, Wong H Poromechanics of compacted earth for building applications. In: Poromechanics VI, pp 664–671
46. Fabbri A, Morel JC (2016) Nonconventional and vernacular construction materials: characterisation, properties and applications. In: Harries KA, Sharma B (eds.) Woodhead publishing.
47. Faria P, Santos T, Aubert J-E (2016) Experimental characterization of an earth eco-efficient plastering mortar. J Mater Civ Eng 28(1):04015085
48. Faria P, LimaJ (2018) Rebocos de terra. Cadernos de Construção com Terra 3. Argumentum. ISBN 978-989-8885-04-3
49. Faria P, Silva V, Pereira C, Rocha M (2012) The monitoring of rammed earth experimental walls and characterization of rammed earth samples. Rammed Earth Conservation pp 91–97
50. Fernandes F, Lourenço PB, Castro F (2010) Ancient clay bricks: manufacture and properties. In: Bostenaru Dan M, Prikryl A, Torok A (eds.) Materials, technologies and practice in historic heritage structures, Springer Science+Business Media B.V
51. Fontaine L, Anger R, Houben H (2009) Some stabilization mechanisms of earth—stabilization of earth by clay-polymer. In: Mediterra 2009—1st Mediterranean conference on earth architecture, Edicom Edition, Cagliari (CD)
52. Fuentes-García R, Valverde-Palacios I, Valverde-Espinosa I (2015) A new procedure to adapt any type of soil for the consolidation and construction of earthen structures: projected earth system. Mater Constr 65(319):063
53. Gallipoli D, Bruno AW (2017) A bounding surface compression model with a unified virgin line for saturated and unsaturated soils. Géotech 67(8):703–712
54. Gerard P, Mahdad M, McCormack AR, Francois B (2015) A unified failure criterion for unstabilized rammed earth materials upon varying relative humidity conditions. Constr Build Mater 95:437–447
55. Gomes MI, Gonçalves TD, Faria P (2016) Hydric behavior of earth materials and the effects of their stabilization with cement or lime: study on repair mortars for historical rammed earth structures. J Mater Civ Eng 28(7):1–11. https://doi.org/10.1061/(ASCE)MT.1943-5533.0001536

56. Gomes MI, Gonçalves TD, Faria P (2017) Earth-based mortars for repair and protection of rammed earth walls. Stabilization with mineral binders and fibers. J Clean Prod https://doi.org/10.1016/j.jclepro.2017.11.170
57. Grossein O (2009) Modélisation et simulation numérique des transferts couplés d'eau, de chaleur et de solutés dans le patrimoine architectural en terre, en relation avec sa dégradation. Ph.D. Thesis, Université Joseph Fourier, Grenoble, France
58. Guettala A, Abibsi A, Houari H (2006) Durability study of stabilized earth concrete under both laboratory and climatic conditions exposure. Constr Build Mater 20(3):119–127
59. Guerrero Baca LF, Soria López FJ (2015) Sustainability of low income dwellings with shed compacted earth (TVC) in México. In: Arquitectura en tierra, patrimonio cultural-XII CIATTI. Congreso de arquitectura en tierra en cuenca de campos. Cátedra juan de Villanueva, Valladolid,Spanish pp 143–152
60. Guerrero Baca LF, Soria J, Garcia B (2010) Lime on earth architecture design and conservation (in Spanish). In: Arquitectura construida en tierra, tradición e innovación. Cátedra juan de Villanueva, Valladolid, pp 177–186
61. Han Q, Qu J, Dong Z, Zu R, Zhang K, Wang H, Xie S (2014) The effect of air density on sand transport structures and the adobe abrasion profile: a field wind-tunnel experiment over a wide range of altitude. Bound-Layer Meteorol 150(2):299–317
62. Heathcote KA (1995) Durability of earthwall buildings. Constr Build Mater 9(3):185–189
63. Houben H, Guillaud H (2008) Earth Construction: a comprehensive guide, 3rd edn. CRATerre-EAG, Intermediate Technology Publication, London, UK
64. Huang P, Peng X (2015) Experimental study on raindrop splash erosion of Fujian earth building rammed earth material. Mater Res Innovations 19(sup8):S8-639
65. ICONTEC NTC 5324 (2004) Bloques de suelo cemento para muros y divisiones. Definiciones. Especificaciones. Métodos de ensayo. Condiciones de entrega. Instituto Colombiano de Normas Técnicas y Certificación, Bogotá
66. Ivanov V, Chu J, Stabnikov V (2014) Iron and calcium-based biogrouts for porous soils. Constr Mater 167:36–41. https://doi.org/10.1680/coma.12.00002
67. Jaquin PA, Augarde CE, Gallipoli D, Toll DG (2009) The strength of unstabilised rammed earth materials. Géotech 59(5):487–490
68. Jessberger HL (1981) A state-of-the-art report. Ground freezing: mechanical properties, processes and design. Eng Geology 18(1–4), 5–30
69. Jones MS (2002) Effects of UV radiation on building materials. In UV workshop, Christchurch
70. Jones J, McMullen MJ, Dougherty J (1987) Toxic smoke inhalation: cyanide poisoning in fire victims. Am J Emerg Med 5(4):317–321
71. Kebao R, Kagi D (2012) Integral admixtures and surface treatments for modern earth buildings. In: Modern earth buildings: materials, engineering, constructions and applications 256
72. Khanduri AC, Stathopoulos T, Bédard C (1998) Wind-induced interference effects on buildings—a review of the state-of-the-art. Eng Struct 20(7):617–630
73. Kinuthia JM (2015) The durability of compressed earth-based masonry blocks. In: Eco-efficient masonry bricks and blocks. Woodhead Publishing, Oxford, pp 393–421
74. Kita Y, Daneels A, Romo de Vivar A (2013) Bitumen as raw earth stabilizer. In: Kalish R, Cetina C (eds) Tecnohistoria—Objectos y Artefactos de Piedra Caliza. Madera y Otros materiales, Universidad Autónoma de Yucatan, Merida, Yucatan, pp 174–193
75. Kodur VKR, Phan L (2007) Critical factors governing the fire performance of high strength concrete systems. Fire Saf J 42(6–7):482–488
76. Krakowiak K (2011) Assessment of the mechanical microstructure of masonry clay brick by nanoindentation, PhD Thesis, University of Minho
77. Laborel-Préneron A, Aubert J-E, Magniont C, Tribout C, Bertron A (2016) Plant aggregates and fibers in earth construction materials: a review. Constr Build Mater 111:719–734
78. Lai BT, Wong H, Fabbri A, Branque D (2016) A new constitutive model of unsaturated soils using bounding surface plasticity (BSP) and a non-associative flow rule. Innov Infrast Solut 1(1):3

79. Laím L, Rodrigues JPC, da Silva LS (2014) Experimental analysis on cold-formed steel beams subjected to fire. Thin-Walled Struct 74:104–117
80. Lian-You L, Shang-Yu G, Pei-Jun S, Xiao-Yan L, Zhi-Bao D (2003) Wind tunnel measurements of adobe abrasion by blown sand: profile characteristics in relation to wind velocity and sand flux. J Arid Environ 53(3):351–363
81. Lima J, Correia D, Faria P (2016) Earth mortars: the influence of adding gypsum and particle size of sand (in Portuguese). In: ARGAMASSAS 2016—II Simpósio de Argamassas e Soluções Térmicas de Revestimento, ITeCons, Coimbra, pp 119–130
82. Lima J, Silva S, Faria P (2016) Earth mortars: influence of linseed oil addition and comparison with conventional mortars (in Portuguese). In: Neves J, Ribeiro A (ed) TEST&E 2016—1° Congresso de Ensaios e Experimentação em Engenharia Civil. IST, Lisbon. https://doi.org/10.5281/zenodo.164637
83. Lincoln weather and climate. Growing season data, 1887 to 2016. 2017 13 May 2017; Available from: http://snr.unl.edu/lincolnweather/data/GrowingSeasonData.asp
84. Liu J, Wang T, Tian Y (2010) Experimental study of the dynamic properties of cement-and lime-modified clay soils subjected to freeze–thaw cycles. Cold Reg Sci Technol 61(1):29–33
85. Lubelli B, Cnudde V, Goncalves TD, Franzoni E, van Hees R, Ioannou I, Menendez B, Nunes C, Siede H, Stefanidou M, Verges-Belmin V, Viles H (2018) Towards a more effective and reliable salt crystallization test for porous building materials: state of the art. Mater Struct 51:55
86. Mamlouk MS, Zaniewski JP (2011) Materials for civil and construction engineers, 3rd edn. Pearson Educational International, USA
87. Maniatidis V, Walker P (2003) A review of rammed earth construction. In: Innovation project "Developing Rammed Earth for UK Housing", Natural building Technology Group, Department of Architecture & Civil Engineering, University of Bath
88. Marston NJ (2002) Effects of UV radiation on building materials
89. Maskell D, Heath A, Walker P (2014) Inorganic stabilisation methods for extruded earth masonry units. Constr Build Mater 71:602–609
90. Melo MJ, Bracci S, Camaiti M, Chiantore O, Piacenti F (1999) Photodegradation of acrylic resins used in the conservation of stone. Polym Degrad Stab 66(1):23–30
91. Millogo Y, Aubert JE, Séré AD, Fabbri A, Morel JC (2016) Earth blocks stabilized by cow-dung. Mater Struct 49(11):4583–4594
92. Minke G (2006) Building with earth: design and technology of a sustainable architecture. 2006, Boston, United State of America: Birkhaeuser-Publishers for Architecture, 198
93. Minke G (2007) Building with earth-30 years of research and development at the University of Kassel. In: International symposium on earthen structures, Bangalore, Interline Publishing
94. Mukherjee A, Dhami NK, Reddy BVV, Reddy MS (2013) Bacterial calcification for enhancing performance of low embodied energy soil-cement bricks. In: 3rd International conference on sustainable construction materials and technology
95. National Centers for Environmental Information (2018) Past weather by Zip Code-data table. U.S, Department of Commerce, United States
96. Nowamooz H, Masrouri F (2008) Hydromechanical behaviour of an expansive bentonite/silt mixture in cyclic suction-controlled drying and wetting tests. Eng Geol 101(3):154–164
97. O'Connor J (1973) The Adobe book. Ancient City Press, Santa Fe, NM
98. Obonyo E, Exelbirt J, Baskaran M (2010) Durability of compressed earth bricks: assessing erosion resistance using the modified spray testing. Sustainability 2(12):3639–3649
99. Oliveira C, Varum H, Costa A (2013) Adobe in art-nouveau constructions in Aveiro. In: Art nouveau and ecology. Historical lab 4-raw materials and art nouveau, 26 January, Aveiro
100. Ouellet-Plamondon CM, Habert G (2016) Self-compacted clay based concrete (SCCC): proof-of-concept. J Clean Prod 117:160–168
101. Parracha J, Santos Silva A, Cotrim M, Faria P (2019) Mineralogical and microstructural characterisation of rammed earth and earthen mortars from 12th century Paderne Castle. J Cult Herit 42:226–239

102. Parracha J, Pereira AS, Velez da Silva R, Almeida N, Faria P (2019) Efficacy of iron-based bioproducts as surface biotreatment for earth-based plastering mortars. J Clean Prod 237:117803
103. Perrot A, Rangeard D, Menasria F, Guiheneuf S (2018) Strategies for optimizing the mechanical strengths of raw earth-based mortars. Constr Build Mater 167:496–504
104. Pontes J, Santos Silva A, Faria P (2013) Evaluation of pozzolanic reactivity of artificial pozzolans. Mater Sci Forum 730:433–438
105. Qu JJ, Cheng GD, Zhang KC, Wang JC, Zu RP, Fang HY (2007) An experimental study of the mechanisms of freeze/thaw and wind erosion of ancient adobe buildings in northwest China. Bull Eng Geol Env 66(2):153–159
106. Quagliarini E, Lenci S, Iorio M (2010) Mechanical properties of adobe walls in a Roman Republican domus at Suasa. J Cult Herit 11:130–137
107. Randriamanana TR, Lavola A, Julkunen-Tiitto R (2015) Interactive effects of supplemental UV-B and temperature in European aspen seedlings: implications for growth, leaf traits, phenolic defense and associated organisms. Plant Physiol Biochem 93:84–93
108. Ricaud E (2014) Architecture en terre aux Etats-Unis: hybridation des techniques précolombiennes et coloniales. Rapport de mission Richard Morris Hunt Priza-Labex AE and CC, Paris, octobre
109. Rossi P, Charron JP, Bastien-Masse M, Tailhan JL, le Maou F, Ramanich S (2014) Tensile basic creep versus compressive basic creep at early ages: comparison between normal strength concrete and a very high strength fibre reinforced concrete. Mater Struct 47:1773–1785
110. RILEM TC 200-HTC (2007). Recommendation of RILEM TC 200-HTC: mechanical concrete properties at high temperatures—modelling and applications: Part 1: introduction-general presentation. Mater Struct 40:841-853. https://doi.org/10.1617/s11527-007-9285-2
111. Russo S, Sciarretta F (2013) Masonry exposed to high temperatures: mechanical behaviour and properties—an overview. Fire Saf J 55:69–86
112. Santos T, Nunes L, Faria P (2017) Production of eco-efficient earth-based plasters: influence of composition on physical performance and bio-susceptibility. J Clean Prod 167:55–67
113. Scarato P, Jeannet J (2015) Cahier d'expert bâti en pisé: Connaissance, analyse, traitement des pathologies du bâti en pisé en Rhône-Alpes et Auvergne. ISBN 2746678756:978274667875
114. Schroeder H (2016) Sustainable building with earth. Springer
115. Shibi T, Kamei T (2014) Effect of freeze–thaw cycles on the strength and physical properties of cement-stabilised soil containing recycled bassanite and coal ash. Cold Reg Sci Technol 106:36–45
116. Shihata SA, Baghdadi ZA (2001) Simplified method to assess freeze-thaw durability of soil cement. J Mater Civ Eng 13(4):243–247
117. Siavichay D, Narváez M (2010) Propuesta de mejoramiento de las características técnicas del Adobe para la aplicación en viviendas unifamiliares emplazadas en el área periurbana de la ciudad de Cuenca (Bachelor's thesis)
118. Simonsen E, Isacsson U (2001) Soil behavior during freezing and thawing using variable and constant confining pressure triaxial tests. Can Geotech J 38(4):863–875
119. Tamrakar SB, Toyosawa Y, Mitachi T, Itoh K (2005) Tensile strength of compacted and saturated soils using newly developed tensile strength measuring apparatus. Soils Found 45(6):103–110
120. Tien C-C, Chang C-H, Liu B-H, Stanley D, Rabb SA, Yu LL, Nguyen T, Sung L (2014) Effects of temperature on surface accumulation and release of silica nanoparticles in an epoxy nanocoating exposed to UV radiation. In: Proceedings nanotech2014, Washington, DC
121. Val D, Faria P, Silva V (2015) Eco-efficient monolithic walls building solution. Contribute for characterization (in Portuguese). In: CONPAT 2015-XIII Congresso Latino-Americano de Patologia da Construção. IST, Lisboa, pp 7343-1–7343-8
122. Velez da Silva R (2017) Bioconsolidation of construction materials—effect on the durability of an eco-efficient earthen plaster. MSc thesis, NOVA University of Lisbon
123. Vissac A, Bourgès A, Gandreau D, Anger R, Fontaine L (2017). Clays and biopolymers—natural stabilizers for earth construction. CRATerre Éditions, Villefontaine

124. Walker PJ (1995) Strength, durability and shrinkage characteristics of cement stabilised soil blocks. Cement Concr Compos 17(4):301–310
125. Walker P (2002) The Australian earth building handbook HB195, Standards Australia
126. Walker PJ (2004) Strength and erosion characteristics of earth blocks and earth block masonry. J Mater Civ Eng 16(5):497–506
127. Warren J (1999) Conservation of earth structures. Butterworth-Heinemann, Oxford
128. Wendt CJ, Cyphers A (2008) How the Olmec used bitumen in ancient Mesoamerica. J Anthropol Archaeol 27(2):175–191
129. Wright GRH (2005) Ancient building technology. Materials Brill Leiden, Boston
130. Zakaria SF, Rosnan SM (2015) Photodegradation of materials: an overview. In: Proceedings of the international symposium on research of arts, design and humanities (ISRADH 2014). Springer, Singapore, pp 171–178

Chapter 7
Codes and Standards on Earth Construction

B. V. Venkatarama Reddy, Jean-Claude Morel, Paulina Faria, P. Fontana, Daniel V. Oliveira, I. Serclerat, P. Walker, and Pascal Maillard

Abstract Earthen structures provide solutions to build green and sustainable buildings. Any engineered construction needs guidelines on the production of materials, construction of the structural elements, quality control methods and design guidance. There is lack of universally accepted standards on the production of earth construction materials and construction methods as compared to the standards available on conventional materials. The paper attempts to review the existing standards and norms on the earth construction, and bring out the need for comprehensive standard codes on earth construction. An analysis of the existing standard codes on earth construction has been provided. There are about 70 standards, but there is lack of coherence among the standards and globally acceptable terminology. The paper highlights the points needing attention while developing comprehensive globally applicable standards on different types of construction methods.

B. V. V. Reddy (✉)
Indian Institute of Science, Bangalore, India
e-mail: venkat@iisc.ac.in

J.-C. Morel
LTDS-ENTPE, CNRS, University of Lyon, Vaulx-en-Velin, France

P. Faria
CERIS and NOVA School of Science and Technology, NOVA University of Lisbon, Caparica, Portugal

P. Fontana
BAM, Stuttgart, Germany

D. V. Oliveira
ISISE & IB-S, University of Minho, Guimarães, Portugal

I. Serclerat
Lafarge Centre de Recherche, Saint-Quentin-Fallavier, France

P. Walker
Universtiy of Bath, Bath, UK

P. Maillard
CTMNC, Research and Development Department of Ceramic, Ester Technopole, Limoges, France

A. Fabbri et al. (eds.), *Testing and Characterisation of Earth-based Building Materials and Elements*, RILEM State-of-the-Art Reports 35,
https://doi.org/10.1007/978-3-030-83297-1_7

Keywords Codes · Standards · Earth construction · Compressed earth · Rammed earth · Cob · Adobe

7.1 Introduction

Earthen structures, especially earthen dwellings assume importance in the context of environment conservation and emission reduction. History of earthen constructions can be traced back to the dawn of civilisation. Earth or soil is a mixture of clay minerals and inert materials such as silt, sand and gravel. In addition, minor quantities of impurities such as organic matter, salts, etc. could be present. The soil characteristics are mainly controlled by the type and quantity of clay minerals present. The clay minerals can be grouped under two broad categories: (a) expansive clays and (b) less-expansive clays (called non-expansive clays). The expansive clays possess high swell-shrink characteristics when compared to the non-expansive clays. Generally, soils with non-expansive clays are used for earth construction. The earth construction finds applications in (a) walls, including wall elements and masonry mortars, (b) floor/roof systems, (c) foundations and (d) renderings and plastering. There are different types of earth construction techniques. The earth construction methods used for the walls can be grouped under (a) monolithic walls and (b) masonry walls.

Any engineered construction demands guidelines for the production of materials, construction of the structural elements, thermal comfort and durability of the structures/buildings. Even though earth construction exists since the dawn of civilisation, there is lack of universally accepted standards on the material production and construction methods as compared to the standards available on conventional materials such as concrete, masonry, steel, etc. There are attempts in the recent past (6–7 decades) to develop codes of practice and normative standards for modern earth construction, which can facilitate the building professional to design and construct modern earth buildings. There are a large number of regional and national standards on earth construction. The specifications and information on the earth construction in these standards differ widely. There is a need to take stock of the existing information on the earth construction standards and propose comprehensive and universally applicable ones.

7.2 Earthen Structures and Construction Techniques

The earth or the soil has been used for the construction of varieties of structures: buildings, retaining structures (dams, bunds, etc.), embankments, canals, roads, etc. The buildings consist of construction of different structural elements (foundations, walls, roofs/floors, etc.). Table 7.1 gives the details of the earth construction techniques for each of the building components and the status of codes or standards available. Also, the Table gives possibilities of different earth construction techniques

Table 7.1 Earth construction techniques for buildings

Building component	Types of construction techniques	Status of codes or standards
Foundation	– Rammed earth – Earth mortars	Few codes exist
Masonry walls	– Adobe – Compressed earth – Extruded earth – Cut earth (sod and laterite)	Codes exist for some of the techniques
Monolithic walls	– Cob – Rammed earth – Poured earth – Flowable earth mix concrete	Codes exist for some of the techniques
Infill walls	– Wattle and daub – Bagged earth	No codes
Floors/roofs	– Earth panels – Precast elements – In-situ techniques	No codes
Plastering and renderings/finishes	– Earth mortars – Plastering/renderings – Clay panels	Few codes exist

for each of these building components. One of the major problems in developing codes/standards is the common terminology for different types of earth construction techniques and earthen materials. There are regional specific terms for some of the modern earth construction techniques and materials. The regional definitions may have to be avoided to suite with the common terminology for the global usage. The common definitions or terminology for earthen materials and earthen structures are as follows.

(a) ***Earth***

Earth is the basic material required for the manufacture of earthen building materials and construction of earthen structures. Another commonly used term for the earth is soil. It is formed due to the weathering process of rocks over a period of millions of years. Earth or soil is a granular material consisting of inert crystalline silica particles and clay minerals. The silica particles are called silt, sand and gravel (depending upon the grain size of the particle). There are different types of clay minerals and the soil or earth characteristics depend upon the type and quantity of clay minerals present in the granular mixture. The clay minerals are responsible for the swell-shrink characteristics of the soil and the strength of the earth or the soil. There are different regional terminologies for the earth or the soil. From the point of view of global standardisation with reference to earthen structures, the term "Earth" can be used.

(b) ***Earth mortar (EM)***

The EM is basically composed of earth and sand. Sometimes fibres (generally natural fibres) are mixed in order to control the shrinkage cracks. The earth mortars find applications in plasters or renders as well as bed joints in the earth block masonry. The earth mortars used for rendering and plastering can contain colouring agents such as natural oils or pigments.

(iii) ***Stabilised earth mortar (SEM)***

The SEM consists of earth, sand and inorganic stabilisers such as cement or lime. Such mortars are used for the construction of masonry using stabilised earth bricks or blocks. The stabilisers and the granular composition of the mix is dictated by the strength and workability characteristics of the mortar.

(iv) ***Earth block (EB)***

The EB is a masonry unit. There are varieties of EB's, such as adobe bricks or blocks, compressed earth blocks (CEB), extruded earth bricks (EEB), cut earth blocks (sod and laterite), etc. Adobe and CEB can be further classified as stabilised adobe and compressed stabilised earth block (CSEB). The stabilised EB's have inorganic binders in addition to the earth based granular mix.

(e) ***Cob (Cb)***

For cob, processed earth mixture in a plastic state is piled up with some tamping to form a monolithic wall. Frequently organic fibres such as straw are mixed while processing the earth mix. Cob wall construction does not require any formwork.

(f) ***Rammed earth (RE)***

RE is a monolithic construction involving compaction of partially saturated soil or earth-aggregate mixture in layers inside a rigid formwork. A layered texture resembling a sedimentary rock is ascertained for the wall. When inorganic stabilisers (such as cement or lime) are added to the earthen mixture it is designated as stabilised rammed earth (SRE).

(g) ***Wattle and daub***

For wattle and daub, processed earth of plastic consistency is applied on either side of a skeleton structure and finished manually to get even surfaces for the wall. Generally, the skeleton structure is made up of unshaped wooden members and wooden sticks. This skeleton is the wattle, which is daubed from both the sides with the processed earth material.

(h) ***Poured earth (PE)***

For poured earth, processed earth in slurry consistency is poured into a formwork and the formwork is stripped after the earth ascertains stiffness. Generally, it is stabilised with an inorganic binder. Alternatively, the mix can fill fibre bags that are piled while the earth/soil is in plastic state to form walls (bagged earth).

(i) ***Flowable earth mix (FEM) concrete***

This is nearly similar to poured earth but for the inclusion of coarse aggregates. Also, this is termed as mud concrete. The FEM concrete is poured into a formwork and can be vibrated like a conventional cement concrete. The stiffness of the formwork used is much less than that used in the rammed earth construction.

(j) ***Earthen or clay panels***

These are thin precast panels (25–30 mm thick) finding application in the partitioning of spaces and cladding the inner surfaces of walls. The composition can be diverse; one of the possibilities is basically the processed earth mixed with the fibres.

Precast earth elements can also be fabricated with some of the previously mentioned techniques.

7.3 Developments in Earth Construction Standardisation

Archaeological excavations and surveys indicate evidence of using earthen structures and dwellings in almost all the past civilisations. Even in the modern era, large chunk of people living in earthen dwellings is evident. The earliest forms of documents in conveying the methods and rules for earth constructions are in the form of sketches and paintings [25, 77]. Such documents depict the block sizes, wall thickness, use of plumb, etc. During the 16–19th century period (industrial revolution) developing some standards and codes on earth construction can be seen [77]. After the emergence of steel and concrete, the earth construction works for dwellings decreased due to the dominance of modern materials. After natural or man-made disasters, the need for building homes is acute, in such situations earth has emerged as the easily available local material to build homes. Construction of earthen dwellings after the World War II is an example of this kind. Even a German standard [23] emerged in the 1950s and was later withdrawn in 1971 [77]. Large scale rehabilitation works can be seen in Asia and other parts of the world evidenced by the articles and documents [26, 34, 83]. A revival of interest in earth construction was noticed in the past 6–7 decades, which can be mainly attributed to the interest on the abatement of emissions, creating healthy indoor living conditions, conservation of ecology and environment.

7.4 Types of Earth Building Standards/Codes and Documents

Earth construction standards can be grouped under global, regional and national level standard/code documents. The global standards are formed by the International Standards Organisation (ISO). There are hardly any ISO standards on earth construction. Nevertheless, many national and regional standards on earth construction exist.

The existing standards address certain issues pertaining to some specific earth-based technologies. Any form of earth construction needs information on the following.

1. Earth or soil composition
2. Moulds and machinery
3. Earthen products production or manufacturing details
4. Testing and quality control
5. Design guidance
6. Construction methodology and construction procedure
7. Earthquake resistant guidelines
8. Durability, limitations and maintenance.

Existing codes/standards may not address all the points mentioned above in a comprehensive manner. Table 7.2 outlines the details of the some of the standards/documents available on earth construction. The type of document, group it belongs to and the information on the above-mentioned points are provided in the Table 7.2. Many of the available standards (70) spreading across all the continents were considered [1–16, 19–24, 28–33, 35–37, 40–76, 78–82]. The information and guidelines needed for stabilised and un-stabilised earth constructions differ widely. The information on the above-mentioned points is discussed in the following sections.

7.4.1 Soil or Earth Composition

Since it is not possible yet, to guarantee the performance of earthen material from its composition, the TC recommend that the assessment of soil or earth suitability should rather be made using a performance-based approach on earthen samples and/or structures. For stabilised earth construction (such as CEB and rammed earth), the soil composition, especially limiting values on clay content, provides satisfactory performance in terms of strength and durability.

Soil or Earth composition includes primarily grain size distribution, especially the type and quantity of the clay mineral in the mix. The clay mineral type and its quantity in the soil/earth mix is indirectly judged based on the Atterberg's limits. There shall be limiting numbers for other parameters such as salts, organic matter and pH. The range of values specified in several standards on earth construction under two categories: with or without inorganic stabilisers, are given in Table 7.3. The limits on different properties of soil or earth given are limited. Not many codes/standards specify the limits explicitly and many a times a wide range of values are specified. There is a need for developing comprehensive codes on each type of earth construction technology where a range of values are specified for the soil or the earth properties, which are universally acceptable.

Table 7.2 Details of earth building standards and documents

S. No.	Standard/document	Region/country	EC type	Information, guidelines and specifications on earth construction							
				Earth	MM	EPP	T&QC	DG	CM	EQ	ML
1	NBDS [64]	Bolivia	Ad	NA	NA	NA	NA	NA	NA	√	NA
2	NBR 8491 [40]	Brazil	CSEB	NA	NA	NA	√	NA	NA	NA	NA
3	NBR 8492 [41]	Brazil	CSEB	NA	NA	NA	√	NA	NA	NA	NA
4	NBR 10833 [42]	Brazil	CSEB	√	√	√	√	NA	NA	NA	NA
5	NBR 10834 [43]	Brazil	CSEB	NA	NA	NA	√	NA	NA	NA	NA
6	NBR 10836 [44]	Brazil	CSEB	NA	NA	NA	√	NA	NA	NA	NA
7	NBR 11798 [45]	Brazil	EFL	NA	NA	NA	√	NA	NA	NA	NA
8	NBR 12023 [46]	Brazil	CSEB	NA	NA	NA	√	NA	NA	NA	NA
9	NBR 12024 [47]	Brazil	CSEB	NA	NA	NA	√	NA	NA	NA	NA
10	NBR 12025 [48]	Brazil	CSEB	NA	NA	NA	√	NA	NA	NA	NA
11	NBR 12253 [49]	Brazil	EFL	NA	NA	NA	√	NA	NA	NA	NA
12	NBR 13553:[52]	Brazil	SRE	NA	NA	NA	√	NA	NA	NA	NA
13	NBR 13554 [51]	Brazil	EC	NA	NA	NA	√	NA	NA	NA	NA
14	NBR 13555 [52]	Brazil	EC	NA	NA	NA	√	NA	NA	NA	NA
15	NBR 16096 [53]	Brazil	EFL	NA	NA	NA	√	NA	NA	NA	NA
16	Nch 3332 [54]	Chile	Ad, RE, WD, EM	NA	NA	NA	NA	√	NA	√	√
18	DIN 18945-08 [20]	Germany	Ad	√	√	√	√	NA	NA	NA	NA
19	DIN 18946-08 [21]	Germany	Ad	√	√	√	√	NA	NA	NA	NA
20	DIN 18947-08 [22]	Germany	Ad	√	√	√	√	NA	NA	NA	NA
21	IS 4332 Part IV [28]	India	SEWT	NA	NA	NA	√	NA	NA	NA	NA

(continued)

Table 7.2 (continued)

S. No.	Standard/document	Region/country	EC type	Information, guidelines and specifications on earth construction							
				Earth	MM	EPP	T&QC	DG	CM	EQ	ML
22	IS 4332 Part–V [29]	India	SEST	NA	NA	NA	√	NA	NA	NA	NA
23	IS 2110 [30]	India	SRE	√	√	√	√	√	√	NA	–
24	IS 13827 [31]	India	Ad, Cb, RE, WD	NA	NA	NA	√	√	√	√	NA
25	IS 1725 [32]	India	CSEB	√	NA	NA	√	√	NA	NA	√
26	IS 17165 [33]	India	CSEB	NA	√	√	NA	NA	√	NA	NA
27	NMX-C-508-ONNCCE [57]	Mexico	CSEB	√	NA	√	√	NA	NA	NA	NA
28	NBC 203 [58]	Nepal	CSEB and others	–	–	–	–	√	√	√	–
29	NBC 204 [59]	Nepal	Cb, Ad, RE,	–	–	–	–	√	√	√	–
30	NZS 4297 [60]	New Zealand	Ad, CEB, RE, PE, SE	NA	NA	NA	√	√	–	√	NA
31	NZS 4298 [61]	New Zealand	Ad, CEB, RE, PE, SE	√	–	–	√	NA	–	√	√
32	NZS 4299 [62]	New Zealand	Ad, CEB, RE, SE	NA			√	√		√	
33	NTE E. 080 [65]	Peru	EC, Ad, RE	√	√	√	√	√	√	√	√

(continued)

Table 7.2 (continued)

S. No.	Standard/document	Region/country	EC type	Information, guidelines and specifications on earth construction							
				Earth	MM	EPP	T&QC	DG	CM	EQ	ML
34	LNEC [37]	Portugal	EC, RE, SRE	√	√	√	√	NA	√	NA	NA
35	UNE 41410 [82]	Spain	CEB, CSEB	√	√	√	√	√	NA	NA	√
36	SLS 1382 Part 1 [74]	Sri Lanka	CSEB	√		√	√	NA			
37	SLS 1382 Part 2 [75]	Sri Lanka	CSEB	√			√	NA			
38	SLS 1382 Part 3 [76]	Sri Lanka	CSEB	√	√	√		√			
39	CID-GCB-NMBC-14.7.4 [19]	USA	All forms of EC				√	√	√	√	
40	ASTM E2392/E2392-10 [16]	USA	All forms of EC	General statements and general guidelines on all aspects							
Earth building documents											
41	EBAA [24]	Australia	Ad, CEB, RE				√		√	√	
42	HB 195 [27]	Australia	Ad, CEB, RE, PE, SE, EFL, EFT	√	√		√	√	√		√
43	Bulletin 5 [18]	Australia	Ad, CEB, RE, SE	√	√	√	√	√	√		√
44	Lehmbau [38]	Germany	Ad, RE, Cb	√	√	√	√	NA	√	NA	√

Earth Earth composition/grading; *EC* Earth Construction; *MM* Moulds and machinery; *EPP* Earthen product production or manufacturing; *T&QC* Testing and quality control; *DG* Structural design guidance; *CM* Construction methodology and procedure; *EQ* Earthquake resistance design; *ML* Maintenance and limitations; *SEWT* Stabilised earth weathering test; *SEST* Stabilised earth strength test; ***SRE*** Stabilised rammed earth; *Ad* Adobe; *Cb* Cob; *RE* Rammed earth; *WD* Wattle and Daub; *CEB* Compressed earth block; *CSEB* Compressed stabilised earth block; ***PE*** Poured earth; *SE* Stabilised Earth; *NA* Not Applicable; *EFL* Earth floors; *EFT* Earth foundations; *EM* Earth mortar; *SEM* Stabilised earth mortar

Table 7.3 Earth composition parameters specified in different standards/codes

Composition parameter	Unstabilised earth						Stabilised earth					
	Adobe		Rammed earth		CEB		Adobe		Rammed earth		CEB	
	Range	References	Range	References	Range	References	Range	References	Range	References	Range	References
Clay content (%) Silt, sand and gravel (%)	10–12 > 70	NTE E.080 [65]	5–15 > 65	SAZS 724 [73]	≥ 10	UNE 41410 [82]			– > 35	IS 2110 [30]	2–18 > 60	IS 1725 [32], PCH-2–87 [71], SLS 1382 - Part 1 [76]
Liquid limit (%) Plasticity Index (%)									27.0 8.5–10.5	IS 2110 – [30]	≤ 45 < 12	NBR 8491 [40], NBR 8492 [41], NBR 10833 [42], IS 1725 [32]
Organic content (%)												
pH											6 – 8.5	IS 1725 [32], SLS 1382 - Part 1 [76]
Salts			≤ 2	CID-GCB-NMBC -14.7.4. [19], SAZS 724 [73]					≤ 3.0	IS 1725 [32], PCH-2-87 [71]		

7.4.2 Moulds/machinery and Manufacturing Earthen Building Products

Generally, the standards/codes do not provide detailed information on the moulds/machinery used in the production of earthen building products. Some of the codes/standards provide generic information on the moulds/machinery in brief. But some of the earth building documents mentioned in the Table 7.2 provide fairly detailed production techniques for compressed earth blocks, rammed earth and adobe.

7.4.3 Testing/evaluation and Quality Control

Most codes/standards touch upon the issues of testing/evaluation procedures for various properties and quality control aspects. Specifications or procedures for testing/evaluation and quality control aspects of earthen building products pertain to the following:

(1) Dimensions and density
(2) Strength (compression, tension and shear)
(3) Water absorption
(4) Durability

   a. Spray erosion
   b. Mass loss in cyclic wet and dry tests
   c. Expansion/shrinkage.

Procedures for some of the above-mentioned tests have evolved well for the stabilised earth products. Apart from some field tests, the quality control issues are addressed mainly through the limiting values for various tests suggested.

7.4.4 Design Guidance

The design guidance shall address: (1) structural design for gravity, wind and seismic loading; (2) fire safety and accidental damage and (3) thermal performance and moisture movement/buffering. Some standards/codes give information on certain aspects of design guidance. The information on design guidance for earthen buildings, available in the codes/standards, is summarised in Table 7.4. The codes/standards addressing the design guidance are limited. Some codes [32], especially on CEB masonry walls, mention referring to the codes on structural design of masonry. There is hardly any information on the structural design aspects of cob and wattle and daub. There are attempts to evolve design guidance for rammed earth walls. Some documents attempt to follow masonry design philosophy for rammed earth. There is a need for R&D into developing comprehensive design guidance for the earthen structures.

Table 7.4 Design guidance on earthen buildings

Design guidance parameters	Type of earth construction			
	Adobe masonry	CEB masonry	Rammed earth	Wattle and Daub
	Code/References	Code/References		
1. Structural design Design for gravity/wind loading and seismic design	IS 13827 [31], NBC 203 [58], NBC 204 [59]	NBC 203 [58], NBC 204 [59], SLS 1382-Part 3 [76]	IS 13827 [31], IS 2110 [30], NBC 203 [58], NBC 204 [59]	IS 13827 [31]

7.4.5 *Construction Methodology and Construction Procedure*

Details on construction techniques and construction procedures are essential for the construction of earthen structures. These aspects have been addressed by a large number of codes/standards, especially on adobe, CEB and rammed earth buildings, as detailed in the Table 7.2. Also, several books and manuals on earth construction detail these aspects [1, 18, 24, 25, 27, 77].

7.4.6 *Earthquake Resistant Guidelines*

There are few codes/standards on earthquake resistant design for earthen buildings [19, 31, 58–62]. Earthquake resistant features common to load bearing masonry buildings are suggested for earthen buildings with load bearing walls. Providing ties at sill, lintel and roof level and strengthening at the corner junctions and openings is commonly suggested as earthquake resistant feature. Another common type of earthquake resistant feature suggested for adobe masonry walls is providing a mesh layer on the two faces of the wall embedded in an earthen or air lime render and plaster. The height of the earthen buildings in severe earthquake zones is restricted to 1–2 storeyed [58, 59].

7.4.7 *Durability, Limitations and Maintenance*

The durability of unstabilised earthen products is evaluated mainly through accelerated spray erosion test (ASET). There are many attempts by different research groups to standardise ASET. The codes/standards which specify ASET include (NZS 4298 [61], SLS 1382 Part 2 [76]) and generally a limit on the depth of erosion is specified. For stabilised earth products, the durability assessment is carried out by a test measuring mass loss after cyclic wet/dry test and the linear expansion on saturation [32, 76]. For earth plasters a dry abrasion test is also proposed [22].

7.5 Conclusion and Guidelines for Comprehensive Code

There are attempts to develop codes/standards on different types of earthen products for building construction, in different regions across the globe. It appears there are a greater number of attempts in developing the codes/standards on stabilised earth (especially stabilised earth blocks and rammed earth), rather than unstabilised earthen building products. Evolving comprehensive global standards on each one of the earthen products listed in Table 7.1, will be more useful for better promotion of the earth construction. Moreover, it is an absolute necessity to develop new codes and standards in order to encourage the designers to build with earth, as well as to convince the regulatory bodies, which are sometimes reluctant to the use of non-conventional materials [39], that earth is relevant for construction.

Currently many national and regional standards on earth construction exist, but they are not comprehensive in addressing all the existing earth construction techniques and the earthen materials. In the absence of proper regulatory support, the building designers have to fend for themselves [17] leading to difficulties in the building approval processes. Apart from design of buildings there is a need for international laboratory standards in testing the earthen materials and the earthen building products. The new comprehensive global standards should address the following generic items:

1. Earth selection, composition/grading
2. Moulds and machinery
3. Production or manufacturing techniques
4. Testing and quality control
5. Structural design guidance including earthquake resistance design
6. Construction methodology and construction procedure
7. Thermal performance, hygroscopicity and moisture buffering
8. Durability, maintenance and limitations
9. Common glossary on earthen products.

References

1. ACP-EU/CRATerre-BASIN Center for the Development of Industry: Compressed Earth Blocks, Series Technologies. Nr. 5 Production equipment (1996); Nr. 11: Standards (1998); Nr. 16 Testing procedures (1998). Brussels (1996–1998)
2. ARS 670 (1996) Standard for terminology. Centre for Development of Industry (CDI), African Regional Organisation for Standardisation ARSO, Nairobi
3. ARS 671 (1996) Standard for definition, classification and designation of Compressed Earth Blocks. Centre for Development of Industry (CDI), African Regional Organisation for Standardisation ARSO, Nairobi
4. ARS 672 (1996) Standard for definition, classification and designation of earth mortars. Centre for Development of Industry (CDI), African Regional Organisation for Standardisation ARSO, Nairobi

5. ARS 673 (1996) Standard for definition, classification and designation of Compressed Earth Block masonry. Centre for Development of Industry (CDI), African Regional Organisation for Standardisation ARSO, Nairobi
6. ARS 674 (1996) Technical specifications for ordinary Compressed Earth Blocks. Centre for Development of Industry (CDI), African Regional Organisation for Standardisation ARSO, Nairobi
7. ARS 675 (1996) Technical specifications for facing Compressed Earth Blocks. Centre for Development of Industry (CDI), African Regional Organisation for Standardisation ARSO, Nairobi
8. ARS 676 (1996) Technical specifications for ordinary mortars. Centre for Development of Industry (CDI), African Regional Organisation for Standardisation ARSO, Nairobi
9. ARS 677 (1996) Technical specifications for facing mortars. Centre for Development of Industry (CDI), African Regional Organisation for Standardisation ARSO, Nairobi
10. ARS 678 (1996). Technical specifications for ordinary Compressed Earth Block masonry. Centre for Development of Industry (CDI), African Regional Organisation for Standardisation ARSO, Nairobi
11. ARS 679 (1996) Technical specifications for facing Compressed Earth Block masonry. Centre for Development of Industry (CDI), African Regional Organisation for Standardisation ARSO, Nairobi
12. ARS 680 (1996) Code of practice for the production of Compressed Earth Block. Centre for Development of Industry (CDI), African Regional Organisation for Standardisation ARSO, Nairobi
13. ARS 681 (1996) Code of practice for the preparation of earth mortars. Centre for Development of Industry (CDI), African Regional Organisation for Standardisation ARSO, Nairobi
14. ARS 682 (1996) Code of practice for the assembly of Compressed Earth Block masonry. Centre for Development of Industry (CDI), African Regional Organisation for Standardisation ARSO, Nairobi
15. ARS 683 (1996) Standard for classification of material identification tests and mechanical tests. Centre for Development of Industry (CDI), African Regional Organisation for Standardisation ARSO, Nairobi
16. ASTM E2392/E2392—10 (2010) Standard guide for design of earthen wall building systems. ASTM International West Conshohocken
17. Ben-Alon L, Loftness V, Harries KA (2017) Integrating earthen building materials and methods into mainstream housing projects throughout design, construction, and commissioning stages, 11
18. Bulletin 5 (1992) Earth wall construction, 4th edn (Middleton GF, Revised by Schneider LM). Commonwealth Scientific and Industrial Research Organisation (Division of Building Construction and Engineering), Australia
19. CID-GCB-NMBC-14.7.4 (2006) Regulation & Licensing Dept., Construction Industries Div., General Constr. Bureau: 2006, New Mexico Earthen Building Materials Code. Santa Fe
20. DIN 18945 (2018) Lehmsteine – Begriffe, Anforderungen, Prüfverfahren (Earth blocks—terms and definitions, requirements, test methods)
21. DIN 18946 (2018) Lehmmauermörtel – Begriffe, Anforderungen, Prüfverfahren (Earth masonry mortar—terms and definitions, requirements, test methods)
22. DIN 18947 (2018) Lehmputzmörtel – Begriffe, Anforderungen, Prüfverfahren (Earth plasters—terms and definitions, requirements, test methods)
23. DIN 18951, Old German standard
24. EBAA (2004). Earth Building Association of Australia: building with earth bricks & rammed earth in Australia. EBAA, Wangaratta
25. Fathy H (1973) Architecture for the poor: an experiment in rural Egypt. Chicago
26. Fitzmaurice RF (1958) Manual on stabilised soil construction for housing, U.N. Technical Assistance Programme, New York
27. HB 195 (2002) Peter Walker and Standards Australia, The Australian Earth Building Handbook

28. IS 4332 Part–IV (1967) Methods of tests for stabilised soils, Wetting and drying, and freezing and thawing tests for compacted soil-cement mixtures. Bureau of Indian Standards, New Delhi, India
29. IS 4332 Part V (1967) Methods of tests for stabilised soils, Determination of unconfined compressive strength of stabilised soils. Bureau of Indian Standards, New Delhi, India
30. IS 2110 (1980) (reaffirmed 1998), Code of practice for in situ construction of walls in buildings with soil-cement (first revision). Bureau of Indian Standards, New Delhi, India
31. IS 13827 (1993) Improving earthquake resistance of earthen buildings—guidelines. Indian Standard, New Delhi
32. IS 1725 (2013) Stabilized soil blocks used in general building construction—specification (2nd revision). Bureau of Indian Standards, New Delhi, India
33. IS 17165 (2020) Manufacture of stabilized soil blocks—guidelines. Bureau of Indian Standards, New Delhi, India
34. Kotak T (2007) Constructing cement stabilised rammed earth houses in Gujarat after 2001 Bhuj earthquake. In: Proceedings international symposium on earthen structures. Interline publisher, Bangalore, pp 62–71
35. KS02-1070 (1993. Specifications for stabilized soil blocks. Kenya Bureau of Standards, Nairobi, Kenya
36. Legge 24 Diciembre n. 378 (2003) Disposizioni per la tutela e la valorizzazione dell'architettura rurale. Gazzetta Ufficiale, n° 13 (2004)
37. LNEC (1953) The use of earth as a building material (in Portuguese). CIT 9, Laboratório Nacional de Engenharia Civil, Lisbon, Portugal
38. Regeln L (2009) Dachverband Lehm e.V. (Hrsg.): Lehmbau Regeln—Begriffe, Baustoffe, Bauteile, 3., überarbeitete Aufl (Rules of earth construction—terms and definitions, building materials, building elements, 3rd revised edn). Vieweg+Teubner/GWV Fachverlage, Wiesbaden
39. MacDougall C (2008) Natural building materials in mainstream construction: lessons from the UK. J Green Build 3:1–14
40. NBR 8491 (2012) Tijolo de solo-cimento - Requisitos. Rio de Janeiro (Earth-cement bricks—requirements). ABNT
41. NBR 8492 (2012) Tijolo de solo-cimento - Análise dimensional, determinação da resistência à compressão e da absorção de água - Método de ensaio (Earth-cement bricks—dimensions, compressive strength and water adsorption—test procedures). ABNT, Rio de Janeiro
42. NBR 10833 (2012) Fabricação de tijolo e bloco de solo-cimento com utilização de prensa manual ou hidráulica - Procedimento (Production of earth-cement bricks and blocks with manual or hydraulic press—procedure). ABNT, Rio de Janeiro
43. NBR 10834 (2012) Bloco de solo-cimento sem função estrutural – Requisitos (Soil-cement block without structural function—requirements). Associação Brasileira de Normas Técnicas (ABNT), Rio de Janeiro
44. NBR 10836 (2013) Bloco de solo-cimento sem função estrutural – Análise dimensional, determinação da resistência à compressão e da absorção de água (Soil-cement block without structural function—dimensional analysis, compressive strength determination and water absorption—test method). ABNT, Río de Janeiro
45. NBR11798 DE 08 (2012) Materiais para base de solo-cimento — Requisitos. ABNT, Río de Janeiro
46. NBR 12023 (2012) Solo-cimento. Ensaio de compactação (Soil-cement. Compaction test method). Rio de Janeiro, ABNT
47. NBR 12024 (2012). Solo-cimento. Moldagem e cura de corpos de prova cilíndricos. Procedimenyo (Soil-cement. Molding and curing of cylindrical specimens. Procedure). Rio de Janeiro, ABNT
48. NBR 12025 (2012) Solo-cimento. Ensaio de compressão simples de corpos de prova cilíndricos. Método de ensaio (Soil-cement. Simple compression test of cylindrical specimens. Method of test). Rio de Janeiro, ABNT

49. NBR 12253 (2012). Soil-cement—mixture for use in pavement layer—procedure. ABNT, Rio de Janeiro
50. NBR 13553 (2012) Materiais para emprego em parede monolítica de solo-cimento sem função estrutural. Requisitos (Soil-cement materials for monolithic walls of soil-cement without structural function. Requirements). ABNT, Rio de Janeiro
51. NBR 13554 (2012). Solo-cimento. Ensaio de durabilidade por molhagem e secagem. Método de ensaio (Soil-cement. Durability test by wetting and drying. Test method). ABNT, Rio de Janeiro
52. NBR 13555 (2012) Solo-cimento. Determinação da absorção de água. Método de ensaio (Soil-cement. Determination of water absorption. Test method). ABNT, Rio de Janeiro
53. NBR 16096 (2012) Solo-cimento. Determinação do grau de pulverização. Método de ensaio (Soil-cement. Determination of pulverization rate. Test method). ABNT, Rio de Janeiro
54. Nch 3332 (2013) Estructuras - Intervención de construcciones patrimoniales de tierra cruda - Requisitos del proyecto structural (Structural design—retrofitting of historic earth buildings—requirements for the structural design planning). Instituto Nacional de Normalización, Santiago de Chile
55. NTC 5324 (2004) Bloques de suelo cemento para muros y divisones. Definiciones. Especificaciones. Métodos de Ensayo. Condiciones de entrega (Earth-cement blocks for structural and partition walls. Definitions. Specifications. Test methods. Delivery conditions). Bogotá, ICONTEC
56. NF XP P13-901 (2001) Blocs de terre comprimée pour murs et cloisons. Definitions, spécifications, méthodes d'essai, conditions de reception. AFNOR, Paris
57. NMX-C-508-ONNCCE (2015) Industria de la construcción - Bloques de tierra comprimida estabilizados con cal - Especificaciones y métodos de ensayo (Construction industry—eEarth blocks—specifications and test methods)
58. NBC 203 (1994). Nepal National Building code, Nepal, Guidelines for earthquake resistant building construction: Low strength masonry
59. NBC 204 (1994) Nepal National Building code, Nepal, Guidelines for earthquake resistant building construction: Earthen Buildings (EB)
60. NZS 4297 (1998) Engineering design of earth buildings. Standards, New Zealand
61. NZS 4298 (1998) Materials and workmanship for earth buildings. Standards, New Zealand
62. NZS 4299 (1998) Earth buildings not requiring specific design. Standards, New Zealand
63. NIS 369 (1997) Standard for stabilized earth bricks, Standards Organisation of Nigeria, Lagos, Nigeria
64. National Building Code (2006) 1st ed. Federal Republic of Nigeria: LexisNexis Butterworths, Cape Town
65. NTE E. 080 (2000) Adobe. Reglamento Nacional de Construcciones (Adobe. National Building Standards of Peru. SENCICO, Technical Building Standard, Lima
66. NTP 331.201 (1978) Elementos de suelo sin cocer: adobe estabilizado con asfalto para muros: Requisitos (Non-fired earth elements: bitumen stabilized adobe for walls. Requirements). Instituto Nacional de Defensa de la Competencia y de la Protección de la Propiedad Intelectual (INDECOPI), Lima
67. NTP 331.202 (1978) Elementos de suelos sin cocer: adobe estabilizado con asfalto para muros: Métodos de ensayo (Non-fired earth elements: bitumen stabilized adobe for walls. Test methods). Instituto Nacional de Defensa de la Competencia y de la Protección de la Propiedad Intelectual (INDECOPI), Lima
68. NTP 331.203 (1978) Elementos de suelos sin cocer: adobe estabilizado con asfalto para muros: Muestra y recepción (Non fired earth elements: bitumen stabilized adobe for walls. Samples and acceptance). Instituto Nacional de Defensa de la Competencia y de la Protección de la Propiedad Intelectual (INDECOPI), Lima
69. NT 21.33 (1996) Blocs de terre comprimée ordinaires - Spécifications techniques (Common compressed earth blocks—technical specifications). Tunis, Institut National de la Normalisation et de la Proprieté. Industrielle (INNOPRI)

70. NT 21.35 (1996) Blocs de terre comprimée. Définition, classification et désignation (Compressed earth blocks. Definition, classification and designation). Tunis, INNOPRI
71. PCH-2-87 (1988) State Building Committee of the Republic of Kyrgyzstan/Gosstroi of Kyrgyzstan: Возведение малоэтжных зданий и сооружений из грунтоцементобетона PCH-2–87 (Building of low storied houses with stabilized rammed earth), Republic Building Norms RBN-2–87, Frunse (Bischkek) Republic of Kyrgyzstan
72. Regione Piemonte L.R. 2/06 (2006) Norme per la valorizzazione delle costruzioni in terra cruda, B.U.R. Piemonte, n° 3
73. SAZS 724 (2001) Standards Association of Zimbabwe: rammed earth structures. Zimbabwe Standard Code of Practice, Harare
74. SLS 1382 Part 1 (2009) Specification for compressed stabilized Earth Blocks, Part 1: Requirements, Sri Lanka Standards Institution, Colombo, Sri Lanka
75. SLS 1382 Part 2 (2009) Specification for compressed stabilized Earth Blocks—Part 2: Test methods. Sri Lanka Standards Institution, Colombo, Sri Lanka
76. SLS 1382-Part 3 (2009) Specification for compressed stabilized Earth Blocks, Part 3: Guidelines on production, design and construction. Sri Lanka Standards Institution, Colombo, Sri Lanka
77. Schroeder H (2012) Modern earth building codes, standards and normative development. Woodhead Publishing, UK, pp 72–106
78. TS 537 (1985) Cement treated adobe bricks. Ankara, Turkish Standard Institution
79. TS 2514 (1997) Adobe blocks and production methods. Ankara. Turkish Standard Institution
80. TS 2515 (1985). Adobe buildings and construction methods. Ankara. Turkish Standard Institution
81. US 849 (2011) Specification for stabilized soil blocks, 1st edn. Uganda Standard, Uganda
82. UNE 41410 (2008) Bloques de tierra comprimida para muros y tabiques. Definiciones, especificaciones y métodos de ensayo (Compressed earth blocks for structural and partition walls. Definitions, specifications and test procedures). AENOR, Madrid
83. Verma PL, Mehra SR (1950) Use of soil-cement in house construction in the Punjab. Indian Concrete J 24(4):91–96

Chapter 8
Environmental Potential of Earth-Based Building Materials: Key Facts and Issues from a Life Cycle Assessment Perspective

Anne Ventura, Claudiane Ouellet-Plamondon, Martin Röck, Torben Hecht, Vincent Roy, Paula Higuera, Thibaut Lecompte, Paulina Faria, Erwan Hamard, Jean-Claude Morel, and Guillaume Habert

Abstract The global challenge of large-scale climate change mitigation requires action also in the building and construction sector. From a life cycle perspective, and considering the mitigation timeframe, the issue of reducing embodied GHG emissions is gaining attention. Effective ways to reduce embodied GHG emissions have been proposed by the use of fast-growing, bio-based materials, due to carbon sequestered in the biomass. Another promising, yet largely under-explored option is to harness the environmental potentials and low embodied GHG emissions of earth-based materials for building construction. Earth construction dates back from

A. Ventura (✉) · P. Higuera · E. Hamard
University Gustave Eiffel MAST/GPEM, Campus of Nantes, Route de Bouaye CS5004, FR-44344 Bouguenais, France
e-mail: anne.ventura@univ-eiffel.fr

C. Ouellet-Plamondon · V. Roy
Département de Génie de La Construction, École de technologie supérieure, 1100, rue Notre-Dame Ouest, Montréal, Qc 3C 1K3, Canada

M. Röck
Faculty of Engineering Science, Architectural Engineering Unit, KU Leuven, Kasteelpark Arenberg 1, 3001 Leuven, Belgium

M. Röck · T. Hecht
Graz University of Technology, Working Group Sustainable Construction, Waagner-Biro-Straße 100, 8020 Graz, Austria

T. Lecompte
University Bretagne Sud, UMR CNRS 6027, IRDL, 56100 Lorient, France

P. Faria
CERIS and NOVA School of Science and Technology, NOVA University of Lisbon, 2829-516 Caparica, Portugal

J.-C. Morel
Faculty of Engineering, Environment and Computing, Coventry University, 3 Gulson Road, Coventry CV1 2JH, UK

G. Habert
Chair of Sustainable Construction, ETH Zurich, Zürich, Switzerland

A. Fabbri et al. (eds.), *Testing and Characterisation of Earth-based Building Materials and Elements*, RILEM State-of-the-Art Reports 35,
https://doi.org/10.1007/978-3-030-83297-1_8

10,000 to 8000 BC and has been derived in many vernacular construction techniques. More recently, some earthen techniques have been modified, using stabilizers, mainly cement and lime, to increase strength and water stability. The objective of this article is to compare existing literature performed on the LCAs applied to various earthen construction techniques and seek for key factors. Transports as well as binder stabilizations are very influent on the results. Climate, nature of local soil, and geographical context are very influent on functionalities of buildings, mix design and transports, themselves influencing environmental impacts. According to design choices and local context, earthen construction is not always better than concrete. This means that no universal solution can be recommended with the LCA of an earthen wall. The solution has to be adapted to the local context. All references comparing walls material to conventional materials at the building scale, find better environmental performances of earthen walls compared to fired brick walls. However, a full comparison between earthen construction and conventional materials should account for the use phase: combining LCA models with thermal and durability models is a key research issue. Finally, it certainly would be useful to seek for solutions with best environmental performances in a local context, accounting for the nature of soil, the building's functional requirements as well as geographical and cultural specificities. Such an approach would ensure to lower environmental impacts but represents a drastic change in current construction practices. Whereas today building materials are standardized in order to fit with construction working practices, this paradigm shift would require to adapt construction working practices to the local material and context. As earthen construction is today, in many countries of the world, a re-emerging technique, and new professional practices are yet to be established, it seems possible to make this paradigm shift happen. Certainly, in the current context of the need to substantially reduce building-related GHG emissions, there is still strong potential in earth construction techniques for both research and building practice.

Keywords Cumulative energy demand · Mix design · Life cycle inventory · Functionality

8.1 Introduction

The construction sector has for long been identified as one of the most contributing sectors to climate change with 30% of total greenhouse gases emissions in the world, mainly due to heating and cooling energy [59]. Moreover, recent studies have highlighted the growing importance of reducing buildings' material-related, embodied GHG emissions for effective climate change mitigation [49].

It is in this context that one can see a growing interest of civil engineering research on earthen construction. Earth construction dates back from 10,000 to 8000 BC [12, 52] and has been derived in many vernacular construction techniques. Earth can be implemented to build monolithic walls (rammed earth and cob techniques), using masonry units (adobe and Compressed Earth Blocks techniques), as infill of timber frame structures (wattle and daub and light earth techniques), as plasters to protect walls or as mortars either for earth and stone masonry units. Adobe are earth molded air-dried masonry units bedding with a mortar in order to build masonry walls. Adobes can have different dimensions and include, or not, plant fibers, namely if the earth clay content is high. If the clay content is low, they can be stabilized with air-slaked lime. The mortar can be of the same earth as the adobe or air lime-based. Compressed Earth Blocks (CEB) are produced by compacting humid earth in a manual or hydraulic press and joint with a mortar in order to build masonry walls. The CEB can have, or not, holes depending on the mold. A low content of binder is added to the earth to produce stabilized CEB [43]. The masonry units are layered with a mortar that can be earth-based or based on the binder that stabilizes the CEB. Cob walls are made by pilling successive portions of earth-plant fiber mixture, commonly without a formwork [29]. Rammed earth walls are made by compacting successive layers of humid earth inside a formwork until completing the formwork; afterwards, the formwork is disassembled and assembled for the adjacent rammed earth parcel and the process repeated [46].

More recently, some earthen techniques have been modified, using stabilizers to increase strength and water stability. Depending on the local availability of resources and of the construction technique, many different additives have been used from biopolymers such as Casein, starch or blood [67] to mineral additives such as bitumen or lime [31]. Currently the most common stabilizer is cement and is used depending on countries between 3 and 15% in mass of earth products [60].

One reason for the renewal of earth construction is the easiness of implementation as they do not require heavy industrial transformation processes. But the other key interest is that they can be used for excavated soils from earthworks which represent around 75% of total inert waste produced in Europe [50] and currently represent a raising problem for disposal around major urban centers. Finally, earthen construction may have lower environmental impacts in comparison with conventional materials such as cement concrete structures or fired brick masonry, which releases fossil CO_2 for their production [8].

Because environmental impacts of a building are not only provoked by the production of materials, but also by the use and end-of-life phases, it is important to estimate environmental impacts using Life Cycle Assessment (LCA) [21]. Environmental policies concerning the building sector have led to new incentives, tools and regulations, many based on LCA such as standards [13] in Europe. Today, to obtain a chance of spreading in the current practices, earthen construction has, among other aspects, to prove its environmental advantages through LCA studies. Moreover, in a long-term vision, new paradigms for construction practices must emerge towards at least minimal environmental impacts or at best environmental benefits, from the building sector. Earthen construction may be one among other possible solutions, especially if environmental innocuity can be reached.

However, the generic term “earthen construction” hides a wide variety of techniques, of dimensions and of mix designs, including or not additional materials, according to various types of soils, climate, and cultures around the World. Furthermore, existing traditional techniques have to adapt to current economic and regulation mechanisms, and evolve to save costs and to respect standard conformity. These adaptations can vary according to location, and they can require additional processes compared to traditional techniques such as the use of binders, of calibrated materials processed in quarries, or additional mechanical equipment, etc. These additional materials and processes often lead to additional environmental impacts. It is thus important to estimate how much environmental impacts earthen construction could generate considering their variety.

From these reasons, some countries developed their regional and national standards on earth construction. Indeed, earth construction is not limited to a specific climate zone. Standards exist for countries in Europe (Germany, France and Spain), Asia (Nepal, India and Sri Lanka), North and South America (USA, Peru, Chile, Bolivia, Brazil and Mexico) and Oceania (Australia and New Zealand) [63], meaning that earth construction can be versatile and has an expansion potential across the globe. These standards cover varied earth construction materials, from adobe to compressed stabilized earth blocks, from mortar to foundations, to floors and plasters, while also covering many technical aspects such as earth composition, molds, manufacturing, testing, structural design, construction methods, earthquake resistance and maintenance, just to name a few. However, there is still a lack of universally accepted standardization on the material production and construction methods as compared to the standards available on conventional materials, such as concrete or steel.

The objective of this article is thus to provide a review of existing literature performed on the LCAs applied to various earthen construction techniques. In the long term, this can help earthen construction actors to minimize their environmental impacts according to existing local conditions. The present review surveys the different construction techniques which have been analyzed and focuses on variability between studies. The article also wants to highlight key issues for LCA of earthen construction and future research to be done in the future.

8.2 Method

The article selection was conducted using Google Scholar as well as references collected from the different co-authors of the present paper. Because the number of available references is quite small, the review is not restricted to peer-reviewed articles but also includes reports and conference papers. The search included the following key words: earth construction (and derivatives such as earthen construction, earth buildings …) and other key words such as "energy", "life cycle", "impact" and "environment". The review concerns 26 references found in the literature, ranging from 2001 to 2019, with among them: 19 peer-reviewed scientific journals, 5 reports (on-line publications and master's thesis), and 2 conference proceedings.

References cover various earthen construction techniques and various countries (see Fig. 8.1) and some cover more than one technique: 2 articles on cob [10, 18],

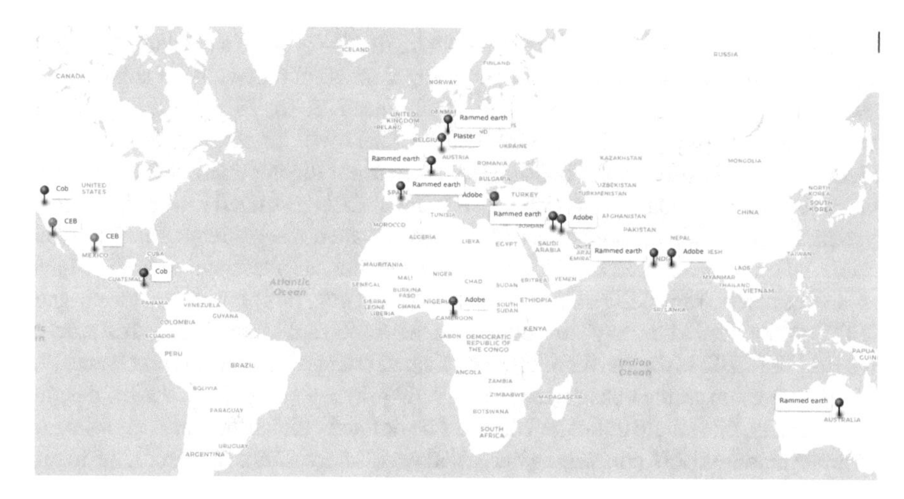

Fig. 8.1 LCA and earthen constructions: locations and techniques found in the corpus of references

4 articles on CEB [16, 20, 25, 51], 4 articles on adobe [2, 15, 17, 56], 8 articles on rammed earth [5, 41, 42, 53–55, 62, 64], one on earth plaster [38] and several other articles on various techniques not based on traditional methods.

References also cover life cycle phases differently. All references consider extraction and manufacturing steps, but only 9 of them consider the use phase. When included, the use phase is exclusively focused on maintenance aspects and does not consider thermal aspects and energy to achieve comfort and indoor air quality. Only 3 references consider the end-of-life phase.

In general, LCA studies in the building sector can have different scales for different purposes. Some aim at comparing different materials and thus collect and provide results at material scale (one block, 1 kg …), others aim at comparing several possibilities of a given building elements (wall, window, roof …), and finally others aim at comparing entire buildings with different solutions including materials, elements, as well as usage scenarios. An increase of scale means an increase of choices and complexity of interpretation, because the number of possible parameters and possible interactions between choices drastically increase, but it also means comparing functions that are more similar. In the present article, the attention is focused on crude earth material used for walls. Thus, in order to allow comparisons between references, results were recalculated at 1 m^2 of the wall surface, when sufficient information was available to do so. Because many articles use different functional units, some assumptions and some calculations were necessary: they are all described in the Appendices of this chapter. The calculation could not be performed for all references because some of them lacked sufficient information. Although some references provide indicators on many LCA impact categories, methods were generally different between references, and it was not possible to compare. Thus, the review focuses on energy. The term Cumulative Energy Demand (CED), is used by some authors [24], whereas others [51] use the term "saturated energy". In both cases, it corresponds to CED, defined by authors as the "energy required along the life cycle of a product, including energy of non-renewable fossil origin, nuclear, biomass or renewable of solar origin, geothermal, wind, and water" [51]. Other references [1, 15, 28, 58, 62, 64] use the term "embodied energy" (EE). In this article, CED will be chosen because EE is a term commonly used in the building sector, but not the appellation for a specific indicator. EE of a product corresponds to a value of CED restricted to a cradle-to-gate system, i.e. use phase and end of life of the product are not considered. Further in the article, when values of CED and EE are compared, the use phase and end of life have been subtracted from CED values when needed. Thus, both terms "cradle-to-gate CED" or "embodied energy" are both used with the same meaning in this article. In LCA, the CED is the most common dedicated energy indicator: it represents the energy harvested in the ecosphere, also called "primary energy". However, despite its popularity, this single indicator can itself be defined quite differently according to existing standards [23], concerning harvested versus harvestable resources, the inclusion or not of renewables, fission and chemical energy sources. Thus, it is important to notice that some uncertainties of further presented results can be due to this lack of uniformity.

8.3 Extraction and Production

8.3.1 Influence of Clay Content in Raw Earth

Venkatarama Reddy and Prasanna Kumar [64] measured compaction energy on experimental rammed earth wallettes and showed an increase from 125 to 150% with an increase of the clay fraction (from 21 to 31.6%). An increase in cement content also increased compaction energy, with a coupled interaction with clay content: a higher clay content required a higher addition of cement, which resulted in an additional increase of compaction energy. However, the importance of compaction energy was very small compared to total energy of the system.

8.3.2 Influence of Binder in Mix Design

Influence of binder content is also interesting to observe. For references that made it possible (enough information provided), cradle to gate CED (or EE) has been calculated for 1 m^2 of wall and plotted versus binder content (see Fig. 8.2). Binders are mainly cement and lime, but when it was different it has been indicated in Fig. 8.2.

The figure shows a cluster of CED values between 0 and 400 MJ/m^2 for binders' contents ranging from 0 to 10%. Inside the cluster, it appears that LCA studies focusing on materials show CEDs values found slightly below studies focusing on wall scale.

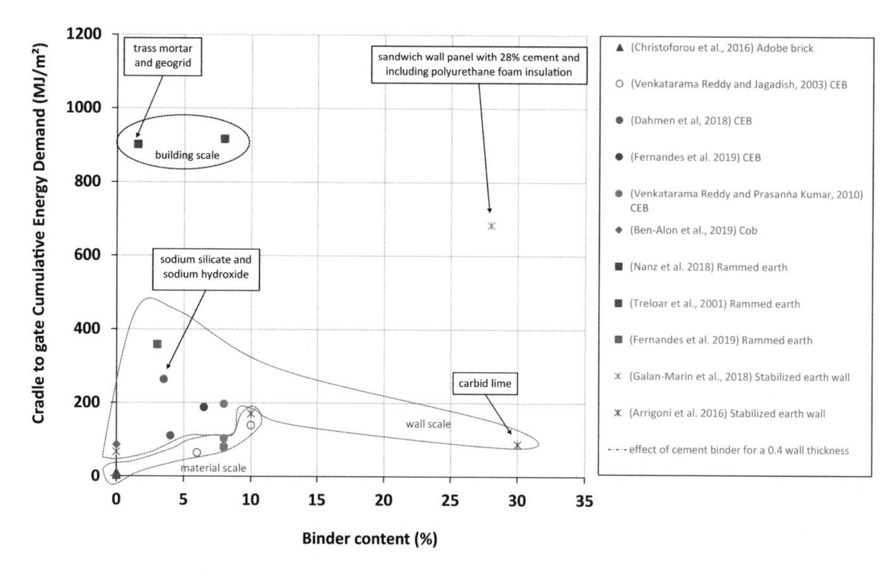

Fig. 8.2 Influence of binder content on the cradle to gate cumulative energy demand—shape is linked to construction technique: Empty circle CEB, Empty square Rammed earth, Empty diamond Cob, * Stabilized earth wall, Empty triangle Adobe—unless indicated otherwise, binders are cement or lime

Outside of this cluster, four outliers are observed. Two outliers provide high results with a low binder content [42, 58]. Both concern studies conducted at building scale. The contribution of transport was found very high for one reference [42] and the binder is not cemented but a mix of trass mortar and geogrid, but no sufficient details are provided to analyze results from Treloar et al. [58]. Another outlier [5] only founds around 90 MJ/m^2 of EE for a 30% binder content. In that specific solution, the binder was composed of flying ashes and carbid lime. The difference can directly be attributed to the type of binder used, because for other results from the same study [5] concerning materials containing cement, the EE was found consistent with the other references. Both flying ashes and carbid lime were obtained from waste valorization and the authors considered them as zero impact inputs. Another system model (end of waste or partition of valorization processes) would probably increase the impacts of that solution. Finally, the last outlier [24] shows a high EE value around 700 MJ/m^2 for a high cement content of 28%. This study is very specific as it does not correspond to a traditional earthen construction technique but to a sandwich panel including a polyurethane foam insulation.

CED values obtained with no binder (for both material and wall scales) are found the lowest. According to details provided by some authors [16], binders are responsible for more than 50% of the total energy consumption, thus it is likely to think that a change in cement content is very influent on CED. To check that idea, the CED of various cement contents of a mix design containing earth and binder in an average wall (thickness 0.4 m and density of earth dry density 2000 kg/m^3) has been calculated, using the ecoinvent 3.3 cut-off database (market process at global scale, CED = 4.2 MJ/kg of Portland cement). It is represented in Fig. 8.2 (dot line). Most of the CED values obtained at material or wall scales and using cement or lime as binders, provide results close to that trend, showing the predominance of these binders to CED in a cradle to gate system. Except the reference using carbid lime [5], two references using others binders, i.e. sodium silicate and sodium hydroxide [16] or trass mortar and geogrid [42], provide CED values largely above the line.

8.3.3 Influence of the Scale of Data Collection

As observed in the previous section, there seem to be a tremendous difference between results when considered scales vary. To further analyze this initial observation, EE per m^2 of the wall have been plotted versus the volume of earth material considered in each studied in Fig. 8.3 (log scale has been used for more convenient representation). There is a clear increase of EE with an increase if scale from material to wall and then to building. This can be explained by the higher complexity of the system studied and the additional materials considered from main material (where only earth is considered) to wall and buildings where many other materials are included.

At wall scale, one could explain the increase by the contribution of bedding mortar as well as construction operations, and possibly renders and plasters.

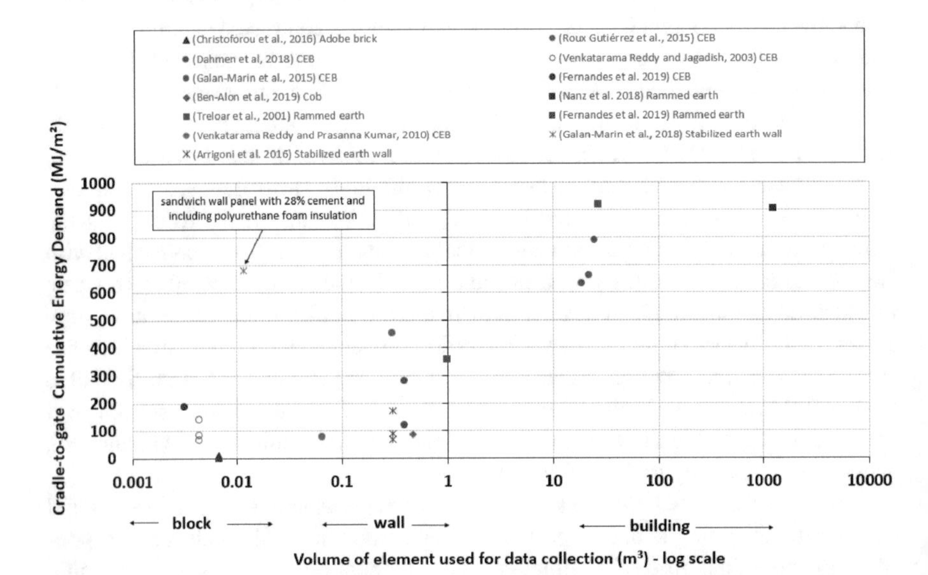

Fig. 8.3 Influence of scale used for data collection—shapes of plots are linked to the construction technique: Empty circle CEB, Empty square Rammed earth, Empty diamond Cob, * Stabilized earth wall, Empty triangle Adobe

At building scale, the number of possible design choices drastically increases compared to material or wall scales. And these choices may interact one another, such as the interaction between the number of floors and wall thickness and/or cement content. In addition, the wall of a building also includes openings, that cannot be subtracted from the results if details are not provided. Furthermore, local conditions and cultural aspects will also play an important role concerning the use of insulators, external and internal coatings. Finally, whereas transport distances have to be assumed for studies at material or wall scales, they are better known at building scale, and may be quite high compared to assumptions. The building scale thus introduces a complexity that is not enough accurately described in existing references, to allow an accurate downscale and comparisons with material or wall scales. However, this complexity represents the actual practice, and highlights that a wall or a building cannot be resumed nor solely characterized by a material, at least in terms of environmental impacts.

8.3.4 Influence of Transports

Various results can be obtained from the corpus of reference. Some results are detailed below for LCAs at wall or building scales.

8.3.4.1 Wall Scale

One study looked in detail about transportation-related impacts on adobe production [15]. They show that compared to soil extraction alone, soil transportation between the soil extraction site and the adobe manufacturing site on a 50 km distance multiplied both CED and GWP100 by a 3 factor [15]. The same study showed that adding a 100 km transport distance between a manufacturing site for adobe and the building site increased these two indicators by around 50% [15]. Including transports, CED for soil extraction was found to be around 4.7 MJ/m^2 for an adobe wall (thickness 0.15 m) [15].

This value can be compared to the CED = 5.5 MJ/m^2 obtained in a previous study [64] for a rammed earth wall (wall thickness 0.2 m, thin in comparison to common rammed earth walls) with soil transported on a 25 km distance. Although distances are different, as well as dry densities and total masses per square meter of the wall, orders of magnitudes are similar for both references. Results on transports are also provided for two cases of rammed earth walls in Germany [42]. For those two cases, transport contribution was found between 55 and 84% of total energy (see Fig. 8.7 in the appendix).

On the contrary, a study concerning two stabilized earth blocks in California [16] found transports representing 22% and 9% of total energy (calculated from Table 8.8 in the appendix). This difference shows clearly that when only earth is used, transport is a predominant parameter to consider for environmental impact assessment. On the contrary, when earth is stabilized with cement, the impact of transport becomes a second order parameter, as it is the case for other industrialized building materials such as steel [27] or concrete [22].

8.3.4.2 Building Scale

The comparison of a cob house to a concrete house located in Nicaragua and Costa Rica [18] also investigated the contribution of transport on climate change indicators. The authors showed that transports contributed to 25% of total GWP100 for the cob house.

Morel et al. [41] provided detailed scenarios of transports for three cases of buildings located in France (see Table 8.9 in appendix). Calculation of energy corresponding to transports was not conducted in the article, only total ton km were provided, but the calculation has been done for the present review from data obtained in the article (see Table 8.10 in appendix). The studied earthen buildings reduced of around 80% the amount energy for transport compared to the current concrete building taken as a reference. Transport only contributed to around 2% of total energy consumption for earthen buildings using local resources when it contributed to around 4% for the reference current concrete building.

The contribution of transports on environmental impacts for two rammed earth façades, one with on-site soil extraction, and the other with off-site soil extraction, was analyzed by Nanz et al. [42]. Two transport operations were distinguished:

- transports between soil extraction and manufacturing sites (A2 stage according to EN15804 [13]), with a distance of 0.61 km/m^2 for the on-site solution, and 7.93 km/m^2 for the off-site one [42],
- transports between the manufacturing and the construction site (A4 stage according to standard EN15804 [13]), with a distance of 0.13 km/m^2 for the on-site solution and null for the off-site one.

For the on-site solution, more than 98% of the materials were transported on a distance under 10 km to the production site.

For the off-site solution, the 1061 tons of soil were excavated from a tunnel construction works, and were transported on a total distance of 9143 km using trucks of 24 tons capacity. The total primary energy demand was found equal to 5200 MJ/m^2. With a transportation credit considering that this soil should have been transported to the nearest landfill 60 km away, the energy was then decreased to 3833 MJ/m^2.

Globally, the total energy consumption of transports accounted for more than 55% for the on-site solution and 84% for the off-site solution [42]. This example shows that it is important to keep results on the A1–A3 phases disaggregated, and also clearly shows that transports have a considerable influence on environmental impacts for building materials with low carbon intensive production processes. On-site soil extraction is indeed an important factor to minimize environmental impacts.

8.3.5 *Influence of the building's Design*

Galan-Marin et al. [25] compared the effect of environmental impacts for different heights of buildings. They found an increase from CED = 630 MJ/m^2 (and GWP = 38.9 kg CO_2 eq/m^2) for one level, to CED = 788 MJ/m^2 (and GWP = 47.9 kg CO_2 eq/m^2) for three levels. More precisely, adding one floor was found to increase both impacts of 4–5% compared to the one-floor building, but adding one more floor was found to increase both impacts of 19% compared to the two-floor building. The necessity to increase mechanical resistance of walls when building with three levels explains this result.

8.4 Use Phase

8.4.1 *Maintenance of Earthen Walls*

The LCA of the maintenance phase of earthen buildings have been included in three LCAs studies [25, 51, 56]. However, none of the studies provide details on maintenance scenarios (i.e. descriptions of types and frequencies of maintenance operations).

Some results are provided on the total of construction and maintenance phase of a CEB wall stabilized with calcium hydroxide and located in Mexico [51] per 1 m^2 of CBE wall (see Table 8.13 in the appendix). Their results showed that both construction and maintenance phases were well below the manufacturing phase, but the detail of maintenance compared to construction, as well as the value of service life considered, are not provided.

The LCA study on an adobe house with a 40-year life span [56] also provided results on maintenance, not detailing maintenance scenarios, but providing amounts of materials necessary to maintain interior and exterior walls. For exterior walls, white cement, samosam and hydrophobizing agent were used for rendering, and for interior walls, painted white cement-samosam plasters were used. This study [56] does not provide impacts of maintenance, only masses of materials, and those are found negligible compared to masses of materials for initial construction (see Fig. 8.8 in the appendix).

8.4.2 *Heating and Cooling Energy: Thermal Aspects*

Thermal properties of materials are a key aspect of the usage phase of any building because they drastically influence the building's energy consumption during its service life. Several physical considerations of the materials have to be considered: thermal conductivity, thermal mass, as well as hygroscopic properties. In addition to materials' properties, the buildings' design and the construction method also play an important role for energy savings. These aspects are detailed below.

8.4.2.1 Material Scale

Thermal conductivity reflects the ability of a material to transfer heat, and it is expressed in W m^{-1} K^{-1}. The thermal resistance of a material is calculated as the ratio between the materials' thickness and its conductivity. A material can be considered as a thermal insulator when its conductivity is at most 0.065 W m^{-1} K^{-1} [40]. For earthen materials, the conductivity increases with the materials' water content. In a plastic physical state, with an important water content, earthen materials' conductivity was found around 2.4 W m^{-1} K^{-1} and it could go down to 0.6 W m^{-1} K^{-1} for a perfectly dried state [57]. Conductivity was linked to density considering that the water content is accounted in the density of an earthen material [30] as resumed in Table 8.1. Thermal resistances of earthen construction walls were found comparable to those of classical materials with an adequate thickness, at least 0.45 m [30].

Thermal inertia represents the ability of a material to resist to a change of temperature. The thermal mass, associated to the thermal conductivity of a material, plays a role in terms of the time necessary for a change of outside temperature to be transferred to the inside temperature, defined as time lag ϕ (Eq. 8.1). The decrement factor f (Eq. 8.2) represents the attenuation of the change of outside temperatures compared to the change of the inside temperatures.

$$\phi = t_{T\text{outside\_max}} - t_{T\text{inside\_max}} \tag{8.1}$$

With ϕ the time lag of temperature wave (h), $t_{T\text{outside\_max}}$ (h) the time of the day at which the outside temperature is minimum, and $t_{T\text{inside\_max}}$ the time of the day (h) at which the inside temperature is maximum.

$$f = \frac{T_{\text{inside\_max}} - T_{\text{inside\_min}}}{T_{\text{outside\_max}} - T_{\text{outside\_min}}} \tag{8.2}$$

Table 8.1 Relationship between the earthen construction technique, thermal conductivity and density [30]

Construction technique	Density (kg/m^3)	Conductivity (W m^{-1} K^{-1})
Cob	1450	0.60
Adobe	1650	0.82
CEB (manual)	1750	0.93
Rammed earth or CEB (mechanical)	2000	1.20

With f the decrement factor (no unit), T_{inside_max} (°C) the maximum inside temperature, T_{inside_min} (°C) the minimum inside temperature, $T_{outside\_max}$ (°C) the maximum outside temperature, and $T_{outside\_min}$ (°C) the minimum outside temperature.

Roux Gutiérrez Rubén et al. [51] measured thermal delays on eight different wall structures out of CEB. The time to reach maximum temperature was further measured on conventional wall structures from concrete blocks and fired bricks. The comparison showed that the time delay took five and a half hours longer with a CEB wall than with other materials. According to Baggs and Mortensen [7], the thermal mass of earthen walls (1740 kJ/(m^3 K) for CEB and 1300 kJ/(m^3 K) for adobe) was comparable to the one of a fired brick wall (1360 kJ/(m^3 K)), below the one of cement concrete (2060 kJ/(m^3 K)) and above the one of an autoclaved aerated concrete block (550 kJ/(m^3 K)). Asan [6] investigated the time lag and decrement factors of several building materials, including clayish earth and pure clay. Table 8.2 provides a part of his results about mineral bulk materials for the building. Thus, the thermal mass effects of clay were found of the same order of magnitude than concrete blocks and bricks, while earth layers were about twice higher regarding time lag and decrement factors [6]. According to Asan [6], it means that, due to thermal inertia, earthen walls buildings are fresher in summer and warmer in winter than conventional building systems.

Hence, beyond thermal inertia, it is important to know if such thermal properties can lead to energy savings. Serrano et al. [55] made an experimental study in summer conditions in Spain using cubicles made of different materials. They compared two kinds of walls: rammed earth and fired brick masonry, with several insulation systems and roofs. They found that the energy consumption with 0.29 m of rammed earth associated with a bio-based insulation material (0.06 m) gave the same cooling consumption than 0.21 m of brick walls insulated with polyurethane (0.03 m). In this study, the theoretical transmittance was 0.563 W/(m^2 K) for the rammed earth wall, and 0.383 W/(m^2 K) for the insulated brick wall. The thermal mass of earthen material thus counterbalanced its lower theoretical transmittance. However, in the same study, rammed earth without insulation (theoretical transmittance of 2.429 W/(m^2 K)) consumed 18–37% more cooling energy than the reference

Table 8.2 Time lag and decrement factors of several mineral building materials, after [6]

Material	Thickness: 0.1 m		Thickness: 0.2 m	
	Time lag ϕ (h)	Decrement factor f	Time lag ϕ (h)	Decrement factor f
Fired brick	2.83	0.343	6.65	0.137
Concrete block	2.88	0.312	6.81	0.118
Sandstone block	2.03	0.519	4.47	0.306
Pure clay layer	2.61	0.396	5.98	0.178
Cement layer	1.89	0.284	5.82	0.128
Earth layer	6.12	0.184	12.08	0.036

brick wall. Thus, if thermal mass of the inner wall certainly is an asset, it is expected not sufficient for energy savings.

Earthen materials are hygroscopic materials, meaning that they tend to adsorb or attract humidity from the air and afterwards desorb or release that moisture. This ability is due to their porosity that allows water and water vapor to circulate into the wall [19, 66]. This hygroscopicity plays a role into thermal behavior of the wall. When an earthen wall is exposed to sun radiations, water contained into pores can evaporate, the water vapor can circulate inside pores towards colder zones, and re-condense. Water condensation will release heat, due to water latent energy, and thus increase temperature. This knowledge on thermal behavior of earthen materials allows to expect energy savings during service life of buildings. It has to be considered on LCA studies.

8.4.2.2 Building Scale

Although materials' thermal properties play an important role for buildings' heating and cooling energy consumptions, many other considerations also influence actual energy consumption. To fully benefit from the interesting thermal properties of earthen materials, buildings' design plays an important role. The actual energy savings due to the thermal mass of earthen constructions depends on the climate as well as on other design choices (buildings' orientation, windows, roof, ground floor).

For houses in New South Wales, Australia, Albayyaa et al. [3] questioned the design strategies in terms of passive solar and thermal mass of the walls. In their case study (transmittance of about 0.3 W/(m^2 K), NSW climate) they found that including thermal mass in the system allowed 35% of energy savings for both heating and cooling. In that specific case study [3], the energy savings due to the use of high thermal mass (brick) instead of low thermal mass (fibro concrete panels) per total floor area was found around 19 kWh/(y m^2) of floor area (to be compared to 68.4 MJ/(y m^2) of floor area for the fibro concrete panels). With a life span of 50 years, it leads to estimate energy consumption of 3.4 GJ/m^2 of floor area, that is drastically more important than the EE of the materials.

The construction technique was also investigated for the building walls in order to allow water vapor to circulate and favor walls' hygroscopic behavior. According to Minke [39], if water vapor cannot be evacuated it would reduce walls' mechanical resistance and favor biological colonization, such as mould. Renders (plaster applied outdoors) protect external walls from rain, but they should not be waterproof so they can be water vapor permeable. For interior walls, direct contact between the wall and indoor air or the use of a porous plaster more permeable than the exterior render, was recommended by Minke [39]. Compared to ancient techniques, recent earthen constructions now use classical concrete foundations and waterproof barriers applied on top of those foundations, that separate the wall from soil, avoiding water to rise by capillarity from the ground into the wall, thus optimizing hygroscopic transfers between interior and exterior.

This complexity certainly explains why no consolidated LCA results can be produced for building walls' service life.

8.5 End of Life

A few references have considered an end-of-life scenario.

The study of a CEB considered inert landfill at the end of life stages and found: GWP100 = 3.4 kg CO_2 eq/m^2 and CED = 48.11 MJ/m^2 [51], that represent 8.2 and 9.4% of the total life cycle, respectively.

On a stabilized CEB building case study [25], the inert landfill was also considered as end-of-life scenario. Results are not provided for 1 m^2 of the wall, but it is possible to estimate, from provided graphs, that the deconstruction and disposal operations contribute around 8–14 and 21–23% of total GWP100 and CED life cycle impacts, respectively.

The study of a stabilized earth façade panel included demolition and disposal phases [24]. From the outer to the inner wall, the façade is composed of a cement mortar render, polyurethane foam as thermal insulator between two layers of stabilized earth, and gypsum plaster inside. The end-of life scenario assumes the final disposal of each of these elements. The climate change indicator GWP100 results for the panel are found equal to 11.466 kg CO_2 eq/m^2 for demolition and 3.106 kg CO_2 eq/m^2 for disposal [24], that represents 38.9% of the total life cycle indicator. The CED results are found equal to 185.731 MJ/m^2 for demolition and 26.637 MJ/m^2 for disposal [24, 42] that represents 42.1% of the total life cycle indicator. Details are also provided for the final disposal of stabilized earth material only [24]: GWP100 = 0.370 kg CO_2 eq/m^2 and CED = 10.242 MJ/m^2 that represent 0.6% and 1.1% of each total life cycle indicator, respectively.

Finally, it has to be noted that landfill impacts associated with earth materials are also considered sometimes as avoided impacts as a growing interest is seen for the use of excavated materials as earth construction products. In this case, the environmental impact associated with earth extraction is allocated to the excavation activities (not related with earth production) and earth production is avoiding an extra landfill impact. This raises the question of allocation [14] but for the moment, earth coming from excavation activities is clearly seen as a waste from the excavation activities.

8.6 Comparisons of Earthen Walls to Other Construction Techniques

In this part, studies that performed comparisons between earthen construction and other more conventional materials are gathered. A distinction is made between studies that were conducted at wall scale to those that were conducted at building scale.

8.6.1 Comparisons at Wall Scale

EE and carbon of several scenarios of adobe have been compared to several other materials [15]: fired clayed brick, concrete blocks and hollow concrete blocks. However, the reference flows are different for the materials (1 kg, 1 brick, or 1 m^3) compared in that reference, and no information is available to recalculate all values for 1 m^2. According to Venkatarama Reddy and Jagadish [61], in the Indian context, the EE of 1 m^3 of an earth-cement block masonry ranges from 646 to 810 MJ/m^3 that is lower than for hollow concrete block masonry (819–971 MJ/m^3), steam cured clayish earth block masonry (1396 MJ/m^3) and fired clay brick masonry (2141 MJ/m^3). However, cubic metre is either not relevant for comparison, as it does not correspond to similar functions.

One study compared various façades designed for similar thermal performance in the Spanish context [24]: a double-sheet façade of stabilized earth panels (SSPF), a double-sheet façade made of ceramic brick masonry (FCBF), a similar double-sheet façade of ceramic brick where the inner sheet is replaced with gypsum plasterboard (PBF), and another double-sheet façade of concrete block masonry (CBF). Although the walls have different total thicknesses (that correspond to different indoor living areas), authors obtained the following results by decreasing order on GWP100 in kg CO_2 eq/m^2: 0.120 for FCBF, 0.103 for CBF, 0.093 for PBF and 0.057 for SSPF [24]. The same ranking between compared solution is obtained for CED in MJ/m^2: 1.615 for FCBF, 1.453 for CBF, 1.241 for PBF and 0.895 for SSPF [24].

In the context of continental USA, another case study compared 4 different exterior load-bearing wall assemblies suitable for up to 2-story construction and having an insulation value meeting or exceeding the requirements of the USA regulation for warm-hot climates [10]: an insulated lightweight sheathed timber platform frame (W), an uninsulated concrete block masonry (CB), an insulated concrete block masonry (ICB) and a cob wall (COB). Authors obtained the following results by decreasing order on GWP100 in kg CO_2 eq/m^2: 74.8 for ICB, 62.7 for CB, 53.1 for W and 13.2 for COB [10]. The same ranking between compared solution is obtained for CED in MJ/m^2: 491 for ICB, 241 for W, 226 for CB and 86.4 for COB [10].

8.6.2 Comparisons at Building Scale

Some studies compared different types of walls for an identical building, thus accounting for their structural functions as well as comparable thermal performance, in order to design the building.

A residential building (one level) made of different wall materials have been compared in the Australian context [58]: rammed earth stabilized with 8% cement, brick veener, and fired brick masonry. EE has been recalculated from the buildings' wall surface for 1 m^2 of wall area (see appendixes) providing: 917, 2460.4 and 2717.4 MJ/m^2, respectively [58].

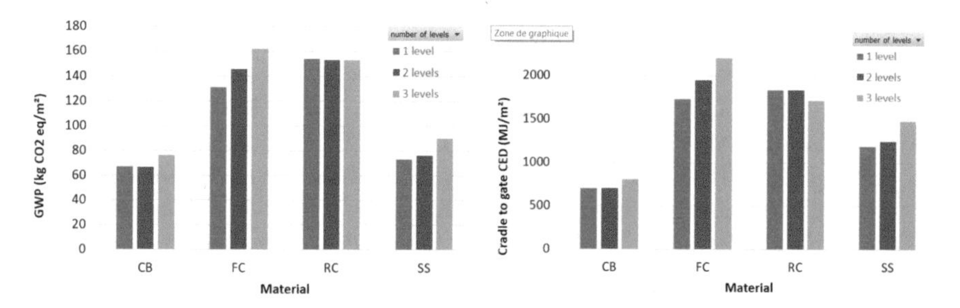

Fig. 8.4 Global warming potential (GWP100) and cumulative energy demand (CED) for the same building, according to the number of levels, after [25]—CB: concrete block masonry, FC: fire clay brick masonry, RC: reinforced concrete and SS: stabilized earth block masonry

The case study of Nanz et al. [42] compared two buildings in the German context, each of them including a comparison between a rammed earth and a fired bricks façade. For the first building, the primary energy of the rammed earth façade is found 150 MJ/m^2 whereas the fired brick façade is found equal to 498 MJ/m^2. For the second building, the primary energy of the rammed earth façade is found 395 MJ/m^2 whereas the brick façade is found equal to 500 MJ/m^2.

In their case study, Galan-Marin et al. [25] considered a building with one, two or three story floors. Their results have been gathered in Fig. 8.4. A stabilized earth wall (SS) is found to have similar GWP but higher CED than a concrete block wall (CB). A fired clay brick wall (FC) and a reinforced concrete wall (RC) are found largely higher for both indicators. Indicators per square metre of wall all increase with the number of floors, except the reinforced concrete wall, for which they remain stable. It is also noticeable that CED obtained by this study is one of the highest EE values of all references considered in the present article, with the highest cement content as previously shown in Fig. 8.2.

8.7 General Discussion

The review highlights some influential aspects on LCA results, mainly the energy consumption indicators, for the three life cycle steps. One could easily conclude from that review that minimum transport and minimum binder content surely improve environmental aspects. Although these are surely key factors that should always be kept in mind by architects and building designers, they have to be further discussed.

Van Damme and Houben [60] modeled CO_2 intensity of earth mix designs by gained resistance (kg CO_2 eq/MPa) as a function of cement content. They showed that the binder addition in earth does not increase the resistance to a sufficient level to make it competitive, from an environmental point of view to cement based concrete: kg CO_2 eq/m^3/MPa seems much higher in stabilized earth construction than for conventional concrete. Thus, the need for using a binder can be questioned as several earthen techniques are available without a binder. However, this study considers that strength is the only function of using cement in earth construction and lack of sufficient consideration for the broader factors needed to make a fair comparison between stabilized earth, unstabilized earth and concrete blocks. In particular the fact that earth stabilization is used for weathering resistance and that earth in general provides wider benefits in terms of indoor comfort which are modified by stabilization [37]. In earthen walls durability should not be assessed only by strength. Weathering simulated tests are also very important [9]. Furthermore, the need for a high strength is very linked to the buildings design, Galan-Marin et al. [25] showing that an increase in the number of floors would change the choice of a material regarding minimum values of GWP100 or CED (see Fig. 8.4).

The functional requirements of building walls are numerous [26] and they are gathered in Table 8.3. Hence, the use of a binder can be required for durability or safety reasons, and this aspect is not considered by Van Damme and Houben [60]. Indeed, in countries with frequent flooding events or frequent pouring rains, binders are useful to avoid penetration of humidity and collapse of walls. The durability

Table 8.3 Possible functions of a building wall—after [26]

Functionality	Description and possible measurable observation
Strength: ability to take up the loads due to its own weight, superimposed loads and lateral pressures	Materials' resistance to compression
	Materials' resistance to rotation
Durability	Wall ability to keep its functionalities in time
	Wall ability to resist current weather events in buildings' location: wind or rain erosion
Thermal performance: ability to preserve desired temperature indoors	Materials' ability to conduct heat flows
	Materials ability to adsorb and desorb water vapor
	Winter comfort: wall ability to preserve comfortable sensation indoors while outdoor temperature is low and ventilation rate is low
	Summer comfort: wall ability to preserve comfortable sensation indoors while outdoor temperature is high
Privacy: ability to preserve intimacy for inhabitants	Sound insulation: wall ability to absorb noise
	Sight insulation
Security: ability to temporary resist to exceptional aggressive events in order to allow safe evacuation	Fire: ability to resist a fire for a certain amount of time
	Seism: ability not to be ruined by a seism
	Water floods: ability not to be ruined by a flood
Safety: ability to be innocuous to health of inhabitants in usual conditions	Chemical inertia towards variable usage conditions

aspects are very dependent on the building's location. As an example, Bui et al. [11] studied the erosion of different rammed earth walls. Over a period of 20 years, the mean erosion depth of the examined walls was found 2 mm (0.5% of wall thickness) for walls stabilized with 5% dry weight of hydraulic lime, and 6.4 mm (1.6% of wall thickness) for unstabilized walls [11]. Thus, service life span would be reduced for unstabilized walls compared to stabilized ones. However, these results are typical of rammed earth walls of a given climate, that is wet continental in that study [11].

The use of wastes as stabilizers instead or at least partially replacing common binders, such as lime or cement replaced by artificial pozzolans, can contribute to reduce the environmental impact of earthen walls. That reduction will be directly correlated to the consumption of those energy intensive binders and the type of binder and, simultaneously, the reduction on waste landfilling [4].

Furthermore, the eventual need to stabilize local earth to use it as building material and water consumption also depend on the building technique. For instance, considering walls with similar thickness, to build an adobe wall consumes much more water in comparison to build a rammed earth wall. An earth with coarse aggregates may be directly used to build a rammed earth wall, while the coarse aggregate needs to be removed, by sieving, before using the earth to produce CEB, adobe or even cob. An earth with relatively low clay content can be used to build unstabilized rammed earth while a binder addition should be needed to use the same earth for CEB, abode or cob. All these aspects should be considered for environmental assessment.

As fully described in a previous part ($0), the thermal aspects are also very complex because saving heating and cooling energy requires to have an overview of correlated aspects. Materials' properties and mix design are important, and may be different according to local soil resources' properties, such as clay content. Wall design has to be considered, with possible additional layers enabling water permeability control and/or additional thermal insulation. The review (see §0) shows big differences in results between LCA studies conducted at masonry unit or material scale and studies at building scale. These differences are probably due to wall designs, generally not considered for studies at masonry unit or material scales. At building scales, wall design generates higher impacts because of additional layers, but that should be balanced with possible energy savings. The building design is also a key aspect especially concerning isolation of walls from foundations, orientation of façades in relationship with local climate. In fact, the application of compatible protective renders and capillary rise barriers can be fundamental for durability but also for thermal performance, depending on the earth technique, exposure and architectural design.

Several references confirm the high hygroscopicity of clayish earth materials as being one of the most advantageous in comparison to other building materials. It may depend on the type and content of clay [36], eventual stabilization [4] and on the surface of the wall. Therefore, earth walls may provide a contribution to passively equilibrate indoor relative humidity and so, reduce the energy consumption to achieve hygrothermal comfort. However, that aspect is not yet quantified on environmental assessment literature.

All references tend to confirm the influence of transport on the environmental performance of earthen construction. Using local material could appear as a very efficient way to lower environmental impacts, and a true added value of earthen construction materials compared to conventional ones. However, the notion of "local" is itself important: as shown by Nanz et al. [42], off-site soil extraction can still be local (distances around 10 km), and will drastically change the results.

Finally, end-of-life phases in existing references, all consider inert landfill scenarios. Clay is a material of natural origin and reintegration of the unsterilized material at the end of life has been described as unproblematic [51]. That is important because earth is not a renewable material. Although the recyclability of clay is documented in several publications [11, 44, 47], current LCAs do not consider such a scenario. Existing LCAs studies considering end-of-life phase all considered the use of stabilizers in the mix design. Stabilizers can be used to improve the properties of CEB [5], but according to Pacheco-Torgal and Jalali [45], Roux Gutiérrez Ruben [51], even if clay has been stabilized with lime or cement, its recyclability is only minimally impaired. The resultant earth product stabilized with lime turns out similar to a clay limestone; however, the same does not happen when it is stabilized with cement. The recycling scenario of unstabilized or air lime stabilized earth could thus be accounted as an alternative to the inert landfill for comparison.

Inert landfill scenarios also fail to highlight the fact that in the context of circular economy, using the earth coming from excavation sites of conventional buildings and infrastructure construction is an economically viable activity [34, 48]. Actually, it becomes more and more difficult due to difficulty for quarry extension to access to natural sand and gravel around cities [32, 33]. Furthermore, landfill costs for excavation materials are increasing due to space limitation and transport distance costs. Both aspects combined raise the interest to use excavation material directly as a building material becomes economically interesting [35]. From an environmental perspective, it means that impacts associated with extraction of earth are allocated to the main excavation activity (construction) and not earth production.

8.8 Conclusion

This review of existing LCAs applied to earthen construction concerns all current techniques: CEB and adobe masonry, cob and rammed earth monolithic walls, some plasters as well as some other particular techniques. This review provides some key points mainly concerning energy demand of earthen construction.

First, it shows that transports, even on small distances, as well as binder stabilizations are very influent on the results. If no cement stabilization is used, the transport of material seems to be the critical parameter. On the opposite, if cement stabilization is used, then the amount becomes the critical driver of environmental impact. However, it should not be concluded that it is necessary to eliminate binder stabilization from earthen techniques. The binder stabilization can prove useful for particular functions (durability or safety) in a given context of use, accounting for local specificities such as the nature of soil and the climate. If reducing transports is a generic advice to lower environmental impacts of earthen construction, and the use of local materials is strongly beneficial, the ability to use local soil, the need to prepare it by sieving and the need to add other materials to the earth mix also depends on the nature of the soil and the building technique.

This leads to the second point concerning local specificities. Climate, nature of local soil, and geographical context are very influent on functionalities of buildings, mix design and transports, themselves influencing environmental impacts. This means that no universal solution can be recommended with the LCA of an earthen wall. The solution has to be adapted to the local context. This also explains the absence of a universal standard for earth construction in favor of regional or national standards.

As a third point, it was not possible to provide a general ranking of different materials among all references, as existing studies use different sets of indicators and lack of sufficient information to enable conversions. Nevertheless, all references comparing wall material to conventional materials at the building scale, find better environmental performances of earthen walls compared to fired brick walls. However, for cement concrete walls, it is not always the case. Then, although one intuitively and commonly assumes that earthen construction has better environmental performances than conventional materials, our analysis shows that according to design choices and local situations earthen construction can have lower performances than concrete.

However, as a fourth point, a full comparison between earthen construction and conventional materials should account for the use and end-of life phases. These are key issues for future researches on LCA of earthen construction. For the use phase, combining LCA models with thermal and durability models is a key issue to enable life cycle performances. Concerning thermal models, there are still research needs to provide models accounting for all particular properties and behaviors of earthen materials as thermal insulators and hygrothermal passive buffers. Concerning durability, some combined approach already on carbonation of reinforced cement concrete [65], and this type of combined models should be extended to all construction materials when relevant. For the end-of-life phase, the existing references only consider inert landfills, and no study considers recyclability of the material or even reuse when the earth is not stabilized.

Finally, it certainly would be useful to seek for solutions with best environmental performances in a local context, accounting for the nature of soil, the building's functional requirements as well as geographical and cultural specificities. Such an approach would ensure to lower environmental impacts but represents a drastic change in current construction practices. Whereas today building materials are standardized in order to fit with construction working practices, this paradigm shift would require to adapt construction working practices to the local material and context. Some countries are paving the way with their standards for earthen construction across various continents. As earthen construction is today, in many countries of the world, a re-emerging technique, and new professional practices are yet to be established, it seems possible to make this paradigm shift happen. Certainly, in the current context of an urgent need to substantially reduce building-related GHG emissions, there is still strong potential in earth construction techniques for both research and building practice.

Appendix 1

Calculations of CED and GWP

Treloar et al. [58]

The LCA study is conducted on a building (Fig. 8.5). To obtain a value for one square meter, the total wall surface is estimated.

Height: 2.4 m, perimeter: $12.3 + 7.2 + 16.5 + 7 + (12.3 - 7) = 41.4$ m.

Total wall surface = 99.36 m^2.

With a global assumption of 10% of openings, the obtained surface is 89.424 m^2.

The article provides an embodied energy of 82 GJ for the building rammed earth walls, that is 917 MJ/m^2.

For other types of walls (brick veneer and hollow brick), the total EE are 220 and 243 GJ, respectively, that is 2460.4 and 2717.4 MJ/m^2, respectively.

Venkatarama Reddy and Jagadish [61]

Earth-cement block

Results are provided for a cubic meter of earth-cement block masonry wall: 646 MJ/m^3 with 6% cement, and 810 MJ/m^3 with 8% cement. Size of blocks are 230 mm × 190 mm × 100 mm (volume = 0.00437 m^3) and height of blocks is assumed to be 100 mm.

The external surface of a block is thus $0.23 \times 0.19 = 0.0437$ m^2. Thus 22.88 blocks are necessary to cover 1 m^2 of the wall surface, that is 0.09998 m^3.

Thus, 1 m^2 of earth-cement block masonry wall requires 64.6 MJ/m^2 with 6% cement, and 81.0 MJ/m^2 with 8% cement.

Lime stabilized steam cured earth blocks

Results are provided for a cubic meter of lime stabilized steam cured earth block walls: 1396 MJ/m^3 with 10% lime. Size of blocks are 230 mm × 190 mm × 100 mm and height of blocks is assumed to be 100 mm.

The external surface of a block is thus $0.23 \times 0.19 = 0.0437$ m^2. Thus 22.88 blocks are necessary to cover 1 m^2 of the wall surface, that is 0.09998 m^3.

Thus, 1 m^2 of lime stabilized steam cured earth block requires 139.6 MJ/m^2.

Venkatarama Reddy and Prasanna Kumar [64]

See Fig. 8.6 and Table 8.4.

Melià et al. [38]

All results are directly provided in the supplementary materials available on the journal's website.

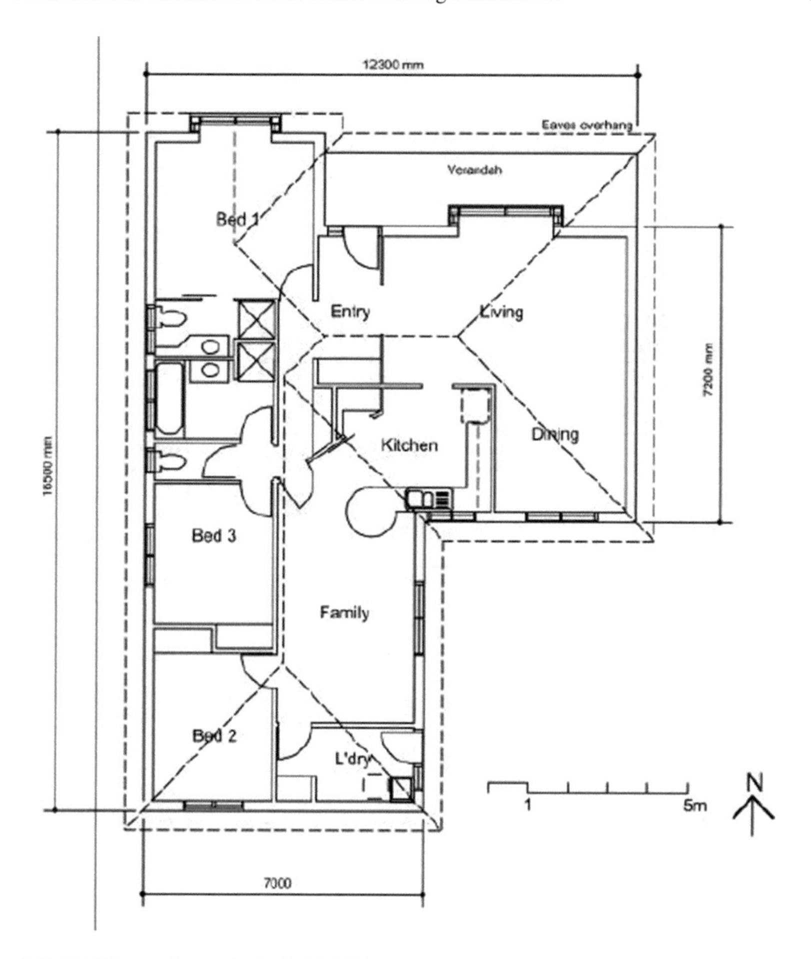

Fig. 8.5 Building under study for LCA [58]

Galan-Marin et al. [25]

See Table 8.5.

Christoforou et al. [15]

Block dimension: 0.30 m × 0.45 m × 0.05 m.

External surface: $0.05 \times 0.3 = 0.015$ m^2.

Number of blocks for 1 m^2: 66.7 blocks/m^2.

Density 1544 kg/m^3 for straw and 1568 kg/m^3 for sawdust.

Table 8.4 Original results [64] and calculations (grey cells) for 1 m^2 of wall

Parameter (unit)	Building A	Building B	Experimental wallette
Clay fraction of the mix (%)	16	12.6	15.8
Moulding water content (%)	10.6	10.8	11
Dry density (kg/m^3)	1800	1800	1800
wall thickness (m)	0.2	0.375	0.15
Compacted in layers of thickness (mm)	35	100	100
Cement content (by weight) (%)	8	8	8
Energy in cement (MJ/m^3)	489.6	489.6	489.6
Compaction energy (MJ/m^3) (animate)	0.174	0.084	0.139
Number of observations	35	8	3
Standard deviation	0.059	0.016	0.009
Energy in mixing (MJ/m^3)	0	7.35	0
Energy in transportation of raw materials (MJ/m^3)	27.5	27.5	27.5
Total energy in rammed earth wall (MJ/m^3)	517.27	524.45	517.24
Surface of the wall (m^2/m^3)	5	2.67	6.67
Soil extraction: energy in transportation of raw materials (MJ/m^2)	5.5	10.31	4.13
Construction: energy for compaction (MJ/m^2)	0.0348	0.0315	0.0209
Construction energy for mixing (MJ/m^2)	0	2.756	0
Construction: energy in cement (MJ/m^2)	97.9	183.6	73.4
Total energy for construction (MJ/m^2)	98.0	186.4	73.5
Total energy in rammed earth wall (MJ/m^2)	103.45	196.67	77.59

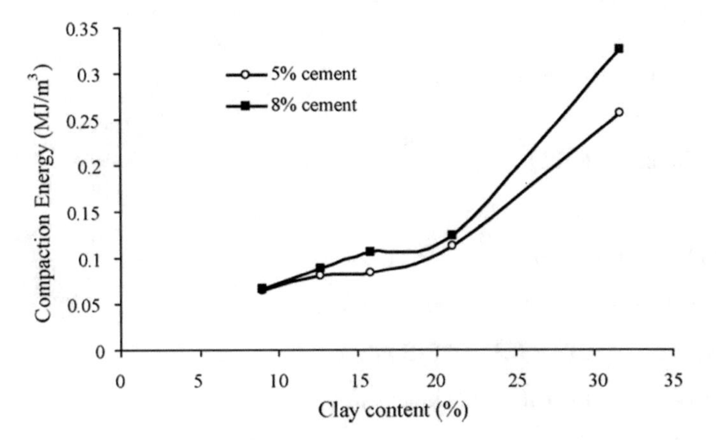

Fig. 8.6 Compaction energy for experimental wallettes according to clay and cement contents [64]

Table 8.5 Original results [25] and calculation or estimations (grey cells) to obtain total volume of walls and values for 1 m^2 of wall

Scenario/unit			1	2	3	4	Average
Span between walls		m	3	3.5	4	4.5	3.75
Floor area		m^2	48	56	64	72	60
Wall thickness		m	0.28	0.3	0.32	0.33	0.3075
Total volume of walls		m^3	18.6	21.6	24.6	26.9	22.9
Density		g/cm^2	1.79	1.79	1.79	1.79	1.79
Total wall mass		kg/m^2		121.83	128.599	142.136	130.855
Length of building		m	16	16	16	16	16
Width of building		m	6	7	8	9	7.5
Height of wall		m	2.4	2.4	2.4	2.4	2.4
Calculated wall surface	L1	m^2	105.6	110.4	115.2	120	112.8
	L2	m^2	211.2	220.8	230.4	240	225.6
	L3	m^2	316.8	331.2	345.6	360	338.4
GWP/m^2 area	SS L1	kg CO_2 eq					4386.23
	SS L2	kg CO_2 eq					9112.18
	SS L3	kg CO_2 eq					16,201.1
GWP/m^2 wall	SS L1	kg CO_2 eq					38.89
	SS L2	kg CO_2 eq					40.39
	SS L3	kg CO_2 eq					47.88
CED/m^2 area	SS L1	MJ					71,145.05
	SS L2	MJ					149,312.04
	SS L3	MJ					266,562.54
CED/m^2 wall	SS L1	MJ					630.72
	SS L2	MJ					661.84
	SS L3	MJ					787.71

See all results in Tables 8.6 and 8.7.

Dahmen et al. [16]

Values are given for one block of which dimension are $0.19 \times 0.19 \times 0.39$ m.

The exposed surface area of one block is thus $0.19 \times 0.39 = 0.0741$ m^2, requiring 13.5 blocks to cover 1 m^2 of the wall. This factor has been applied to provided LCA results.

One block of stabilized soil is 0.00839 m^3, with a 2100 kg/m^3 density, thus a mass of 17.619 kg. Its cement content has been calculated: 0.71 kg cement/block that is 4%.

See all results in Table 8.8.

Fernandes et al. [20]

Results are given for 1 m^3 of the wall.

Table 8.6 Intermediate calculated results for energy from [15]

Scenario (unit)		1	2	3	4	5	6
Diesel fuel Soil extraction	(kWh)	0.00728	0.00728	0.00728	0.00716	0.00716	0.00716
Diesel fuel soil transportation	(kWh)	e	0.0124	0.0124	e	0.0122	0.0122
Diesel fuel Straw/Sawdust transportation	(kWh)	0.000312	0.000312	0.000312	0.000580	0.000580	0.000580
Diesel fuel Adobe brick transportation	(kWh)	e	e	0.0255	e	e	0.0255
Electricity Well water supply	(kWh)	6.08E−5	6.08E−5	6.08E−5	5.98E−5	5.98E−5	5.98E−5
Electricity Straw pre-mixing treatment	(kWh)	0.000122	0.000122	0.000122	e	e	e
Electricity Mixing	(kWh)	0.00141	0.00141	0.00141	0.00139	0.00139	0.00139
Total soil extraction	(MJ)	0.026208	0.070848	0.070848	0.025776	0.069696	0.069696
Total wall construction	(MJ)	0.00685728	0.00685728	0.09865728	0.00730728	0.00730728	0.09910728

Table 8.7 Calculated results for energy and GWP from [15]

Scenario	GWP		Energy soil extraction (MJ)		Energy wall construction (MJ)	
	Results from article (1 block)	Results converted to 1 m^2	Results from article (1 block)	Results converted to 1 m^2	Results from article (1 block)	Results converted to 1 m^2
1	1.76E−03	0.117	0.026	1.748	0.007	0.457
2	5.41E−03	0.360	0.071	4.726	0.007	0.457
3	1.29E−02	0.860	0.071	4.726	0.099	6.580
4	1.70E−03	0.113	0.026	1.719	0.007	0.487
5	5.3E−03	0.353	0.070	4.649	0.007	0.487
6	1.28E−02	0.854	0.070	4.649	0.099	6.610

Table 8.8 Calculated results for energy [16]

Resources (MJ)	Values for one masonry units		Values for 1 m^2	
	Stabilized earth block	Alkali activated block	Stabilized earth block	Alkali activated block
Transportation	2.0	1.9	27.0	25.6
Manufacture	3.3	5.8	44.5	78.3
Cement	2.5	0.0	33.7	0.0
Fine aggregate	1.0	0.8	13.5	10.8
Sodium silicate		2.7	0.0	36.4
Sodium hydroxide		9.5	0.0	128.2
Total	8.7	20.7	117.4	279.3

CEB:

One block is sized 300 × 150 × 70 mm, with thus a volume of 0.00315 m^3.

The external surface of one block is 0.30 × 0.07 = 0.021 m^2.

For 1 m^2 surface area 47.62 blocks are required.

The binder content (lime) is 6.5% of mass.

Rammed earth:

One cubic meter of dried wall weights 1127.36 kg.

The wall thickness is 0.6 m, thus 1 m^2 surface area is 0.6 m^3.

Results have to be multiplied by 0.6.

The binder content (lime) is 3%.

Appendix 2

Available Results Concerning Transport

Morel et al. [41]

Assuming that transportation would occur in France, it is possible to calculate energy, i.e. around 1.5 MJ/(ton km).

- For building A in stone masonry with earth mortar, total transport of 1390 ton km (Table 8.9) is found to be 2.1 GJ.
- For building B in stone masonry with earth mortar and rammed earth, total transport of 1041 ton km (Table 8.9) is found to be 1.6 GJ.
- For building C in concrete, total transport of 6707 ton km (Table 8.9) is found to be 10.2 GJ (Table 8.9).

With these results, a proportion of transport compared to total energy consumed for construction is obtained (Table 8.10). The energy reduction due to the use of local materials can be calculated for buildings A and B compared to building C (Table 8.10).

Estrada [18]

See Tables 8.11 and 8.12.

For a cob house, contribution of transport is around 25% of total emission, whereas for concrete house it is 2.7%.

Nanz et al. [42]

See Fig. 8.7.

Variant A: total PE = 902 MJ/m^2, A2 + A4 = 500 MJ/m^2, transport = 55%

Variant B: total PE = 4,537 MJ/m^2, A2 + A4 = 3833 MJ/m^2, transport = 84%.

Appendix 3

Available Results Concerning the Maintenance Phase

See Fig. 8.8 and Table 8.13.

Table 8.9 Available information concerning transport of 3 types of houses [41]

		Stone masonry with soil mortar	Base: stone masonry with sail mortar and rammed soil	Concrete
Earthworks				
Excavated volume (m^3)	Total	100	100	65
	Stone	16	16	
	Organic soil	44	44	10
	Soil	40	40	
	Transp (t km)	0	0	413[a]
Vertical masonry and timber frame or concrete				
Cement	Mass (t)	7	8	20
	Energy (GJ)	36	41	103
	Transp (t km)	357[b+e]	408[b+e]	1440[d]
Aggregates	Mass (t)	0	0	66
	Energy (GJ)	0	0	27
	Transp (t km)	0	0	4752[d]
Stone	Mass (t)	120	40	0
	Energy (GJ)	48	16	0
	Transp (t km)	600[c]	200[c]	0
Timber	Mass (m^3)	7.5	7.5	0
	Energy (GJ)	3	3	0
	Transp (t km)	431[f]	431[f]	0
Steel	Mass (t)	0.21	0.21	2.0
	Energy (GJ)	10	10	95
	Transp (t km)	1[b]	1[b]	12[b]
Thermal insulation				
Mineral wool	Volume (m^3)	0	0	10
	Energy (GJ)	0	0	8
	Transp (t km)	0	0	30[b]
Baked bricks	Mass (t)	0	0	10
	Energy (GJ)	0	0	6
	Transp (t km)	0	0	60[b]
Total				
	Energy (GJ)	97	70	239

(continued)

Table 8.9 (continued)

		Stone masonry with soil mortar	Base: stone masonry with sail mortar and rammed soil	Concrete
	Transp (t km)	1390	1041	6707

[a]Distance from the building site to dump 5 km
[b]Distance from the building site to the material seller 6 km
[c]Distance from the building site to the stone quarry 5 km
[d]Distance from the agregate quarry to building site 72 km
[e]Distance from the cement quarry to the material seller 45 km
[f]Distance from the forest to the building site 115 km

Table 8.10 Contribution of transport calculated after [41]

Building	Description	Transport/total energ (%)	Transport energy compared to building C (%)
A	Stone masonry with earth mortar	2.1	– 79
B	Stone masonry with earth mortar and rammed earth	2.2	– 84
C	Concrete	4.1	100

Table 8.11 Results of CO_2 emissions [18] for a cob house

Material group	Weight	CO_2 emissions		
		Materials	Transport	Total
Unit	kg	$kgCO_2$	$kgCO_2$	$kgCO_2$
Solvent/adhesive paints	51	1778	2	1781
Wood	5670	0	27	27
Cooked earth	10,500	2271	203	2474
Metals	101	440	1	441
Glass	162	97	2	99
Other building materials	186,680	1656	1863	3519
Total	203,163	6242	2099	8340

Table 8.12 Results of CO_2 emissions [18] for a concrete house

Material group	Weight	CO_2 emissions		
		Materials	Transport	Total
Unit	kg	$kgCO_2$	$kgCO_2$	$kgCO_2$
Solvent/adhesive paints	10	350	0	350
Wood	30	0	0	0
Cooked earth	0	0	0	0
Metal	721	7823	8	7831
Glass	162	97	2	99
Other building materials	52,645	9740	505	10,245
Total	53,567	18,010	515	18,525

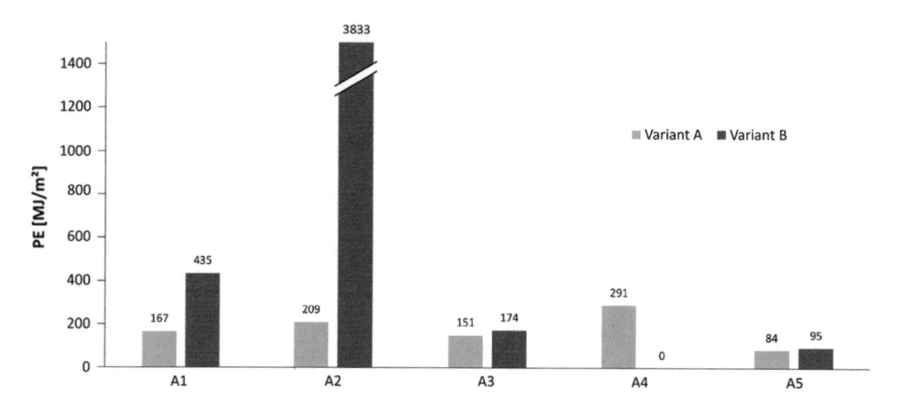

Fig. 8.7 Primary energy demand for the two variants studied—transports correspond to A2 and A4 stages [42]

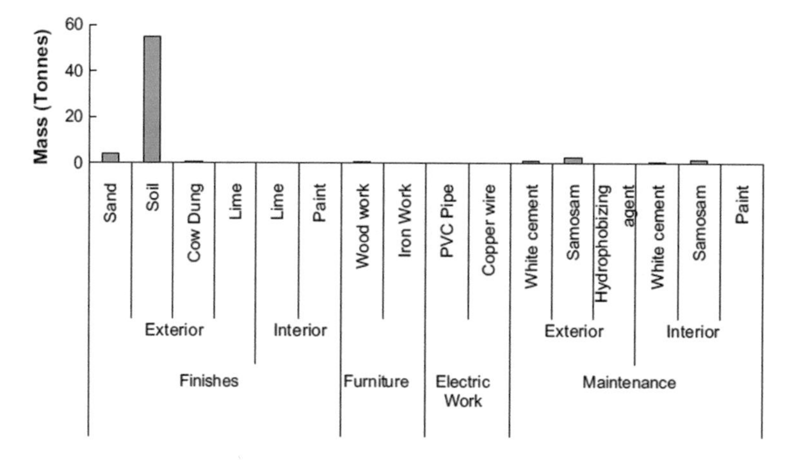

Fig. 8.8 Masses of materials for life phases of an adobe house in India [56]

Table 8.13 LCA results for 1 m^2 of the CEB wall stabilized with calcium hydroxide [51]

Impact category	MP and manufacturing	Construction and maintenance	End purpose	Total
Thinning of the ozone layer (kg CFC-11 eq)	1.08E−06	1.04E−07	1.32E−08	1.20E−06
Climate change (kg CO_2 eq)	35.74	2.04	3.40	41.18
Photochemical oxidants—smog (kg O_3 eq)	6.34	0.07	1.01	7.42
Acidification (mol H + eq)	12.24	0.18	1.86	14.28
Eutrophication (kg N eq)	1.50E−02	2.47E−04	1.97E−03	1.72E−02
Human health: carcinogens (HTU)	2.88E−07	1.41E−09	4.73E−08	3.37E−07
SH: non-carcinogens (HTU)	3.49E−06	7.37E−08	4.56E−07	4.02E−06
SH: respiratory effects (kg MP10 eq)	1.83E−02	6.62E−04	2.57E−03	2.15E−02
Ecotoxicity (ETU)	50.89	0.04	8.73	59.66
Use of agricultural and urban soil (m^2a)	11.21	0.00	0.02	11.24
Water resources' depletion (m^3)	0.41	1.43	0.00	1.84
Mineral resources' depletion (kg Fe eq)	3.51E−02	3.07E−03	4.70E−06	3.82E−02
Fossil resources' depletion (kg Petroleum eq)	8.85	0.23	1.14	10.22
Saturated energy (MJ)	451.14	11.85	48.11	511.10

HTU human toxicity unit (Toxicity cases/$kg_{emission}$)
ETU ecosystems' toxicity unit (PAF m^3 day/kg emitted)
PAF potentially affected fraction of species

References

1. Abanda H, Nkeng GE, Tah JHM, Ohandja ENF, Majia MB (2014a) Embodied energy and CO2 analyses of mud-brick and cement-block houses. AIMS Energy 2:18–40. https://doi.org/10.3934/energy.2014.1.18
2. Abanda H, Tah JHM, Elambo Nkeng G (2014b) Earth-block versus sandcrete-block houses: embodied energy and CO2 assessment. Embodied energy and CO2 assessment. Eco-Effic Mason Bricks Blocks Prop Durab 481–514. https://doi.org/10.1016/B978-1-78242-305-8.00022-X
3. Albayyaa H, Hagare D, Saha S (2019) Energy conservation in residential buildings by incorporating passive solar and energy efficiency design strategies and higher thermal mass. Energy Build 182:205–213. https://doi.org/10.1016/j.enbuild.2018.09.036
4. Arrigoni A, Beckett C, Ciancio D, Dotelli G (2017) Life cycle analysis of environmental impact vs. durability of stabilised rammed earth. Constr Build Mater 142:128–136. https://doi.org/10.1016/j.conbuildmat.2017.03.066
5. Arrigoni A, Ciancio D, Beckett CTS, Dotelli G (2016) Improving rammed earth buildings' sustainability through life cyle assessment (LC). In: Expanding boundaries: systems thinking for the built environment. Presented at the sustainable built environment, Zurich, p 6
6. Asan H (2006) Numerical computation of time lags and decrement factors for different building materials. Build Environ 41:615–620. https://doi.org/10.1016/j.buildenv.2005.02.020
7. Baggs D, Mortensen N (2006) Thermal mass in building design. Environ Des Guide 1–9
8. Bajželj B, Allwood JM, Cullen JM (2013) Designing climate change mitigation plans that add up. Environ Sci Technol 47:8062–8069. https://doi.org/10.1021/es400399h
9. Beckett CTS, Jaquin PA, Morel J-C (2020) Weathering the storm: a framework to assess the resistance of earthen structures to water damage. Constr Build Mater 242:118098. https://doi.org/10.1016/j.conbuildmat.2020.118098
10. Ben-Alon L, Loftness V, Harries KA, DiPietro G, Hameen EC (2019) Cradle to site life cycle assessment (LCA) of natural vs conventional building materials: a case study on cob earthen material. Build Environ 160:1–10. https://doi.org/10.1016/j.buildenv.2019.05.028
11. Bui QB, Morel JC, Venkatarama Reddy BV, Ghayad W (2009) Durability of rammed earth walls exposed for 20 years to natural weathering. Build Environ 44:912–919. https://doi.org/10.1016/j.buildenv.2008.07.001
12. Cauvin J (2013) Naissance des divinités, naissance de l'agriculture. Cnrs, Paris
13. CEN (2014) EN15804:A1—Sustainability of construction works—environmental product declarations—core rules for the product category of construction products
14. Chen C, Habert G, Bouzidi Y, Jullien A, Ventura A (2010) LCA allocation procedure used as an incitative method for waste recycling: an application to mineral additions in concrete. Resour Conserv Recycl 54:1231–1240. https://doi.org/10.1016/j.resconrec.2010.04.001
15. Christoforou E, Kylili A, Fokaides PA, Ioannou I (2016) Cradle to site life cycle assessment (LCA) of adobe bricks. J Clean Prod 112:443–452. https://doi.org/10.1016/j.jclepro.2015.09.016
16. Dahmen J, Kim J, Ouellet-Plamondon CM (2018) Life cycle assessment of emergent masonry blocks. J Clean Prod 171:1622–1637. https://doi.org/10.1016/j.jclepro.2017.10.044
17. De Wolf C, Cerezo C, Murtadhawi Z, Hajiah A, Al Mumin A, Ochsendorf J, Reinhart C (2017) Life cycle building impact of a Middle Eastern residential neighborhood. Energy 134:336–348. https://doi.org/10.1016/j.energy.2017.06.026
18. Estrada M (2013) A case study of cob earth based building technique in Matagalpa. Nicaragua—LCA perspective and rate of adoption (Master's Thesis No. International Master's Programme in Ecotechnology and Sustainable Development). Mid Sweden University
19. Fabbri A, Morel J-C, Gallipoli D (2018) Assessing the performance of earth building materials: a review of recent developments. RILEM Tech Lett 3:46–58. https://doi.org/10.21809/rilemtechlett.2018.71

20. Fernandes J, Peixoto M, Mateus R, Gervásio H (2019) Life cycle analysis of environmental impacts of earthen materials in the Portuguese context: rammed earth and compressed earth blocks. J Clean Prod 241:118286. https://doi.org/10.1016/j.jclepro.2019.118286
21. Floissac L, Marcom A, Colas A-S, Bui Q.-B, Morel J-C (2009) How to assess the sustainability of building construction. In: Presented at the fifth urban research symposium, Marseilles (France), p 18
22. Flower DJM, Sanjayan JG (2007) Green house gas emissions due to concrete manufacture. Int J Life Cycle Assess 12:282. https://doi.org/10.1065/lca2007.05.327
23. Frischknecht R, Wyss F, Büsser Knöpfel S, Lützkendorf T, Balouktsi M (2015) Cumulative energy demand in LCA: the energy harvested approach. Int J Life Cycle Assess 20:957–969. https://doi.org/10.1007/s11367-015-0897-4
24. Galan-Marin C, Martínez-Rocamora A, Solís-Guzmán J, Rivera-Gómez C (2018) Natural stabilized earth panels versus conventional façade systems. Econ Enviro Impact Assess Sustain 10:1020. https://doi.org/10.3390/su10041020
25. Galan-Marin C, Rivera-Gomez C, Garcia-Martinez A (2015) Embodied energy of conventional load-bearing walls versus natural stabilized earth blocks. Energy Build 97:146–154
26. Gobin C (2003) Analyse fonctionnelle et construction. Tech Ing 19
27. Gomes F, Brière R, Feraille A, Habert G, Lasvaux S, Tessier C (2013) Adaptation of environmental data to national and sectorial context: application for reinforcing steel sold on the French market. Int J Life Cycle Assess 18:926–938. https://doi.org/10.1007/s11367-013-0558-4
28. Habert G, Castillo E, Vincens E, Morel JC (2012) Power: a new paradigm for energy use in sustainable construction. Ecol Indic 23:109–115. https://doi.org/10.1016/j.ecolind.2012.03.016
29. Hamard E, Cazacliu B, Razakamanantsoa A, Morel J-C (2016) Cob, a vernacular earth construction process in the context of modern sustainable building. Build Environ 106:103–119. https://doi.org/10.1016/j.buildenv.2016.06.009
30. Heathcote K (2011) The thermal performance of earth buildings. Inf Constr 63:117–126. https://doi.org/10.3989/ic.10.024
31. Houben H, Guillard H (1989) Earth construction: a comprehensive guide. Practical Action Publishing, Rugby, Warwickshire, United Kingdom. https://doi.org/10.3362/9781780444826
32. Ioannidou D, Meylan G, Sonnemann G, Habert G (2017) Is gravel becoming scarce? Evaluating the local criticality of construction aggregates. Resour Conserv Recycl 126:25–33. https://doi.org/10.1016/j.resconrec.2017.07.016
33. Ioannidou D, Nikias V, Brière R, Zerbi S, Habert G (2015) Land-cover-based indicator to assess the accessibility of resources used in the construction sector. Resour Conserv Recycl 94:80–91. https://doi.org/10.1016/j.resconrec.2014.11.006
34. Kindermans M (2017) Sevran recycle ses terres. Echos 1
35. Lefebvre P (2018) BC architects and studies: the act of building—biennale architettura 2018. Exhibitions International, Antwerpen
36. Lima J, Faria P, Silva AS (2020) Earth plasters: the influence of clay mineralogy in the plasters' properties. Int J Archit Herit 0:1–16. https://doi.org/10.1080/15583058.2020.1727064
37. Marsh ATM, Heath A, Walker P, Reddy BVV, Habert G (2020) Discussion of "EARTH concrete. Stabilization revisited." Cem Concr Res 130:105991. https://doi.org/10.1016/j.cemconres.2020.105991
38. Melià P, Ruggieri G, Sabbadini S, Dotelli G (2014) Environmental impacts of natural and conventional building materials: a case study on earth plasters. J Clean Prod 80:179–186. https://doi.org/10.1016/j.jclepro.2014.05.073
39. Minke G (2012) Building with Earth: design and technology of a sustainable architecture. Walter de Gruyter
40. Moevus M, Anger R, Fontaine L (2012) Hygro-thermo-mechanical properties of earthen materials for construction: a literature review. Terra 2012. Lima, Peru
41. Morel JC, Mesbah A, Oggero M, Walker P (2001) Building houses with local materials: means to drastically reduce the environmental impact of construction. Build Environ 36:1119–1126. https://doi.org/10.1016/S0360-1323(00)00054-8

42. Nanz L, Rauch M, Honermann T, Auer T (2018) Impacts on the embodied energy of rammed earth façades during production and construction stages. J Facade Des Eng 7:75–88. https://doi.org/10.7480/jfde.2019.1.2786
43. Ouedraogo KAJ, Aubert J-E, Tribout C, Escadeillas G (2020) Is stabilization of earth bricks using low cement or lime contents relevant? Constr Build Mater 236:117578. https://doi.org/10.1016/j.conbuildmat.2019.117578
44. Pacheco-Torgal F (2013) Handbook of recycled concrete and demolition waste, Woodhead Publishing Series in Civil and Structural Engineering. Woodhead Publishing, Oxford (UK). https://doi.org/10.1533/9780857096906.1
45. Pacheco-Torgal F, Jalali S (2012) Earth construction: lessons from the past for future eco-efficient construction. Constr Build Mater 29:512–519. https://doi.org/10.1016/j.conbuildmat.2011.10.054
46. Parracha JL, Lima J, Freire MT, Ferreira M, Faria P (2019) Vernacular Earthen buildings from Leiria, Portugal—material characterization. Int J Archit Herit 0:1–16. https://doi.org/10.1080/15583058.2019.1668986
47. Picuno P (2016) Use of traditional material in farm buildings for a sustainable rural environment. Int J Sustain Built Environ 5:451–460. https://doi.org/10.1016/j.ijsbe.2016.05.005
48. Poupeau T (2019) Les remblais du supermétro seront transformés... en logements! Le Parisien 1
49. Röck M, Saade MRM, Balouktsi M, Rasmussen FN, Birgisdottir H, Frischknecht R, Habert G, Lützkendorf T, Passer A (2020) Embodied GHG emissions of buildings—the hidden challenge for effective climate change mitigation. Appl Energy 258:114107. https://doi.org/10.1016/j.apenergy.2019.114107
50. Rouvreau L, Michel P, Vaxelaire S, Villeneuve J, Jayr E, Vernus E, Buclet N, Renault V, de Cazenove A, Vedrine H (2010) Déchets BTP: revue de l'existant (tâche 1) (rapport du projet de recherche ANR ASURET No. BRGM/RP-58935-FR). BRGM, CSTB, INSA Valor, UTT CREID, 13 dévelopement
51. Roux Gutiérrez Rubén S, Velazquez Lozano J, Rodríguez Deytz H (2015) Compressed earth blocks, their thermal delay and environmental impact. In: Energy efficiency. Presented at the National congress on sustainable construction and eco-efficient solutions, Seville, p 12
52. Sauvage M (2009) Les débuts de l'architecture au Proche-Orient. In: Mediterra 2009. Presented at the 1ère conférence méditerranéenne sur l'architecture de terre
53. Serrano S, Barreneche C, Rincón L, Boer D, Cabeza LF (2013) Optimization of three new compositions of stabilized rammed earth incorporating PCM: thermal properties characterization and LCA. Constr Build Mater 47:872–878. https://doi.org/10.1016/j.conbuildmat.2013.05.018
54. Serrano S, Barreneche C, Rincón L, Boer D, Cabeza LF (2012) Stabilized rammed earth incorporating PCM: optimization and improvement of thermal properties and life cycle assessment. Energy Procedia 30:461–470. https://doi.org/10.1016/j.egypro.2012.11.055
55. Serrano S, de Gracia A, Cabeza LF (2016) Adaptation of rammed earth to modern construction systems: comparative study of thermal behavior under summer conditions. Appl Energy 175:180–188. https://doi.org/10.1016/j.apenergy.2016.05.010
56. Shukla A, Tiwari GN, Sodha MS (2009) Embodied energy analysis of adobe house. Renew Energy 34:755–761. https://doi.org/10.1016/j.renene.2008.04.002
57. Soudani L, Woloszyn M, Fabbri A, Morel J-C, Grillet A-C (2017) Energy evaluation of rammed earth walls using long term in-situ measurements. Sol Energy 141:70–80. https://doi.org/10.1016/j.solener.2016.11.002
58. Treloar GJ, Owen C, Fay R (2001) Environmental assessment of rammed earth construction systems. Struct Surv 19:99–106. https://doi.org/10.1108/02630800110393680
59. UNEP Sustainable Building Initiative (2009) Buildings and climate change—ummary for decision makers, sustainable United Nations
60. Van Damme H, Houben H (2018) Earth concrete . Stabilization revisited. Cem Concr Res 114:90–102. https://doi.org/10.1016/j.cemconres.2017.02.035

61. Venkatarama Reddy BV, Jagadish KS (2003) Embodied energy of common and alternative building materials and technologies. Energy Build 35:129–137. https://doi.org/10.1016/S0378-7788(01)00141-4
62. Venkatarama Reddy BV, Leuzinger G, Sreeram VS (2014) Low embodied energy cement stabilised rammed earth building—a case study. Energy Build. 68:541–546. https://doi.org/10.1016/j.enbuild.2013.09.051
63. Venkatarama Reddy BV, Morel J-C, Faria P, Fontana P, Oliveira D, Serclerat I (2020) Codes and standards on earth construction—a review. In: Report of Rilem technical committee 274: testing and characterisation of Earth-based building materials and elements
64. Venkatarama Reddy BV, Prasanna Kumar P (2010) Embodied energy in cement stabilised rammed earth walls. Energy Build 42:380–385. https://doi.org/10.1016/j.enbuild.2009.10.005
65. Ventura A, Ta V-L, Kiessé TS, Bonnet S (2020) Design of concrete: setting a new basis for improving both durability and environmental performance. J Ind Ecol. https://doi.org/10.1111/jiec.13059
66. Vinceslas T (2019) Caractérisation d'éco-matériaux terre-chanvre en prenant en compte la variabilité des ressources disponibles localement
67. Vissac A, Bourgès A, Gandreau D, Anger R, Fontaine L (2017) Argiles et biopolymères: les stabilisants naturels pour la construction en terre. CRAterre, Villefontaine

Printed in the United States
by Baker & Taylor Publisher Services